ELEMENTS OF
SYMBOLIC LOGIC

ELEMENTS OF SYMBOLIC LOGIC

HANS REICHENBACH

Dover Publications, Inc.
New York

Published in Canada by General Publishing Company, Ltd.,
30 Lesmill Road, Don Mills, Toronto, Ontario.
Published in the United Kingdom by Constable and Company, Ltd., 10 Orange Street, London WC2H 7EG.

This Dover edition, first published in 1980, is an unabridged
republication of the work originally published by The Macmillan Company in 1947.

International Standard Book Number: 0-486-24004-5
Library of Congress Catalog Card Number: 80-65530

Manufactured in the United States of America
Dover Publications, Inc.
180 Varick Street
New York, N.Y. 10014

Preface

Symbolic logic has so far been a domain of mathematicians. It grew from the soil of mathematics, it found its first successful applications in mathematics, and it remained accessible only to those who were trained in the mathematical technique.

The present textbook of symbolic logic is written on the assumption that the new logic has a wider meaning, that it is on the march to replace the traditional Aristotelian logic in all fields, and that it can be taught to students who have no special mathematical training. I came to this opinion when I saw that an analysis of science and a general theory of knowledge demand the use of the methods developed in symbolic logic as much as does the analysis of mathematics; and I found overwhelming evidence for it in more than twenty years of academic teaching, which showed me both that symbolic logic is the best initiation to a scientific philosophy and that it can be taught to all who seriously desire to learn it. This book, in fact, was written in close connection with the courses I gave, and the presentation was continuously adjusted to the needs of the students until a point was reached that seemed to justify publication.

I may add the remark that my personal fortunes were of great help in the writing of this book. The migration of intellectuals which followed the disastrous political developments in Germany has greatly contributed to an exchange of the various standards of civilization; and I for one cannot but be grateful to a fate which led me into various countries, not as a traveler, but as a teacher and collaborator in the education of youth. I thus have taught logic and scientific methods in various countries and various languages; and I have studied the reaction of students of many nationalities to an instruction in a scientific logic. I have seen that the logistic approach to philosophy is not bound to a certain type of mind, or of milieu or of educational system, but represents a most successful clarification of ideas, by which all forms of scientific pursuit will profit. At

v

the same time, the necessity of teaching in several languages led me to try to adapt the methods of symbolic logic to the study of conversational language; and I thus undertook an inquiry which turned out to be useful for the understanding both of logic and of language. The present book is the first systematic presentation of such dual use of the logistic symbolism.

The requirements of teaching and the desire to connect logic with the actual use of language have determined the structure of the book. Throughout, emphasis is laid on the applicability of the logistic symbolism to the meanings of conversational language. From the very beginning, the propositional operations are introduced in two interpretations, distinguished as the adjunctive and the connective interpretation, and it is shown that the first, useful as it is for the calculus, cannot exhaust the meaning of the corresponding terms of conversational language until it is supplemented by another which can account for the feeling of connection that is associated, for instance, with the usual notion of implication. It is shown that such connective operations can be constructed from the adjunctive ones by the help of the metalanguage. Though mathematicians are inclined to disregard this problem as irrelevant for their purposes, a logic which claims to be the logic of conversational language and of scientific thought cannot overemphasize it. To prove that a satisfactory solution can be attained without abandoning the principles of what has been called an extensional logic constitutes one of the major objectives of the book.

This line of inquiry is continued by an analysis of grammatical categories aimed at the construction of a logistic grammar, at least the outlines of which are now traceable. The now generally accepted notion of propositional function is used as the major tool of this analysis; to carry it through, however, a number of special investigations had to be made, the results of which, I hope, will be of interest to both the logician and the linguist. I should like to mention in this respect the analysis of event arguments, of token-reflexive symbols, of the tenses of verbs, of the nature of the adjective and the adverb, and of logical terms. I should be glad if the philologist could make use of these contributions by a logician who advances no claim to be an expert in philology, but who feels that the state of traditional grammar is hopelessly muddled by its two-millennial ties to a logic that cannot account even for the simplest linguistic forms.

The interest in practical applications cannot make an analysis of the foundations dispensable. In fact, whoever has taught knows that a clear insight into the meanings of fundamental notions and methods is a prerequisite for successful teaching. In an introductory chapter as well as in repeated nontechnical inserts I have tried to clarify the notions and methods used. I was thus led to an analysis of the nature of deductive methods and a discussion of the proofs of consistency, and I arrived at results that may interest the mathematician who feels disturbed by doubts about the value of deductive proof. It appears to me that, although logic and mathematics are nonempirical sciences, statements about the reliability of logical and mathematical statements belong in an empirical metalanguage; there is no need, therefore, to claim absolute certainty for them.

Among the technical features of the book, what may be interesting for the logician is the extension of the truth-table method to the calculus of functions, the introduction of the higher calculus of functions by a method which permits the derivation of the formulas of this calculus from those of the simple calculus, and the definition of connective operations aforementioned, which is accompanied by a definition of the modalities. An exposition of the foundations of mathematics, on the other hand, is not included in the plan of this book. I trust, however, that the mathematician who has studied symbolic logic from my book will find it easy to follow the presentation in other books dealing with this subject.

The notation I use is a simplified version of Russell's symbolism, with the exception of the sign for the negation, for which a horizontal line on top of the formula offers great advantages. Furthermore, I have abandoned Russell's system of dots and have used parentheses, since they are so much easier to read.

A few words may be added advising the student how to read this book. Those who wish to study symbolic logic thoroughly will do best to follow the presentation in the order given. Those, however, who are chiefly interested in the linguistic applications of symbolic logic may make a short cut by omitting chapters IV and VI, with the exception of §§ 39–40, which should be read. Such readers will be sufficiently prepared to understand, on the whole, the analysis of conversational language given in chapter VII; and they may even read chapter VIII successfully. All students, however, are invited to test their abilities with the exercises given in the appendix for all sections marked at the end by the symbol '(Ex.)'. I have found that an understanding of the methods and the value of symbolic logic will never be reached until a sufficient familiarity with the technique of the symbolization is attained. For the benefit of those who feel uncertain about their achievements, the solutions of all these little problems are given in a second appendix. Should the reader feel unable to do a certain group of problems, it might be helpful to consult the solution of the first one of the group. Teachers who use this book in classes will find it easy to construct, along the lines given, additional exercises of their own.

It goes without saying that the ideas gathered in a book of this kind represent a selection from the contributions of many writers, and that it is not always possible to trace an individual discovery back to its author. Among the logicians to whom I am indebted, however, two men stand out who have shaped modern logic in its essential lines. The first is Bertrand Russell, of whose work I will single out here only the theories of propositional functions and of descriptions, the use of what he calls material implication, and the elaboration of a practical notation including operators and bound variables. The second is David Hilbert, of whose contributions to logic I mention here, in particular, the program of a complete formalization of the object language and of a proof of consistency. I had the good fortune to learn from both men, not only by reading their publications but also by personal contact: from Hilbert when I was a student in the University of Göttingen; from Russell when, some decades later, he was my colleague in the University of California at Los Angeles, long after I had studied his books. Among those who made their way into logic at the same time as I did, I should like to mention my friends Rudolf Carnap, whose insistence on a separation of language and metalanguage I took over, using it to extend his theory of tautological implication to a general theory of nomological statements, and Charles Morris, whose general theory of signs I used for the exposition of the sign nature of language. I cannot give here an account of the contemporary development of symbolic logic, and so I may be excused if I do not name the many others whose opinions

have influenced me through their writings or through personal discussions. The ideas of symbolic logic, in fact, constitute today an
atmosphere common to a large group of philosophers for whom the
name *logical empiricists* has been coined; without my personal par-
ticipation in the activities of this group I would not have been able
to write this book. Let me express the hope that the present book
will help to solidify the conceptions adhered to by the group, and
help to win further adherents for its ideas.

Of those who helped me in the writing of the manuscript I
should like to name my former assistants in Los Angeles: Dr. Norman C. Dalkey, Dr. Abraham Kaplan, and Mrs. Elinor Kesting
Charney. Many detailed formulations were developed by the help
of their criticisms, and a great many of the exercises were constructed by them. I thank, furthermore, Professor Rudolf Carnap
and Professor Charles Morris for the suggestions of improvements
they made after reading the manuscript. My gratitude also goes
to my wife, Maria Reichenbach, for the various forms of help and
advice she afforded me in the writing of this book.

My warmest thanks, however, I wish to express to all the students of
my courses in logic, in Germany, in Turkey, and in the United States,
for their active interest in the subject. The never ending flow of
their questions for the meaning of the terms and methods used
impelled me to make my ideas clearer and clearer, and to develop
methods of exposition which I never would have found in the soli-
tude of the philosopher's study. Let me hope that this book will
show what can become of a philosophy when it is worked out and
experienced in the close community of teacher and students, when
it is the live subject of a common inquiry into the nature of thought.

HANS REICHENBACH

University of California
  at Los Angeles
October 1946

Table of Contents

VIII. CONNECTIVE OPERATIONS AND MODALI-TIES

The sign ' [Ex.] ' at the end of a section indicates
that there are exercises for this section in the
appendix.

ELEMENTS OF
SYMBOLIC LOGIC

I. *Introduction*

§ 1. Logic and language

Logic has often been defined as the science that deals with the laws of thought. This is an ambiguous characterization unless we distinguish between psychological and logical laws of thought. The actual process of thinking evades distinct analysis; it is in part logically determined, in part automatic, in part erratic; and what we observe as its constituents are isolated crystallizations of largely subconscious currents hidden below a haze of emotional processes. As far as there are any laws observable they are formulated in psychology; they include laws both of correct and of incorrect thinking, since the tendency to commit certain fallacies must be considered a psychological law in the same sense as the more fortunate habits of correct thinking. This distinction itself, the distinction between valid and invalid thinking, cannot be made within psychological analysis.

If we want to say that logic deals with thinking, we had better say that logic teaches us how thinking *should* proceed and not how it *does* proceed. This formulation, however, is susceptible of another misunderstanding. It would be very unreasonable to believe that we could improve our thinking by forcing it into the straitjacket of logically ordered operations. We know very well that productive thinking is bound to follow its own dark ways, and that efficiency cannot be secured by prescriptions controlling the paths from the known to the unknown. It is rather the results of thinking, not the thinking processes themselves, that are controlled by logic. Logic is the touchstone of thinking, not its propelling force, a regulative of thought more than a motive; it formulates the laws by which we judge thought products to be correct, not laws that we want to impose upon thinking. Creative thought processes, even of the trained mind, do not move along prepared paths but follow a method of trial and error, in which logic separates the right from the wrong results. If it is true that to a certain extent we can improve our thinking by studying logic, the fact is to be explained as a conditioning of our thought operations in such a way that the relative number of right results is increased.

When we call logic *analysis of thought* the expression should be interpreted so as to leave no doubt that it is not actual thought which we pretend to analyze. It is rather a substitute for thinking processes, their *rational reconstruction*,[1] which constitutes the basis of logical analysis. Once a result of thinking is obtained, we can reorder our thoughts in a cogent way, constructing a chain of thoughts between point of departure and point of arrival; it is this rational reconstruction of thinking that is controlled by logic, and whose analysis reveals those rules which we call logical laws. The two realms of analysis to be distinguished may be called *context of discovery* and *context of justification*. The context of discovery is left to psychological analysis, whereas logic is concerned with the context of justification, i.e., with the analysis of ordered series of thought operations so constructed that they make the results of thought justifiable. We speak of a justification when we possess a proof which shows that we have good grounds to rely upon those results.

It has been questioned whether all thinking processes are accompanied by linguistic utterances, and behavioristic theories stating that thinking *consists* in linguistic utterances have been attacked by other psychologists. We need not enter into this controversy here for the very reason that we connect logical analysis, not with actual thinking, but with thinking in the form of its rational reconstruction. There can be no doubt that this reconstruction is bound to linguistic form; this is the reason that logic is so closely connected with language. Only after thinking processes have been cast into linguistic form do they attain the precision that makes them accessible to logical tests; logical validity is therefore a predicate of linguistic forms. Considerations of this kind have led to the contention that logic is *analysis of language*, and that the term 'logical laws' should be replaced by the term 'rules of language'. Thus in the

---

[1] This term was introduced by Carnap. As to further explanations concerning this notion, cf. the author's *Experience and Prediction*, University of Chicago Press, Chicago, 1938, pp. 5–7.

theory of deduction we study the rules leading from true linguistic utterances to other true linguistic utterances. This terminology appears admissible when it is made clear that the term 'rules of language' is not synonymous with 'arbitrary rules'. Not all rules of language are arbitrary; for instance, the rules of deduction are not, but are determined by the postulate that they must lead from true sentences to true sentences.

It is the value of such an analysis of language that it makes thought processes clear, that it distinguishes meanings and the relations between meanings from the blurred background of psychological motives and intentions. The student of logic will find that an essential instrument for such clarification is supplied by the method of symbolization, which has given its name to the modern form of logic. It is true that simple logical operations can be performed without the help of symbolic representation; but the structure of complicated relations cannot be seen without the aid of symbolism. The reason is that the symbolism eliminates the specific meanings of words and expresses the general structure which controls these words, allotting to them their places within comprehensive relations. The great advantage of modern logic over the older forms of the science results from the fact that this logic is able to analyze structures that traditional logic never has understood, and that it is able to solve problems of whose existence the older logic has never been aware.

We said that logic cannot claim to replace creative thought. This limitation includes symbolic logic; we do not wish to say that the methods of symbolic logic will make unnecessary the imaginative forms of thought used in all domains of life, and it certainly would be a misunderstanding to believe that symbolic logic represents a sort of slide-rule technique by which all problems can be solved. The practical value of a new scientific technique is always a secondary question. Logic is primarily a theoretical science; and it proceeds by giving a determinate form to notions that until then had been employed without a clear understanding of their nature. Whoever has had such an insight into the structure of thought, whoever has experienced in his own mind the great clarification process which logical analysis accomplishes, will know what logic can achieve.

§ 2. Language

If logic is analysis of language, we must begin our inquiry by a consideration of language.

Language consists of *signs*. We should not forget, however, that signs are physical things: ink marks on paper, chalk marks on a board, sound waves produced in a human throat. What makes them signs is the intermediary position they occupy between an object and a sign user, i.e., a person. The person, in the presence of a sign, takes account of an object; the sign therefore appears as the substitute for the object with respect to the sign user. The three-place relation holding between the sign, the object, and the person has therefore been called by Charles Morris[1] the relation of *mediated-taking-account-of*. To simplify the terminology we shall speak of the *sign relation*, or *relation of denotation*.

The notion of *sign* is wider than that of language. Not all signs represent a language. Smoke may appear to us a sign of fire; but smoke does not represent a linguistic utterance. Signs of this kind, which acquire their sign function through a causal connection between the object and the sign, are called *indexical signs*. In other cases something will be the sign of an object because of a similarity in appearance, such as a photograph; such signs are named *iconic signs*. In the third case, that of language, the coordination of the sign to the object is purely conventional; we speak here of *conventional signs*, or *symbols*.[2] The use of the symbols is determined by a set of rules which we call the *rules of language*.

For practical purposes, linguistic signs must be *reproducible* since we use different individual signs for the same logical functions. The individual sign is called a *token*. Thus in the two sentences 'Los Angeles is a city' and 'Los Angeles is situated in California' we have the same word 'Los Angeles', but appearing in two different tokens; and now in making the explanation a third token of this word has been used. Different tokens of the same word have the

---

[1] Charles Morris, 'Foundations of the Theory of Signs', *International Encyclopedia of Unified Science*, Vol. I, No. 2; and *Signs, Language, and Behavior*, Prentice-Hall, New York, 1946.

[2] This distinction originates from Charles Peirce; cf. his *Collected Papers*, Harvard University Press, Cambridge, 1934, Vol. V, p. 50 (originally published 1903). Historically the introduction of symbols may be reducible to iconic signs; thus the hieroglyphics were originally pictures of objects. Furthermore, the onomatopoetic words, occurring in all languages, are iconic signs.

same meaning, or are *equisignificant*. In part, equisignificance is given by geometrical similarity of the tokens, as in our example; but we have also equisignificance between printed and handwritten tokens, and between these and spoken tokens. The coordination of these different kinds of tokens to one another is of course a matter of convention. To have a convenient term we shall call these different kinds of tokens *similar* to each other, using the word 'similar' in a somewhat wider sense. The class of similar tokens is called a *symbol*. Saying 'the same symbol occurs in different places' means 'tokens of the same symbol-class appear in different places'.[1]

The most important unit among signs is the proposition.[2] It is usually composed of several words. The grammatical definition of a word as a group of letters separated from others by an interval is not satisfactory: what is one word in one language may be expressed by several words in another, whereas propositions are always translated into propositions. The German language is noted for its long compound nouns, such as 'Eisenbahnknotenpunkt' which means 'railroad junction' in English. The rule stating that the words 'rail' and 'road' are written in one word, whereas 'junction' is written separately, is merely conventional. Likewise the distinction between words and suffixes, which usually are considered parts of words, is without logical significance. The Turkish language, which uses many suffixes, sometimes expresses a whole sentence in one word. Thus the Turkish word 'alabĭleceğim' is a sentence and means 'I shall be able to buy'.[3] We should not forget that in all languages the division into words more or less disappears as soon as we speak, since in talking we make no intervals between most words; in the French language this habit leads to the *liaison* of spoken words, i.e., the merging of words into one by pronouncing otherwise silent end consonants. We shall later inquire whether a logically satisfactory definition of words, in the sense of sentential parts, can be given.

---

[1] The distinction between tokens and the class of tokens applies also to nonlinguistic signs.

[2] We do not distinguish between 'proposition', 'sentence', and 'statement', and shall therefore use these terms interchangeably.

[3] The two first letters 'al' of the Turkish word constitute the verb root meaning 'to buy'; the three letters 'bĭl' represent the root of a verb meaning 'to be able'. The rest consists of suffixes.

What makes a proposition the fundamental unit is the fact that only a whole proposition can be *true* or *false* — that, as we say, it has a *truth-value*. An isolated word like 'table' is not true or false. Only if a word stands for a sentence, as an abbreviation, can we speak of its truth or falsehood, as, for instance, if a child points at a table and says 'table', when the complete statement would be 'this is a table'. Likewise the property of having a *meaning* is originally restricted to whole sentences. If we want to communicate meanings to other persons we speak in sentences; a word does not communicate anything unless it stands for a sentence.

If none the less we sometimes speak of the meaning of words, the usage is to be interpreted in the following way. The same word may occur in different sentences, and we say that we understand the meaning of a word if we know how to use it in sentences of different meanings. It seems advisable to distinguish the two terms, if necessary, by speaking of *sentence-meaning* and *word-meaning*. It is clear from the given definition that sentence-meaning is logically prior to word-meaning, i.e., that the expression 'word-meaning' is defined in terms of the expression 'sentence-meaning'.

The properties of having a truth-value and a meaning are confined to signs and do not apply to physical things which have no sign function. In the often vague usage of conversational language this rule is not always followed; we speak of a true friendship, or of true facts; we say that the refusal of an ultimatum by a government means war, etc. Sometimes this usage may be excused by the existence of indexical sign relations, as in the last example, or in such expressions as 'smoke means fire', 'smoke is a true sign of fire'. In other instances we may have iconic sign relations, as when we speak of a true portrait. It appears advisable, however, to restrict the predicates *meaning* and *truth* to linguistic signs, or symbols, since a complete interpretation of these terms can be given only within a system of rules constituting a language. Incidentally, it is easy to replace these ambiguous words by more suitable terms; thus we may speak of a genuine friendship, of smoke as indicating fire, etc.

Not all combinations of meaningful words are meaningful. The word sequence 'The emperor of and is' is a *meaningless* set of signs. In this particular example the lack of meaning is indicated by the violation of grammatical rules. However, the observance of grammatical rules is not a guarantee of meaning. The word sequence 'Caesar is a prime number' is also meaningless, although this combination of signs satisfies the rules of grammar. The reason for this further restriction as to meaning will be explained later (§ 40). It is important to realize that a meaningless set of signs does not become meaningful if we take the negation of it. Thus the set of signs 'the emperor of and is not' is as meaningless as the set of signs that does not contain the word 'not'. Neither is the set of signs 'Caesar is not a prime number' meaningful.

We shall use the term 'proposition' in such a way that it refers only to meaningful sets of symbols. We thereby exclude from the domain of logic certain combinations of words which otherwise would lead to serious difficulties, the so-called *antinomies*. The introduction of rules restricting meaning beyond the restrictions of grammatical rules is one of the most important advances made by modern logic; the credit for having recognized the necessity of such rules, formulated in his so-called theory of types (cf. § 40), belongs to Bertrand Russell.

Although the theory of types formulates some *necessary* requirements of meaning, it leaves open the question of *sufficient* requirements. In other words, it leaves unanswered the general question, 'when is a set of signs meaningful?' The answer to this question constitutes an important and much-discussed chapter of modern epistemology. We cannot enter upon this inquiry here and shall merely report that the answer has been given through the *verifiability theory of meaning*, which, in its simplest form, is expressed by the two principles:

1. A proposition is meaningful only if it is verifiable as true or false.
2. Two propositions have the same meaning if they obtain the same verification, as true or false, for all possible observations.

A discussion of these two principles must be left to other expositions of the subject. Let us mention only that the second represents the modern form of a principle that has played a leading role in the history of philosophy. It emerged in the philosophy of *nominalism* and was formulated by William of Occam as the rule — known under the name of *Occam's razor* — that entities should not be multiplied except if necessary. It was brought to the fore by G. W. Leibniz as the principle of the identity of indiscernibles, and it has acquired a particular significance through its application in Einstein's theory of relativity. It constitutes the nucleus of the theories of meaning developed in *pragmatism* and *logical positivism*, conceptions which have been united in the philosophical movement that carries the name of *logical empiricism*.

Analysis in our day has shown that the two principles stated above require some correction, and that the terms 'true' and 'false' should be replaced by the continuous scale of probability.[1] We shall not refer to this modified conception of verification but shall continue to consider propositions as two-valued, i.e., as being true or false. Two-valued logic is the mother of all other logics; further, it can always be carried through, in the sense of an approximation, even if refined analysis demands a probability logic.

The preceding remarks will make it clear why we shall base the system of logic on propositions as fundamental units. Propositions may be called the *atoms of language*, and, just as a piece of matter will always consist of a whole number of atoms, a meaningful speech or article will always consist of a whole number of propositions. This analogy can be carried further. Atoms combine into molecules; in a similar way *atomic propositions* combine into *molecular propositions*. On the other hand, the fact that atoms constitute the fundamental units of matter does not preclude them from being themselves composed of subunits, from having an inner structure accessible to investigation. Similarly, the inner structure of propositions can be investigated. These considerations led modern logicians to a general division of logic into two parts. The first part, the *calculus of propositions*, deals with the operations combining propositions as wholes; the second part, the *calculus of functions*, treats the inner structure of propositions, relating this analysis to the results of the first part.

We cannot explain here the notion of function, which has given its name to the second part, and must refer the reader to § 17 for

---

[1] For a presentation of these ideas and a general exposition of the verifiability theory of meaning, cf. the author's *Experience and Prediction*, University of Chicago Press, Chicago, 1938, chapter I. We refer to this book also for an analysis of the meaning of 'truth'.

an explanation of the term. But we may add that the notion of function possesses a parallel in the notion of class and that thus another interpretation of the second part can be constructed, which is usually added to the given classification as a third part under the name of *calculus of classes*.

We shall present the three parts in the order given.

§ 3. Different levels of language

We said that signs are physical things coordinated to other things by certain rules. The process of coordination can be repeated, and we may introduce signs referring to signs. This iteration of the coordination process is not an invention of logicians; conversational language possesses a great many terms of this kind. Thus the word 'word' refers to signs; so do the words 'sentence', 'clause', 'phrase', 'name'. We say that signs of signs constitute a language of a higher level, which we call *metalanguage;* the ordinary language then is called *object language*. From the metalanguage we proceed to meta-languages of higher levels by introducing signs denoting signs of signs.

A means of indicating the transition from a sign to a sign for a sign is offered by quotation marks. Thus 'California' is a sign that denotes California; 'California' is written with ten letters, whereas California grows oranges. The transition may be further repeated. Thus ''California'' is the name of a sign, namely, of 'California', but is not the name of California. In writing quotes we have to watch that the sign combination occurring in our sentence is always one level higher than the object to which it refers. Thus ''California'' is written with one pair of quotes, and 'California' is written without quotes.[1] It would be difficult to add quotes to California; we then would have to construct huge quotes and put those of the left end into the Pacific Ocean, those of the right end into Nevada.

Let us add here the remark that the function of quotes is some-times expressed by other linguistic devices. Thus italics may assume the function of quotes. Furthermore, a formula written on a sepa-rate line is to be conceived as equivalent to being presented in quotes.

---

[1] We assume the usual meaning of the phrase 'is written with', namely, the mean-ing of 'contains as a part of its writing'. On the other hand, it would be correct to say that, in the following expression, 'California' stands between quotes.

Quotes or similar linguistic devices do not represent the only means of introducing signs of signs. We may also introduce independent words as signs of signs, and the mentioned words like 'word' and 'sentence' are such signs of signs not involving quotes. The signs of signs constructed by means of quotes are of a very peculiar kind. In them the object is employed as its own sign, and the function of the quotes consists in indicating this unusual usage. We might introduce a similar usage for the names of other physical objects; thus we might, whenever we write something about sand, put some sand in the place otherwise occupied by the word 'sand'. In order to indicate that this is not an undesired sand spot on our paper, but a part of our language and the name of sand, we should have to put quotes left and right of the sand spot. Unfortunately such a practice, although perhaps suitable for sand, would often lead to serious difficulties, for instance if we wanted to use this method for denoting lions and tigers. It is for these technical reasons that the quotes method is restricted to the introduction of signs of signs.

However, there are some other restrictions to the quotes method which are of a logical nature. What we introduce by quotes are names of signs, i.e., *words* of the metalanguage; but we cannot introduce propositions of the metalanguage in this way. It would not help us to start from propositions of the object language; by adding quotes to a proposition we obtain, not a proposition of the metalanguage, but a word of the metalanguage. In order to construct a proposition in the metalanguage we have to use its words and to combine them in a meaningful way. Thus ''snow' is an English word' and ''snow is white' is true' are propositions of the metalanguage. Many of the words of the metalanguage will correspond to words of the object language, such as the words 'and', 'is'; each has a meaning similar to that of the corresponding word of the object language. But here the transition to the metalanguage cannot be achieved by the addition of quotes either; thus the word 'is' of the metalanguage is not obtainable by adding quotes to the word 'is' of the object language. Words which occur in different languages in similar meanings are called 'ambiguous as to level of language'. In another conception these words are regarded as identical with those of the object language; the metalanguage then is conceived as a mixture of words of the first and the second level. This conception appears preferable because such mixed sentences cannot be completely dispensed with, as is shown by a sentence like "'Peter' denotes Peter", where the second word 'Peter' belongs to the object language.[1]

Whereas signs of signs belong to a language of a higher level, *abbreviations* do not. An abbreviation does not *denote* a sign; it *stands for* a sign. Thus the abbreviation 'U.S.A.' stands for 'United States of America' and belongs to the same language as that sign. An abbreviation consists merely in the introduction, by a convention, of a new kind of token. Thus the tokens of the symbol-class 'U.S.A.' are considered as equisignificant to the tokens of the symbol-class 'United States of America'. Analogously, when letters 'a' and 'b' are used for propositions, they must be regarded as abbreviations, not as names of the propositions.

With the consideration of letters standing for propositions we come to a new distinction. The letter may be an abbreviation for an individual proposition; thus 'a_1' may stand for 'snow is white'. We then call 'a_1' a *propositional constant* and indicate this character by a subscript. Secondly, we may introduce letters as *propositional variables;* we then use letters without subscripts, as 'a' and 'b'. A propositional variable is a sign which does not stand for any individual proposition but which occupies a place that can be filled by any individual proposition. We say individual propositions are special values of propositional variables. Conversational language has no particular signs for propositional variables; these signs are used only within a scientific analysis of language in order to express structural properties considered for all propositions. Thus in order to symbolize the relation of implication we write '$a \supset b$'. (The arc sign is the sign of implication.) Here 'a' and 'b' are propositional variables whose place may be occupied by any special propositions; whether the implication holds will then depend on the special propositions. We see that a formula containing propositional variables is not true or false, in general, but will become so only after specialization. Exception is to be made for formulas that hold for all values of the propositional variables, including the formulas that form the very subject of logic (cf. § 8, § 12). The use of variables in logic

[1] Such sentences belong to semantics; cf. p. 15.

thus serves the same purpose as the corresponding practice in mathematics. A numerical equation containing the variable 'x' will hold only for special numerical values of 'x'. On the other hand, a so-called identical equation, such as '$(a + b)^2 = a^2 + 2ab + b^2$', will hold for all values of the variables.

The use of variables is connected with a specific condition. We said that for a propositional variable we may substitute any special proposition; if the variable occurs in several places within the same context, however, such substitution is permissible only when the same proposition is substituted for the variable in all its places. We therefore say that substitutions for variables are subject to a *coupling condition*. Thus when we have '$a \supset a$' we can put for the two letters 'a' any special proposition; but it must be the same for both. This coupling condition constitutes the basis for the use of variables. It is applied, for instance, when we say that the expressions '$a \supset b$' and '$a \supset a$' are different. Since in both expressions any proposition can be substituted for the letters, the only difference is that the second expression is connected with a coupling condition for the substitution, whereas the first is not.

Propositional variables belong to the same language as the propositions which are their special values. Of course propositional variables are not bound to the language of lowest level; they may be introduced in any higher language. Like propositions, the propositional variables of the higher language cannot be obtained by the addition of quotes to signs of the lower language. We may use letters of different alphabets: for instance, Latin letters for propositions and propositional variables of the object language, Greek letters for propositions and propositional variables of the metalanguage. Special symbols for the metalanguage are necessary, however, only when the metalanguage is formalized; we therefore shall not make much use of such symbols in this book.

When we combine the quotes with propositional variables the resulting expressions are of a peculiar nature. Substitutions made for the variable within the quotes will transform the whole expression into names of various propositions; and the expression consisting of the variable and the quotes will therefore represent a new variable. The new variable, however, is in the metalanguage when the propositional variable is in the object language. We shall call this new variable a *sentence name variable* because its special values are names of propositions. Thus when we say

if 'a.b' is true, 'a' is true

(the period stands for the word 'and') we allow for a substitution of any special propositions for the letters '*a*' and '*b*', subject only to the coupling condition for '*a*'. The expressions containing the quotes and the variables, therefore, represent metalinguistic variables which we employ in order to express generality in the metalanguage.

Quotes used in this form will be called *variable-quotes*. They result from ordinary quotes when a variable is put in quotes, and, in addition, substitutions are regarded as admissible for the letter variable. Sometimes substitutions are not admitted; then quotes put around a variable are ordinary quotes. Usually this distinction is expressed by a term prefixed to the expression in quotes to indicate the range of the substitutions admitted. Thus we say 'the variable '*a*''; the term 'variable' indicates here that the quotes around '*a*' are variable-quotes. When we say 'the letter '*a*'', we mean ordinary quotes around '*a*'. Only in exceptional cases is this notation ambiguous; then the meaning of the quotes must be understood from the context in which they are used. Thus in the sentence 'the variable '*a*' in the statement '*a* ⊃ *b*' is expressed by the first letter of the alphabet', we see from the context that we must not substitute other letters for '*a*', and that therefore the quotes are ordinary quotes. For these reasons it appears unnecessary to introduce special symbols for variable-quotes. They can be dispensed with also for the reason that, when a special proposition is substituted for the variable, the resulting quotes are ordinary quotes. This is clear because a special proposition does not admit of substitutions. Furthermore, quotes used with variables will practically always be variable-quotes; and we shall therefore frequently omit a prefixed range term, the meaning of the quotes being clear from the context.[1]

[1] Variable-quotes were first analyzed by A. Tarski, under the name of 'Anführungs-funktion', in 'Der Wahrheitsbegriff', *Studia Philosophica*, Leopoli, 1935, p. 14. Tarski has pointed out that the use of such quotes requires certain precautions. It is possible to eliminate variable-quotes by the use of special signs for sentence name variables, say Gothic letters (Carnap). This is necessary, however, only when the metalanguage is formalized. For the informal use of the metalanguage in this book we prefer the variable-quotes because they simplify the notation considerably.

In conversational language, variable-quotes are often evaded by means of the phrase 'such as'. We say, for instance, 'a sentence combination such as 'snow is white and water is wet''. These quotes are the ordinary quotes. Using variable-quotes we write the same expression in the form 'a sentence combination '$a.b$''.

On the other hand, the usage of ordinary quotes may be interpreted as involving variable-quotes of a restricted nature. We said in § 2 that a symbol is defined by the use of an equisignificance relation; this relation is given by geometrical similarity, but also by such relations as consist in the correspondence of small and capital letters, or of written and spoken tokens. Now it is in the nature of the equisignificance relation that it admits the replacement of a token by another one without any coupling condition. Thus in the formula '$a \supset a$' we can erase the first token and replace it by a similar one, leaving the second token unaltered. This is usually not admissible for a transition, for instance, to spoken tokens. We would not call it a meaningful expression if we wrote only the first 'a' of the formula and the implication sign, and then pronounced an 'a' for the second token of this letter. We allow a transition only from the written formula to the spoken formula. This restriction represents a coupling condition, and thus indicates that the transition constitutes a substitution rather than a replacement, the word 'replacement' being used for a transition to other signs that is not bound to a coupling condition. Therefore ordinary quotes should be interpreted as variable-quotes of a restricted range of substitution. The range is indicated here also by a prefixed term, such as 'the spoken letter', 'the written letter', 'the capital letter', 'the letter' (including spoken and written tokens and small and capital letters).

Let us summarize the given distinctions of levels of language by the table on the following page.

We consider physical objects as constituting the zero level; they may be called objects in the absolute sense, and the corresponding object language then is the absolute object language. Sometimes other kinds of objects may be considered; for instance numbers. We then say that they are objects in a relative sense, and that the corresponding object language is a relative object language. Physical objects divide into *things*, such as individual human beings,

Objects	Object language	Metalanguage
	bird thing	'bird' word, name
	the bird flies a_1 situation	'the bird flies' 'a_1' sentence, proposition
	a	'a' propositional variable
		'the bird flies' is true α_1
		α

tables, atoms, and *situations*, also called *states of affairs*, which constitute the denotata of sentences. Thus the sentence 'the battleship *Bismarck* was sunk' denotes a situation; the ship itself is a thing.[1]

The metalanguage is divided into three parts, corresponding to the three arguments of the sign relation. The first part, *syntax*, deals with relations between signs only and therefore concerns structural properties of the object language. A syntactical statement is, for instance: 'the sentence 'if water is heated it expands' is an implication'. The second part, *semantics*, refers to both signs and objects; in particular, it therefore includes statements concerning the truth-value of propositions, since truth is a relation between signs and objects. Of this kind is the statement: 'the sentence 'if water is heated it expands' is not always true'. The third part, *pragmatics*, adds a reference to persons; it therefore refers to things,

[1] The word 'proposition' is occasionally used as synonymous, not with 'sentence', but with our term 'situation'; it is thus used by R. Carnap, *Introduction to Semantics*, Cambridge, Harvard University Press, 1942, p. 18. This notation does not seem advisable because it departs too far from established usage. When some logicians thought it necessary to distinguish between 'proposition' and 'sentence' they did so because they believed that there was a third thing between the sentence, i.e., the linguistic expression, and the situation. Such a third thing is certainly unnecessary, and we shall therefore identify sentence and proposition. Only occasionally may it appear useful to regard 'sentence' as a somewhat narrower term; thus we may say that 'Peter drives the car' and 'the car is driven by Peter' are different sentences but represent the same proposition. The term 'proposition', then, is used with a wider equisignificance relation.

signs, and persons. Of this kind is the statement: 'I consider this sentence true', or 'this sentence is a law of physics', as the latter proposition states that physicists consider the sentence true.[1]

Conversational language is a mixture of object language and metalanguage, including all three parts of the metalanguage. The occurrence of words like 'word', 'proposition', indicates the use of the metalanguage. Words like 'conclusion', 'derivable', belong to syntax; words like 'true', 'likely', 'perhaps', belong to semantics; and words like 'assertion', 'incredible', 'of course', belong to pragmatics.

The rules that we need to define a language repeat the trichotomy of the metalanguage. First we must give *formation rules*, which tell us under what conditions a set of signs is meaningful. Of this kind are grammatical rules; they are insufficient, however, and an ideal language would possess rules of formation showing immediately that expressions of the kind 'Caesar is a prime number' are meaningless. Second, we must give *truth rules*, i.e., rules that tell us what kind of truth-values a proposition can have, and how these truth-values determine the truth of compound propositions. Thus two-valued logic and probability logic differ as to their truth rules. Third, we must give *derivation rules*, which tell us ways of deriving new propositions from given propositions; of this kind is the rule of inference. Since the derivation of new propositions from given propositions is needed for practical purposes by persons who use the language, this third kind of rule corresponds to pragmatics.

The rules of derivation are of two kinds. The first lead from true propositions to true propositions; they are called rules of *deduction*. The part of logic that they establish is called *deductive logic*. The second sort of rules lead from true propositions to propositions that are maintained only as *posits*, i.e., as substitutes for true propositions where truth is not knowable and is replaced by a *probability;* [2] they are the rules of *induction*. The part of logic that includes

---

[1] The suggestion to construct a hierarchy of languages corresponding to the theory of types was first made by B. Russell in his introduction to Wittgenstein's *Tractatus Logico-Philosophicus*, London, 1922. The elaboration of this theory of levels of language is chiefly due to R. Carnap and A. Tarski, whose work is concentrated on syntax and semantics. The study of pragmatics has been advanced by C. Morris.

[2] Cf. p. 187 and the author's *Experience and Prediction*, University of Chicago Press, Chicago, 1938, § 34.

inductive rules is called *inductive logic;* it comprises both deductive and inductive derivations and deals with the theory of *indirect evidence.* The exposition of the present book will be restricted to deductive logic. [Ex.]

§ 4. Instrumental usage of language

So far we have spoken only of a *cognitive usage* of language, i.e., a usage of language for the purpose of expressing true statements. Cognitive usage of language belongs to semantics, since truth is a relation between signs and objects. In addition, language is mostly used simultaneously as an instrument for attaining certain other aims. When we speak of *instrumental usage* of language we refer, in particular, to the aim of influencing the listener, or reader, for certain purposes intended by the speaker, or writer. Instrumental usage, therefore, falls into pragmatics.

The first form of instrumental usage consists in the intention of informing the listener. We speak here of the *communicative usage* of language. It should be realized that communication always represents a form of influencing the listener, since it is the speaker's intention, not only to convey a meaning, but to make the listener believe the sentences uttered. Every report given by one person to another is of this kind. When we do not succeed in arousing belief in the listener, the aim of communication has not been attained. When I tell Mr. A that Mr. B has committed an embezzlement, I wish Mr. A to believe what I say. It may be that Mr. A refuses to believe my statement; then I have not succeeded in communicating my knowledge to Mr. A. What I have achieved is only the communicating to Mr. A that *I believe* that Mr. B has committed an embezzlement; this result, namely, is reached when I have succeeded at least in making Mr. A believe this second statement. The raising of such belief is included in the aim of the communication. Maybe Mr. A refuses to believe even in the bona fide intent of my utterance, assuming that, contrary to better knowledge, I wish but to bring Mr. B into bad repute. In that case I have raised in Mr. A the belief that I have the intention of defaming Mr. B; but since it was not the aim of my utterance to raise such belief the incident cannot be called a communication.

The intention of making the listener believe what we say is

usually expressed through a specific intonation which distinguishes a sentence we claim as true from, for instance, a sentence uttered as a question. For written language the acoustical criterion is replaced by a special sign of assertion, namely, the period at the end of the sentence; the period expresses the claim by the writer that the sentence is true, and his intention of making the reader believe the sentence. It is an interesting fact that we cannot completely dispense with such a special symbolism of assertion; it can be omitted only for certain statements, not for all. Thus when we first make a statement 'a' without an indication of a truth claim and then add the statement "'a' is true', or 'I believe 'a' to be true', the latter addition will help us only if it is equipped with the specific indication of assertion. This shows that the indication of assertion, be it made by a sign or by accentuation, is of a pragmatic nature and cannot be completely translated into a statement; it can only be translated into a statement to which an indication of assertion is added. Moreover, the indication of assertion would be redundant if it did not perform its function of making the listener believe, either the statement uttered, or the statement that the speaker believes the statement uttered, or something else; i.e., the indication of assertion exhausts its function in the production of a certain belief in the listener.

As the second form of instrumental usage we consider the *suggestive usage* of language. By this term we understand the use of language with the intention of arousing in the listener certain *emotions*, or of determining him to assume certain *volitional attitudes*. It is well known that the generation of emotions by means of language is one of the main purposes of poetry. But conversational language, too, is seldom free from this usage since even in everyday language we frequently attempt to make the listener approve, disapprove, appreciate, dislike what is said; or to cheer him up and make him our friend; or to make him feel critical of himself. Language is one of the most efficient instruments for interfering with a person's emotional life. From the generation of emotions there is only one step to the initiation of volitional decisions. By means of language we attempt to create in a person certain decisions as to what is good and what is bad. Thus language is the instrument of religious prophets and ethical leaders, of educators, of politicians;

and it is equally the instrument by which every man influences the opinions of his friends or opponents.

The possibility of suggestive usage is derived from the fact that we all have been conditioned since early childhood to associate certain emotional attitudes with certain words. These emotional attitudes do not coincide with the meanings of the words. Two sentences having the same meaning may produce very different emotional effects. Translate a poem into everyday language, or into the clumsy precision of juristic language, and the emotional effects are gone. This fact shows that the suggestive effect of language is entirely in pragmatics.

The third form of instrumental usage is found in the use of language for the purpose of inducing the listener to perform certain actions. We speak here of the *promotive usage* of language. To this category belong commands like 'shut the door', or 'thou shalt not steal'. These examples use the imperative form to indicate promotive usage; but the indicative form can also be used. Thus, sitting in a car we may say to the driver: 'the speedometer points to 85 miles'; by the utterance of this indicative sentence we intend to induce the driver to slow down to a more reasonable speed, and thus we may reach our aim better than by means of an impolite imperative. A famous example of promotive usage of language in seemingly communicative form is given by Mark Antony's oration on Caesar's death. The line of demarcation between suggestive usage and promotive usage is not strictly drawn. The generation of certain feelings in the listener may lead, and may be intended to lead, to actions of a specific type. Political speeches, with their emotional appeal and their intention of inducing the listener to put a certain slip of paper into a ballot box, constitute examples of such combined usage.

It should be clearly seen that the instrumental usage of language falls into a category to which the predicates 'true' and 'false' do not apply. These predicates express a semantical relation, namely, a relation between signs and objects; but since instrumental usage is in pragmatics, i.e., includes the sign user, it cannot be judged as true or false. If we want to introduce, instead of pragmatical relations, statements that are true or false, we can do so by making statements including the sign user. Thus to the imperative 'shut the door' we can coordinate the indicative sentence 'Mr. A wishes the door to be shut'. This sentence is true or false. Or instead of reading a line of poetry we may state that we wish to arouse such and such emotions in the listener. In this way it is possible to coordinate to every instrumental usage a true statement referring to this usage. Frequently the statements so introduced will not have the instrumental function to which they refer.

When we wish to judge, not a coordinated statement, but the instrumental usage itself, we must follow another path; we then shall support, or oppose, the expressed instrumental function by a corresponding instrumental function. This office is performed by the words 'right' and 'wrong', which are themselves imperatives. Thus the command 'thou shalt not steal' will be judged by us as right; this means that we make the command our own, we are willing to enforce it. Similarly, we may share in a statement 'this picture is beautiful' with the remark 'that is right'; we then express the desire to join the efforts of the speaker in inducing other persons to feel the same emotional attitude. But the alternative 'right-wrong' must be clearly distinguished from the alternative 'true-false'. The merging of these two alternatives, or the attempt to regard them as of the same logical nature, has been the root of mischievous confusions, in which the history of philosophy has abounded since the time of Plato. But this is a problem with which we cannot deal in this book; it has its place within an exposition of the theory of value judgments. In this book we shall restrict ourselves to the analysis of the cognitive usage of language.

§ 5. Definitions

The introduction of new terms as a function of known terms is called a definition. It is important to realize that, because of its reference to terms, a definition belongs to the metalanguage, at least in its original meaning. If we say, for instance, 'a submarine is a ship which can go under water' we define, not a submarine, but the term 'submarine'. We can build a submarine, but not define it. The correct form of the definition therefore would be: 'by 'submarine' we denote a ship which can go under water'; or: 'the term 'submarine' is to have the same meaning as the term 'ship capable of going under water''.

It is customary, however, to write definitions as sentences of the object language, as in the example given. This usage is without danger if we recognize it as a simplified procedure which for purposes of logical inquiry has to be translated into sentences of the metalanguage. In symbolic language we use here the sign '$=_{Df}$', meaning 'equal by definition'. Thus we may write

$$\text{submarine} =_{Df} \text{ship capable of going under water}$$

We shall say that the sign '$=_{Df}$' is not a proper object term. Sentences in which this sign occurs are *improper object sentences;* they are introduced by a shifting of the level of language.[1]

In accordance with the nomenclature of traditional logic, we call the term on the left-hand side of the definition the *definiendum*, the term on the right-hand side the *definiens*.

The relation of equality by definition can be considered a special case of the relation of equisignificance, i.e., of having the same meaning. It constitutes the case where the equality of meaning is not derived from other statements but is introduced by a volitional decision with reference to the introduction of a new sign. The question whether an equisignificance is demonstrable or a matter of definition will therefore be answered differently according as the system of language is constructed. It is possible to construct the same system in different ways, varying with the starting points; what is a definition in one construction may be a demonstrable equisignificance in another.

The relation of equisignificance between symbols can be reduced to the relation of equisignificance between tokens. Two symbols are equisignificant if every token of one symbol is equisignificant to every token of the other symbol.[2] This reduction explains the possibility of making definitions. Every definition will occur once for the first time; it then may be interpreted as the convention that every token similar to the token of the definiendum is to be considered equisignificant to every token similar to the token of the definiens.

The example cited is a definition of the special form considered

---

[1] The term 'shifted language' was introduced by R. Carnap, *Logical Syntax of Language*, Harcourt, Brace, New York, 1937.

[2] For convenience we use the same word for the relation of equisignificance between symbols and between tokens, although these relations are of different types (cf. § 40).

in traditional logic, a definition by determination of *genus proximum* and *differentia specifica*. Modern logic has recognized that this is a very limited form of definition and that in scientific as well as in conversational language we ordinarily use other forms of definitions. In general the sign ' $=_{Df}$' stands between sentences, not between words or phrases. Thus we define the sentence 'the metabolism of a person is normal' by a set of sentences about the percentage of certain substances contained in his blood. The general form of such a definition is therefore

$$a =_{Df} [b, c, \ldots]$$

where the brackets indicate a certain combination of sentences 'b', 'c', ... In this combination the sentences may be combined by various logical operations, such as expressed through the words 'and', 'or', 'implies'. We call this a *definition by coordination of propositions.*

If we introduce new terms in this way we do not define them *explicitly* but *implicitly*, i.e., in combination with other terms. Thus we define not the term 'metabolism' but statements of the kind: 'the metabolism is normal', 'the metabolism is subnormal'. Such procedure is permissible because actually we do not use the isolated term 'metabolism' but only certain sentences in which the term occurs; to know the meaning of these sentences is sufficient for all practical purposes. On account of their reference to usage such definitions are often called *definitions in use.*

The importance of these definitions lies in the fact that they allow us to define an abstract term by reference to concrete terms, whereas the scholastic definition through higher genus and specific difference determines an abstract term by reference to more abstract terms. This is the reason that the traditional form of definitions cannot represent the actual procedure of knowledge, in which a verification of abstract statements is always given by verification of statements about directly observable things, such as thermometers, gauges, or small particles seen in a microscope.

II. The Calculus of Propositions

§ 6. Propositional operations

The fundamental operations used to construct molecular sentences out of atomic sentences are expressed by the words 'not', 'or', 'and', 'implies', 'equivalent'. We call them *propositional operations*. In conversational language some of these words are sometimes used to combine, not sentences, but words, such as in 'Peter or William will go with you'. We consider this sentence form an abbreviation standing for 'Peter will go with you or William will go with you'. The operation of negation differs from the others in that it refers to only one sentence and does not combine two sentences; for convenience we shall include sentences containing the word 'not' among molecular sentences. The negation is, therefore, a *monary* operation, whereas the others are *binary* operations.

For propositional variables (cf. § 3) we use letters 'a', 'b', and so forth. For the operations mentioned we use the following symbolic notation, to which we add the name of the operation.

$\bar{a}$	not a	negation
$a \vee b$	a or b	disjunction, sum
$a.b$	a and b	conjunction, product
$a \supset b$	a implies b	implication
$a \equiv b$	a is equivalent to b	equivalence

By 'or' we understand the *inclusive* '*or*', which means that at least one of the two sentences is true, leaving it open whether both are true. This 'or' is to be distinguished from the *exclusive* '*or*', which excludes the case that both 'a' and 'b' are true, and which language expresses by 'either . . . or' or by emphasis on the word 'or'. Conversational language, however, does not always clearly distinguish the two operations, both of which are frequently used; in juristic language the phrase 'and/or' is sometimes used for the inclusive 'or'. We shall later introduce a symbol for the exclusive 'or' (cf. § 8).

23

II. The Calculus of Propositions

The names 'sum' and 'product' for 'disjunction' and 'conjunction' refer to an analogy between these logical operations and the arithmetical operations of addition and multiplication. Writing expressions like

$$7 \text{ men} + 5 \text{ women} = 12 \text{ persons}$$

we see that the addition of numbers is accompanied by an *or-combination* of the coordinated concepts, since 'persons' means the same as 'men or women'.[1] Similar expressions involving multiplication are used in physics, as in

$$3 \text{ meters} \times 4 \text{ kilograms} = 12 \text{ kilogram-meters}$$

The physical term 'kilogram-meter', denoting the unit of energy, may be conceived as a kind of *and-combination* of the two terms 'kilogram' and 'meter', i.e., as determined by the couple formed by these two concepts. It should be noticed that in grammar the word 'conjunction' has a second meaning; namely, it denotes the whole category of words expressing propositional operations, such as 'and', 'or', 'if'.

The expressions between which the sign of the 'or' stands are called the *terms* of the disjunction. The expressions combined by the sign of the 'and' are called the *factors* of the conjunction. In an implication, the expression to the left of the implication sign is called *implicans;* the expression to the right of the implication sign is called *implicate*. In an equivalence the expressions on the two sides of the equivalence sign are called the *terms* of the equivalence. In conversational language the implication is usually expressed by the conjunction 'if . . . then'; the equivalence by the conjunction 'only if . . . then', or, more precisely, 'if and only if . . . then'. The particle 'then' is often omitted. Thus we say: 'if it rains I shall stay at home'; and 'only if nonintersecting straight lines lie in the same plane are they parallel'.

We now must state the *formation rules* holding for the operations introduced. Every propositional expression consists of a combination of elementary propositions or propositional variables, and we therefore use *elementary* propositional expressions, i.e., expressions

---

[1] The deeper reasons of this analogy are revealed in Russell's theory of numbers which shows that addition can be defined in terms of an or-operation between classes; cf. B. Russell and A. N. Whitehead, *Principia Mathematica*, Vol. I, Macmillan, Cambridge, England, second edition, 1924, p. 302.

which do not contain propositional operations, as the starting point
of the rules. We say:

1. Every elementary proposition or propositional variable is a
   propositional expression.
2. When a negation line is drawn over a propositional expression,
   the resulting expression is a propositional expression.
3. Every combination of two propositional expressions by means
   of one of the binary operations is a propositional expression.

Our definition of propositional expressions is a *recursive definition*.
It names in its first rule the simplest propositional expressions and
constructs all other such expressions out of the elementary expressions by the use of propositional operations. Since the rules represent conventions about what we wish to include in the term 'propositional expression', they do not represent cognitive statements but
volitional decisions. We shall always use the term 'rule' in this
sense of a *directive*, i.e., as expressing a command or a permission
stated in the metalanguage and directing the use of names or the
manipulation with terms. The formation rules, in particular,
delimit the domain of *meaning;* i.e., they determine what expressions we wish to consider as having *sentence-meaning*.

It follows from the formation rules that every propositional
expression must be constructed in steps and must therefore consist
of units, divided into subunits, and so on. Examples of propositional expressions are:

$$(a \vee \bar{b}) \supset c \tag{1}$$

$$\overline{(a.b \supset c)} \supset \bar{d} \tag{2}$$

$$\overline{[(a \vee \bar{c}).b] \supset d} \tag{3}$$

The expression above which a negation line extends is called the
scope of the negation; for reasons of expediency, we include the
negation line itself in the scope. In (2), the expression '$\overline{a.b}$' is
the scope of the negation; in (3), the whole expression constitutes
the scope. Expression (1) may be illustrated by the sentence: 'if
a stone is thrown into the air or if it is not supported it will
fall down'. The other expressions are so complicated that conversational language uses phrases like 'it is true' or 'it is certain'
to express them. Thus (2) may be expressed in the statement: 'if
it is true that, in case not both shores of the straits are occupied by

the army, enemy warships can pass the straits, it is certain that no troop transports will be sent across the straits'.

It is a consequence of the step structure that every compound propositional expression, i.e., an expression which contains propositional operations, has a *major operation*. This is the operation introduced in the last step. In (1) the implication is the major operation; in (2) the second implication is the major operation; in (3) the major operation is the negation. If the major operation is binary, as in (1) and (2), it divides the whole expression into two parts which are called the *major terms* of the expression. If the major operation is the negation, we have only one major term, given by the expression below the long negation line.

In order to indicate the units and subunits of which an expression is composed, we use parentheses. We prefer this familiar device to Russell's system of dots.[1] To reduce the number of parentheses we introduce a rule concerning the binding force of the different symbols, similar to the rules used in algebra with respect to the signs of addition, multiplication, etc. We list the operations in the order of binding strength as follows:

strongest binding force $. \lor \supset \equiv$ weakest binding force

A long negation line binds like parentheses.

For instance, the expression

$$a.b \lor c \tag{4}$$

means that 'a' and 'b' are first combined by an 'and', and then this unit is combined by the 'or' with 'c'. In the expression

$$a.(b \lor c) \tag{5}$$

on the contrary, 'b' and 'c' constitute the first unit, combined by the 'or'; this unit is combined with 'a' by an 'and'. According to our rule of binding force we can drop the parentheses in (1), and the brackets in (3).

The expression

$$a \supset b \supset c \tag{6}$$

is meaningless because it has no major operation.[2]

---

[1] B. Russell and A. N. Whitehead, *Principia Mathematica*, Vol. I, second edition, p. 9.

[2] Similar expressions containing the 'or' or the 'and' are meaningful because these operations are associative; cf. § 8.

§ 7. Truth tables

We now turn to the *truth rules* of logic. They are formulated by means of *truth tables*. Like the formation rules, truth rules are directives. By stating the truth-values coordinated to certain combinations of propositions by propositional operations, these rules define the meaning of the operations.

This procedure is based on the following consideration. Propositional operations establish a relation between the truth-value of the molecular sentence and the truth-values of the atomic (or elementary) sentences. For instance, the sentence 'Peter or William will go with you' will be false if it turns out that 'Peter will go with you' is false and also that 'William will go with you' is false. We therefore can define the meaning of a propositional operation by formulating the *truth relations* holding between the elementary sentences and their combinations. These relations are expressed in the following truth tables, in which 'T' means 'truth' and 'F' means 'falsehood'.[1]

TRUTH TABLES

TABLE Ia

a	$\bar{a}$
T	F
F	T

TABLE Ib

a	b	$a \vee b$	$a.b$	$a \supset b$	$a \equiv b$
T	T	T	T	T	T
T	F	T	F	F	F
F	T	T	F	T	F
F	F	F	F	T	T

The truth tables can be read in two directions. The first is from right to left, i.e., from the statement containing the operation, or compound statement, to the elementary propositions. They then state that, if the compound statement is true, one of its T-cases is true. (By a T-case of a compound proposition we understand any of those combinations of the elementary propositions to which a

---

[1] Truth tables were used by L. Wittgenstein, *Tractatus Logico-Philosophicus*, Harcourt, Brace, New York, 1922, p. 93, and by E. L. Post, *Amer. Journal of Math.*, XLIII, 1921, p. 163. Materially, the definition of propositional operations in terms of truth and falsehood was used earlier, for instance, in B. Russell and A. N. Whitehead, *Principia Mathematica*, Vol. I, 1910, p. 6–8. Furthermore, C. S. Peirce employed this definition; cf. footnote on p. 30.

'T' is coordinated in the column of the compound proposition.) For instance, if '$a \vee b$' is true, we know that either 'a' is true and 'b' is true, or 'a' is true and 'b' is false, or 'a' is false and 'b' is true. The second direction is from left to right, i.e., from the elementary proposition to the compound statement. The tables then state that, if one of the T-cases is true, the corresponding operation holds. For instance, if we know that 'a' is true and 'b' is false we shall say that '$a \vee b$' is true.

We shall call the interpretation in which the truth tables are read in both directions the *adjunctive interpretation* of the truth tables. The interpretation in which the tables are read only in the first direction, i.e., from right to left, will be called the *connective interpretation* of the truth tables.

To make the difference clear let us present an example. A surgeon who is treating a serious case admits that he does not know whether the patient can be saved; but he adds: 'The patient will have to be operated upon or he will die'. Let us assume that the operation is made and that the patient does not die. Does this prove that the surgeon's statement was true? If we use the adjunctive interpretation it would indeed be proved, since one of the T-cases of the 'or' has occurred. Using the connective interpretation, however, we would say that what has happened merely conforms with the surgeon's statement without proving its truth. In order to prove the truth we would have to show that no other possibility of recovery was left, that if the patient had not been operated on he would have died. In this interpretation the statement establishes a *connection* between the T-cases such that, if one of the T-cases does not happen, one of the other T-cases is bound to happen. It is obvious therefore that one observation cannot prove the truth of the statement since it shows only one of the T-cases to hold, without informing us about other possibilities. Only a disproof of the statement could be given by one observation; if for instance the patient is not operated on and does not die, the surgeon's statement was certainly false.

Correspondingly we speak of *adjunctive operations* and *connective operations*. An adjunctive operation can be *verified* by one observation; a connective operation can only be *falsified* by one observation, whereas in the positive case one observation can only *conform with* it. We may also say that the connective operation can be *confirmed* by one observation, if by 'confirming' we understand 'conforming with', i.e., 'not contradicting'. The word 'confirm', however, is a dangerous one to use as it has a second meaning that is not always clearly separated from the first, namely, the meaning of 'verifying to a certain degree'. A discussion of this confirmatory evidence belongs in inductive logic, since its use is based on a special form of inductive inference; we shall not enter into this discussion here, and therefore shall not inquire into the conditions under which an observed T-case would be considered confirmatory evidence.[1] The word 'confirm' will always be used here to mean the same as 'conform with'. For the present the question of how to verify a connective operation is left open, and this discussion is postponed to a later section.

The distinction between adjunctive and connective interpretation, which we explained for the or-operation, holds similarly for the other operations. An operation used mostly in the connective interpretation is *implication*. Here the mere coincidence, i.e., the case that 'a' is true and 'b' is true, is usually not taken as proof of the implication. It recently happened in Los Angeles that, while the screen of a movie theater was showing a blasting of lumber jammed in a river, an earthquake shook the theater. The implication 'the blasting of the lumber on the screen implied the shaking of the theater' was then true in the adjunctive sense, whereas it was false in the connective interpretation. Adjunctive implication is scarcely ever used in conversational language. Similarly, equivalence (and also the exclusive 'or') are used mostly in the connective sense. The inclusive 'or' is used in both interpretations; the 'and' mostly in the adjunctive interpretation. The 'and', unlike the other operations, has only one T-case; we therefore cannot speak here of a connection between T-cases. However, there is also a connective interpretation of the 'and', meaning that no other case can happen. Like the 'and', negation is mostly used adjunctively; in the connective interpretation, '$\bar{a}$' means: "'a' is necessarily false'. Incidentally, we see from these considerations that the question of connective operations is closely related to the analysis of the *modalities*, i.e., the terms 'necessary', 'possible', 'impossible'.

[1] Cf. footnote on p. 69.

The duality of interpretation justifies the adoption of the same name for both kinds of operation. Thus adjunctive implication has the same truth tables as connective implication; these operations differ only in the interpretation of the T-cases, which in the first interpretation mean verification, in the second, conformation. The adjunctive operations are the *adjunctive correlates* of the connective operations.

The deviations of adjunctive implication from the connective usage have sometimes been called the 'paradoxes of implication'. It follows from the adjunctive interpretation that a false proposition implies every proposition, and that a true proposition is implied by every proposition. Thus we have 'snow is black implies there will be an earthquake tomorrow', and 'there is an earthquake implies sugar is sweet'. There is of course nothing paradoxical in these statements. We must realize that the word 'implies' here has not the same meaning as in conversational language; the implication in this case simply *adjoins* one statement to the other without *connecting* the statements. Adjunctive implication has a wider meaning than connective implication; if a connective implication holds, there exists also an adjunctive implication, but not vice versa. The struggle between the adherents of adjunctive and connective operations, which goes back to antiquity,[1] has sometimes been confused by the claim that one type of operation is the correct one. Such pretensions indicate a misunderstanding of the situation. In introducing adjunctive implication into our language we make use of the right of the scientist to construct his own simplified concepts. This does not preclude the use of connective implication in other propositions; we should demand only that it be clear which kind of operation is meant.

Since adjunctive operations are verified or falsified by the truth-values of the propositions they combine, these operations are also called truth-functions. Connective operations are not truth-functional in this sense; but they can be called truth-functional in a wider sense if the statements in which they occur are translatable into other statements that contain only adjunctive operations, for instance, such as hold for repeated observation of events of the

[1] Cf. the interesting historical remarks by C. S. Peirce, reported in his *Collected Papers*, Harvard University Press, Cambridge, 1933, Vol. III, sections 441–444.

same kind. The question whether the connective operations of conversational language are truth-functional in this sense constitutes an important issue in the discussion of the foundations of logic; we intend to show in later sections (§ 9, and chapter VIII) that this question is to be answered in the affirmative, although with the qualification that the meaning of connective operations includes statements of the metalanguage.[1]

Although, as we said, adjunctive implication and equivalence are little used in conversational language, there is a rudimentary usage of such operations in certain expressions employed for rhetorical rather than logical purposes. For instance, we may emphasize the truth of a statement by making it the implicate of a certainly true sentence whose meaning has no relation to it, as in the exclamation of the attorney, 'if two times two is four, this man is the rightful heir'. The implication is here asserted as true because the speaker regards both implicans and implicate as true; by this oratorical device the certainty of the implicans is, so to speak, transferred to the implicate. Similar forms of speech are given by assertions like 'as I live and breathe, this man is innocent'. The 'as' indicates here an adjunctive equivalence; it is used to indicate that the two sentences have the same truth-value.

In other forms of this kind an adjunctive implication is used to indicate that, whereas a certain statement 'b' is regarded as true, a statement 'a' is neither asserted nor denied by the speaker. Consider the sentence, 'if this man is not the murderer, he certainly is involved in the crime'. Here the implicate is regarded as true, and since a true sentence is adjunctively implied by every sentence, the assertion of the implication leaves the truth of the implicans open. That, on the other hand, the implicate is asserted is shown by the fact that this sentence would follow from the implicans also when the implicans is false, i.e., when the man is the murderer.

In another version, such adjunctive implications are used in the

---

[1] For the word 'adjunctive', or 'truth-functional' in the narrower sense, the word 'extensional' is frequently used; often, however, the term 'extensional' is also used in the meaning of our term 'truth-functional' in the wider sense. The term 'intensional', which in discussions on logic is frequently opposed to 'extensional', is used sometimes as equivalent to our term 'connective', but at other times as equivalent to our term 'analytic connective', which we shall introduce later. To avoid ambiguities we introduce the terms 'adjunctive' and 'connective' in the meanings defined. Our term 'adjunctive implication' also replaces Russell's 'material implication'.

form of understatements. Imagine that the speaker at a wedding party introduces himself by the following remark: 'If I am not the father of the bride, I am at least her uncle'. What he means is 'I am not the father, but the uncle'. Instead, he states an implication which leaves the truth of the two clauses open but is so constructed that it would entitle him in any case to make the speech. This implication is obviously adjunctive, since it is asserted for the only reason that both implicans and implicate are known to be true. In such flourishes of speech and related rhetorical forms the truth-functional character of the propositional operations becomes recognizable beneath the veil of psychological connotations.[1]

Let us add some remarks concerning the question in what sense the truth tables may be considered definitions of the propositional operations. We may read the definition of the 'or', for instance, as follows:

$$\text{`}a \vee b\text{' is true} =_{Df} \begin{cases} \text{`}a\text{' is true and `}b\text{' is true} \\ \text{or `}a\text{' is true and `}b\text{' is false} \\ \text{or `}a\text{' is false and `}b\text{' is true} \end{cases} \qquad (1)$$

$$\text{`}a \vee b\text{' is false} =_{Df} \text{`}a\text{' is false and `}b\text{' is false} \qquad (2)$$

This is a definition in use, given in semantic language; it contains, not in quotes, the words 'or' and 'and'. It therefore presupposes the operations denoted by these words as known. The definitions of other operations would read similarly and would contain also the words 'or' and 'and'. All that is reached by our definition is therefore a reduction of the operations of the object language to two of the corresponding operations of the metalanguage. This reduction is not entirely useless, as it shows that we can define the operations between any sentences whatsoever if we only know the meaning of the operations between sentences of the form 'a sentence is true' and 'a sentence is false'. Calling propositional operations of this form *semantical operations*, we may say that the truth tables furnish a definition of the propositional operations of the object language in terms of semantical operations.

The duality of adjunctive and connective operations holds also for semantical operations. It is important to realize that, in the

---

[1] An implication which is partly adjunctive and partly connective will be discussed in § 64.

definition of the operations of the object language, the semantical 'and' is adjunctive, whereas the 'or' can be interpreted both as adjunctive and as connective: if the 'or' is adjunctively interpreted we obtain the adjunctive operations of the object language; if it is connectively interpreted we obtain the connective operations of the object language. We see that even the difference of the terms 'connective' and 'adjunctive' cannot be defined unless it is already known for the semantical operations.

This consideration shows that in order to define the fundamentals of logic we must know certain *primitive terms*, i.e., terms whose meanings we understand without definition. Fortunately, it is not necessary that these terms be as clearly conceived as the terms with which we operate in the elaborate system of logic. A definition can be understood even when the meaning of the primitive terms is not distinctly known; and it is possible to define other terms as functions of indistinct terms in such a way that these other terms are not vague, but precisely determined. Thus even a precise meaning of the primitive terms can be defined by the use of their vague meanings originally known. As an example, let us consider the 'or' used on the right-hand side of relation (1). We leave it open whether this 'or' is the inclusive or the exclusive 'or'; and we may do so because, whatever meaning we use, the 'or' defined on the left-hand side will be the inclusive 'or'. Similarly, in order to understand the general characterization of connective operations given we need not be able to state clearly what a connective 'or' is; and we shall later see that it is possible to give a strict definition of connective operations that covers also the meaning of the primitive term. A relation of this kind holds generally between metalanguage and object language. The metalanguage is often vague and unprecise; in spite of this deficiency we can use it for the construction of strictly and unambiguously determined concepts of the object language. Logical methods are like the precision instruments of the physicist, which even when their dials are inexactly read furnish results whose exactness widely surpasses that of observations made without the instruments. Such exactness is possible because the instruments are so built that small differences between the measured objects produce great differences between the readings on the dials. The methods of symbolic logic follow a similar pattern, projecting small differences between the notions to be defined in the object language into great differences between the unprecise notions that we use in the metalanguage. [Ex.]

§ 8. Survey of possible operations. Tautologies

It is the advantage of adjunctive operations that they permit the construction of a very simple calculus. A calculus of connective operations involves technical difficulties; although these difficulties are not insurmountable, we shall not develop such a calculus in this book, as it can be shown that such a calculus can be dispensed with. If the meaning of connective operations is expressed with the help of the metalanguage (cf. § 9 and chapter VIII), all manipulations with these operations can be reduced to manipulations with adjunctive operations. Postponing the discussion of these questions to the later sections mentioned, we now begin with the exposition of the calculus of adjunctive operations.

The negation is the only monary operation. This is demonstrated as follows. There are only four possible arrangements of the letters 'T' and 'F' in the second column of table I a. One arrangement consists of two letters 'T', another of two letters 'F'; the interpretation of these arrangements, which do not define operations, will be given later. Then there is the arrangement 'T–F', which reproduces the first column and therefore states merely the identity of the resulting proposition with 'a'. The fourth arrangement is the one of table I a, i.e., the negation.

It is different with the binary operations. The four operations that we defined in the truth tables are not the only possible operations between two propositions. Other arrangements of the letters 'T' and 'F' in a vertical column are possible, and each such arrangement will define an operation. We see easily that $2^4 = 16$ such arrangements can be made, and that therefore 16 operations between two propositions can be defined.

Among the 16 possible distributions of the letters 'T' and 'F', one consists of letters 'T' only and another one consists of letters 'F' only. We said that these two extreme cases do not define operations and will be dealt with later. Then there is one distribution such that the 'T' corresponds to the 'T' of proposition 'a', and another one such that the 'T' corresponds to the 'T' of 'b';

similarly there are two distributions such that the 'T' corresponds to the 'F' of 'a' or of 'b'. These four are degenerate cases, as the truth of the operation is determined by one of the elementary propositions alone. There remain ten distributions of the letters 'T' and 'F'. Of these operations one half contains the negations of the operations of the other half; thus there remain only five operations. These may be represented by the four operations introduced in table I b and, in addition, by the operation '$b \supset a$'. As '$b \supset a$' is structurally of the same kind as '$a \supset b$', there exist only four proper operations between two propositions, as which we choose the operations of table I b, namely, the operations of disjunction, conjunction, implication, and equivalence.[1] This is the reason that the adjunctive calculus is usually constructed in the symbols of these operations only.

Occasionally two of the negations of the defined operations are denoted by special symbols. These are the exclusive 'or', which we denote by '$\wedge$', and which is the negation of the equivalence; and the stroke, denoted by '$/$', which is the negation of the 'and' and may be read 'not both'. The truth tables of these operations are presented in table I c.

TABLE I c

a	b	$a \wedge b$	a/b
T	T	F	F
T	F	T	T
F	T	T	T
F	F	F	T

We now turn to the consideration of an arrangement of the vertical column consisting of the letter 'T' only. In order to construct a formula having these truth-values, let us first consider another case. The formula

$$\overline{a \vee b} \tag{1}$$

has a truth-value which is a function of the truth-values of 'a' and 'b'; similarly other and more complicated combinations will have

---

[1] Although we have shown that there are only four proper operations, it remains undetermined which operations we should choose as such. Thus instead of choosing the operations of table I b we might replace the equivalence by the exclusive 'or', or the implication '$a \supset b$' by the implication '$b \supset a$'.

truth-values determined by repeated application of the truth tables
and varying with the truth-values of the elementary propositions.
If we now regard the formula

$$\overline{a \vee b} \equiv \bar{a}.\bar{b} \tag{2}$$

and determine its truth-value, we find that this truth-value is
always T for every possible choice of truth-values for 'a' and 'b'.

We construct the proof by *case analysis*, which is carried out as
follows. Let us first assume that 'a' is true and 'b' is true. Then
'$a \vee b$' is true (first horizontal line of the truth tables). '$\overline{a \vee b}$' is
false (truth table of negation). '$\bar{a}$' and '$\bar{b}$' are false; therefore '$\bar{a}.\bar{b}$'
is false (fourth horizontal line of the and-operation). Both sides
of (2) are false, therefore the formula is true (fourth line of the
equivalence operation). Thus in the first case the formula is true.
The second case would be given by assuming 'a' as true and 'b' as
false, and we would prove similarly that the whole formula is true.
In this way we prove its truth for all four cases.

We call such a formula a *tautology* and define it thus:[1]

*Definition of tautologies. A tautology is a formula that is true what-
ever be the truth-values of the elementary propositions of which it is
composed.*

The other case among the 16 possible arrangements of the vertical column to be considered here is that in which we have letter
'F' only. This case is called a *contradiction;* a formula of this kind
is always false. We obtain such formulas by taking the negation of
a tautology; for instance, the formula

$$\overline{\overline{a \vee b} \equiv \bar{a}.\bar{b}} \tag{3}$$

is a contradiction. Not all contradictions show immediately their
character of being the negation of a tautology; this fact is illustrated
by the formula

$$\overline{a \supset b . b \supset a} \tag{4}$$

which by case analysis can easily be proved to be a contradiction.

Tautologies are also called *analytic;* formulas that are neither
tautologies nor contradictions are called *synthetic.* Thus all formulas
may be classified into *analytic, synthetic,* and *contradictory* formulas.

---

[1] This definition of a tautology was used by C. S. Peirce; cf. his *Collected Papers*,
Vol. 3, Harvard University Press, Cambridge, 1933, p. 224 (the original was published
in 1885).

Tautologies constitute the domain of *logical formulas*. They are absolutely reliable because they cannot be false. This property of being necessarily true, characteristic of logical formulas, is explained through the structure of the tautology, which is of such a kind that the truth-value of the formula is *truth* for all possible truth-values of the elementary propositions. Synthetic formulas do not share this property, and so they can be false. On the other hand, synthetic formulas tell us something, namely, that one of the *T*-cases holds, and not one of the *F*-cases; thus they inform us about restrictions for the truth-values of the elementary propositions. In contradistinction to synthetic formulas, tautologies do not tell us anything, since they include no restrictions for the elementary propositions; thus we are left entirely uninformed as to the truth-values of the latter. Tautologies therefore are *empty*. The concept *empty*, however, must be distinguished from the concept *meaningless;* tautologies as well as contradictions possess determinate truth-values and are meaningful, although empty.

The conception of tautologies as empty formulas has been attacked by some logicians who thought that logic would be useless if it were to consist of empty formulas. This is a misunderstanding. Although a tautology is empty, the statement that a certain formula is a tautology is not empty. The latter statement, though easily proved for a simple formula like (2), may require elaborate substantiation in respect to complicated formulas. It is the great task of logic to point out those sign combinations which are empty, in the sense of *analytic;* in these formulas the science of logic presents us with a specific instrument of thought operations necessary in all sciences. In order to know the full bearing of its assumptions, every science must use analytic transformations, which do not add anything to the meaning of the assumptions; it is for this purpose that we need the empty formulas of logic, whose addition to any scientific system is permissible because with them nothing is added to the empirical content of the system.

In our list of logical formulas (p. 38), which refers only to the operations of tables I a and I b, we present a selection of tautologies useful for practical applications. All these formulas can be proved by reference to the truth tables, with the use of case analysis. Those formulas which contain three propositional variables '*a*', '*b*',

'c' must be analyzed for each of the $2^3 = 8$ possible combinations of truth-values of 'a', 'b', 'c'. We classify the formulas with respect to the number of propositional variables or to the kind of operations. The first group referring to only one propositional variable contains formulas presented in traditional logic as the laws of thought, such as the law of identity or the law of contradiction; we see that we are concerned here only with some special formulas out of a long list of other equally important laws of thought. The explicit formulation of these other formulas, few of which were known in traditional logic, is due to the work of the first logisticians such as de Morgan, Boole, Schröder, Peirce, Russell, and Whitehead.

We add some remarks concerning the equivalence relation. Though not all tautologies are equivalences, equivalences play an important role because their function in logic corresponds to the function of equations in mathematics. Most of our formulas are therefore equivalences. Because of formula 7a we obtain from every formula containing an equivalence another formula which states, instead, an implication; thus from 1b we get

$$a \lor a \supset a \tag{5}$$

from 6c we get

$$(a \supset b) \supset (\bar{b} \supset \bar{a}) \tag{6}$$

We do not include such formulas in our list because they can easily be obtained; we rather follow the practice of writing an equivalence sign instead of an implication wherever it is possible. In group 8 we collect those formulas for which an equivalence sign would have been false, calling them 'one-sided implications'. [Ex.]

TAUTOLOGIES IN THE CALCULUS OF PROPOSITIONS

Concerning one proposition:

1a. $a \equiv a$		
1b. $a \lor a \equiv a$		rule of identity
1c. $a.a \equiv a$		
1d. $\bar{\bar{a}} \equiv a$		rule of double negation
1e. $a \lor \bar{a}$		tertium non datur
1f. $\overline{a.\bar{a}}$		rule of contradiction
1g. $a \supset \bar{a} \equiv \bar{a}$		reductio ad absurdum

Sum:

2a. $a \lor b \equiv b \lor a$	commutativity of 'or'
2b. $a \lor (b \lor c) \equiv (a \lor b) \lor c \equiv a \lor b \lor c$	associativity of 'or'

Product:

3a. $a.b \equiv b.a$	commutativity of 'and'
3b. $a.(b.c) \equiv (a.b).c \equiv a.b.c$	associativity of 'and'

Sum and product:

4a. $a.(b \vee c) \equiv a.b \vee a.c$	1st distributive rule
4b. $a \vee b.c \equiv (a \vee b).(a \vee c)$	2nd distributive rule
4c. $(a \vee b).(c \vee d) \equiv a.c \vee b.c \vee a.d \vee b.d$	
4d. $a.b \vee c.d \equiv (a \vee c).(b \vee c).(a \vee d).(b \vee d)$	twofold distribution
4e. $a.(a \vee b) \equiv a \vee a.b \equiv a$	redundance of a term

Negation, product, sum:

5a. $\overline{a.b} \equiv \bar{a} \vee \bar{b}$	breaking of negation line
5b. $\overline{a \vee b} \equiv \bar{a}.\bar{b}$	
5c. $a.(b \vee \bar{b}) \equiv a$	dropping of an always true factor
5d. $a \vee b.\bar{b} \equiv a$	dropping of an always false term
5e. $a \vee \bar{a}.b \equiv a \vee b$	redundance of a negation

Implication, negation, product, sum:

6a. $a \supset b \equiv \bar{a} \vee b$	dissolution of implication
6b. $a \supset b \equiv \overline{a.\bar{b}}$	
6c. $a \supset b \equiv \bar{b} \supset \bar{a}$	contraposition
6d. $a \supset (b \supset c) \equiv b \supset (a \supset c) \equiv a.b \supset c$	symmetry of premises
6e. $(a \supset b).(a \supset c) \equiv a \supset b.c$	
6f. $(a \supset c).(b \supset c) \equiv a \vee b \supset c$	
6g. $(a \supset b) \vee (a \supset c) \equiv a \supset b \vee c$	merging of implications
6h. $(a \supset c) \vee (b \supset c) \equiv a.b \supset c$	

Equivalence, implication, negation, product, sum:

7a. $(a \equiv b) \equiv (a \supset b).(b \supset a)$	dissolution of equivalence
7b. $(a \equiv b) \equiv a.b \vee \bar{a}.\bar{b}$	
7c. $\overline{a \equiv b} \equiv (a \equiv \bar{b})$	negation of equivalence
7d. $(a \equiv b) \equiv (\bar{a} \equiv \bar{b})$	negation of equivalent terms

One-sided implications:

8a. $a \supset a \vee b$	addition of an arbitrary term
8b. $a.b \supset a$	implication from both to any
8c. $a \supset (b \supset a)$	arbitrary addition of an implica-
8d. $\bar{a} \supset (a \supset b)$	tion
8e. $a.(a \supset b) \supset b$	inferential implication
8f. $(a \supset b) \supset (a \supset b \vee c)$	addition of a term in the implicate
8g. $(a \supset b) \supset (a.c \supset b)$	addition of a factor in the impli- cans
8h. $(a \vee c \supset b) \supset (a \supset b)$	dropping of a term in the impli- cans
8i. $(a \supset b.c) \supset (a \supset b)$	dropping of a factor in the impli- cate
8j. $(a \supset b).(c \supset d) \supset (a.c \supset b.d)$	derivation of a merged implica-
8k. $(a \supset b).(c \supset d) \supset (a \vee c \supset b \vee d)$	tion
8l. $(a \supset b).(b \supset c) \supset (a \supset c)$	transitivity of implication
8m. $(a \equiv b).(b \equiv c) \supset (a \equiv c)$	transitivity of equivalence

§ 9. Interpretation of tautological operations as connective operations

The definition of a tautology as an always-true formula has been provided by the use of adjunctive operations. It turns out, however, that tautological relations admit of an interpretation as connective operations.

This can be shown as follows. When a synthetic formula, like '$a \supset b$', has been asserted and we then observe that 'a' and 'b' are true, we do not regard this observation as a proof of a connective implication because we do not know whether in future cases of a similar kind the case '$a.\bar{b}$', which contradicts the implication, will be excluded. It is different when a tautological formula has been asserted. Since such a formula has no F-case, it is impossible ever to make an observation which contradicts the formula; therefore we can say that it is true in the connective sense. In doing so we do not use an individual observation for the verification of the formula; we rather follow the principle that a statement with which all possible observations must conform should be regarded as true.

Let us illustrate this consideration by an example. In a tautology like

$$\overline{a \lor b} \equiv \bar{a}.\bar{b} \tag{1}$$

we have an equivalence operation which holds for all possible truth-values of 'a' and 'b'. The two sides of the equivalence contain the adjunctive operations of disjunction, conjunction, and negation. Now it follows from the tautological character of (1) that, whenever observations make the left-hand side of (1) true, they will also make the right-hand side true, and vice versa; and whenever there are observations making the left-hand side false, they will also make the right-hand side false, and vice versa. It therefore is impossible ever to make observations which falsify the relation (1); all possible observations must conform with it. This allows us to regard the equivalence in (1) as a necessary connection and thus as a connective operation. Its verification is based, not on empirical observations, but on the structure of the formula.

This result is also required by the verifiability theory of meaning. What follows from the tautological character of (1) is that it is impossible ever to have experiences that discriminate between the two sides of the relation, i.e., that make one side true and the other false. Applying the second principle of the verifiability theory of meaning (§ 2) we come to the result that the expressions to the left and right of the equivalence sign have the same meaning. We thus have found a logistic interpretation of the relation of equisignificance; this connective relation is interpretable as a tautological equivalence.

We may therefore replace the sign of the adjunctive equivalence in (1) by a sign for the corresponding connective operation. In order to distinguish connective from adjunctive operations we shall use an accent above the sign of the operation. We then can write, instead of (1), the formula

$$\overline{a \vee b} \overset{\wedge}{=} \bar{a}.\bar{b} \tag{2}$$

The connective character of this equivalence is seen from the fact that it is a necessary equivalence.

In a similar way we can also introduce a connective implication. For instance, in (8g, § 8):

$$(a \supset b) \supset (a.c \supset b) \tag{3}$$

we can replace the major implication sign by the sign of a connective implication, and write the formula as follows:

$$(a \supset b) \overset{\supset}{\supset} (a.c \supset b) \tag{4}$$

This is possible for the same reason as in the previous example; the validity of the major implication in (3) is not established by an empirical observation of the truth-values of the terms on the two sides of the sign, but by showing that, whatever these truth-values may be, the implication must hold.

The same cannot be said of the two other implication signs in (3); they must be left unchanged as adjunctive operations, the verification of which we leave to observation. Only the major operation admits of a connective interpretation. For the introduction of connective operations, and thus of the accent sign, we therefore set up the rule:

The major operation in a tautology can be replaced by the corresponding connective operation.

In order to see that the operations so introduced correspond to our former definition of connective operations, let us take an example. Consider the implication: 'neither Caesar nor Napoleon reached the age of 60 years implies Napoleon did not reach the age of 60 years'. Looking into a reference book we find that Caesar died at the age of 56 and Napoleon at the age of 54; thus we find that both sides of the implication are true. The truth tables then tell us that we have here an implication in the adjunctive sense, whereas the ascertained truth-values only *conform* with a connective implication, leaving open the question of its truth. However, we feel that this is a 'better' implication; it is impossible that the statement about Napoleon is false if the statement concerning Caesar and Napoleon is true. Writing the structure of our example we see that it corresponds to the formula

$$\overline{a \vee b \supset \bar{b}} \tag{5}$$

which is a tautology. It therefore was not necessary to look into a reference book in order to prove that our implication holds; the implication is verified, not by the particular truth-values of the statements combined, but by structural properties of the corresponding general formula. The implication in (5) can therefore be replaced by an accent implication.

We call the major operation of a tautology a *tautological operation*. It follows from the definition of tautologies that tautological operations are adjunctive, and not connective; but we see from the considerations presented that they perform functions similar to those of connective operations. We say that tautological operations *can be interpreted* as connective operations, meaning that they can be replaced by connective operations.

Later (in chapter VIII) we shall analyze the logical nature of connective operations; we shall show that the meaning of these operations can be defined only in the metalanguage.[1] On the other hand, we already see that we can dispense with a certain kind of connective operations if we use tautological operations as their substitutes. This is the reason that we shall not develop a calculus containing operations with the accent sign; everything we need can be expressed in the adjunctive calculus.

Let us add that connective operations introduced in the place of tautological operations do not include all connective operations;

---

[1] The recognition that a tautological implication can often be used to express a 'reasonable' implication, and that we therefore must use the metalanguage for the characterization of such an operation, is due to R. Carnap. Cf. his *Logical Syntax of Language*, Harcourt, Brace, New York, 1937.

they comprise only the *analytic connective operations*. Later we shall consider a second kind, the *synthetic connective operations*, and we shall show that they, too, are reducible to adjunctive operations. The accent sign will be used for both kinds of connective operations.

§ 10. Reduction of operations to other operations

We do not include in our list on page 38 the two operations of table I c. These operations can be expressed in terms of other operations in various ways, for instance, by the following relation:

$$a/b \equiv \overline{a.b} \equiv \bar{a} \vee \bar{b} \tag{1}$$

$$a \wedge b \equiv (a \vee b).\overline{a.b} \equiv (a \vee b).(\bar{a} \vee \bar{b}) \equiv \overline{a \equiv b} \equiv (a \equiv \bar{b}) \tag{2}$$

Since these equivalences are tautologies, as can easily be proved by case analysis, we can say that the meaning of the expressions on the left-hand sides is given by each of the coordinated expressions on the right-hand sides. We can therefore use these relations as *definitions* of the operations of table I c, and can write, for instance:

$$a/b =_{Df} \overline{a.b} \tag{3}$$

$$a \wedge b =_{Df} (a \vee b).\overline{a.b} \tag{4}$$

The symbols on the left-hand sides then have the nature of abbreviations. We say that these symbols are *reducible* to the symbols of tables I a and I b.

This method of reduction can be extended to the symbols of tables I a and I b in such a way that some of these symbols can be defined in terms of the others. Such definitions can be constructed in various ways. For instance, we can consider disjunction and negation as *primitive operations;* the other operations then can be defined as follows:

$$a.b =_{Df} \overline{\bar{a} \vee \bar{b}} \tag{5}$$

$$a \supset b =_{Df} \bar{a} \vee b \tag{6}$$

$$a \equiv b =_{Df} a.b \vee \bar{a}.\bar{b} \tag{7}$$

We use here the equivalences stated in formulas (5a, 6a, 7b, § 8). Similar formulas can be constructed if conjunction and negation are chosen as primitive operations. It is particularly interesting that, as Sheffer [1] has shown, all operations can be reduced to *one*

---

[1] *Trans. Am. Math. Soc.* Vol. 14, 1913, p. 481. The first to see that all propositional operations can be reduced to one was C. S. Peirce, who set forth this result in a paper written in 1880 but not published until it was included in C. S. Peirce's *Collected Papers*, Vol. IV, Harvard University Press, Cambridge, 1933, p. 13. Peirce used the operation $\overline{a \vee b}$ as primitive; it has properties similar to those of the stroke.

primitive operation if we choose the stroke as primitive. We then have:

$$\bar{a} =_{Df} a/a \tag{8}$$
$$a.b =_{Df} (a/b)/(a/b) \tag{9}$$
$$a \vee b =_{Df} (a/a)/(b/b) \tag{10}$$

Formula (8) follows from table I c when we put 'a' for 'b'. Then only the first and the fourth horizontal lines are possible, and these lines furnish, for this case, the column of the negation. Formula (9) follows from the first equivalence in (1) when we express the negation by (8). Formula (10) is a transcription of the relation

$$a \vee b \equiv \overline{\bar{a}.\bar{b}} \tag{11}$$

which follows from (5b, § 8) when we take the negation on both sides. Using the relations (6) and (7), we then can express also implication and equivalence in terms of the stroke.

The reducibility of all other operations to the stroke operation has, of course, only a theoretical interest. The resulting formulas are so complicated that the use of the other signs is preferable. It has been found that for most practical purposes the operations of tables I a and I b are sufficient.

We may add a word on *ternary* operations. Such operations can be introduced by truth tables with three columns as argument columns, such that, for every combination of truth-values of three propositions 'a', 'b', 'c', a truth-value of the operation is defined. It can be shown, however, that ternary operations can be dispensed with because all such operations can be expressed in terms of monary and binary operations. For associative operations, such as conjunction and disjunction, this reduction is very simple. Thus the formulas

$$a \vee b \vee c \tag{12}$$
$$a.b.c \tag{13}$$

which are written as ternary operations, can be reduced by means of formulas (2b and 3b, § 8) to the corresponding binary operations. Therefore expressions (12) and (13) are meaningful although they are not written in a form showing a major operation; this notation is permissible because we can choose either of the or-signs in (12), or either of the and-signs in (13), as the major operation. Furthermore, these two operations have the advantage that the meaning of the ternary operations corresponds to that of the binary operations. Thus (12) means 'at least one of the three propositions is true', and (13) means 'all three propositions are true'.

It is different with the exclusive 'or'. Although this operation is associative (it is even commutative and distributive) and therefore the ternary formula

$$a \wedge b \wedge c \qquad (14)$$

has a meaning expressible as $(a \wedge b) \wedge c$, or as $a \wedge (b \wedge c)$, the meaning of (14) does not correspond to the meaning of the binary exclusive 'or', since (14) is true if all three propositions are true. In other words, (14) does not express the meaning: one and only one of the three propositions is true. This fact is one of the reasons that the use of the exclusive 'or' in the calculus is not expedient.

A disjunction in which one and only one term is true is called *complete* and *exclusive*. By *complete* we mean that at least one of the terms of the disjunction is true, or, in other words, that the disjunction is true. By *exclusive* we mean that at most one term of the disjunction is true. There are various ways of writing a disjunction of three terms which is both complete and exclusive. We can write, for instance:

$$(a \wedge b \wedge c).\overline{a.b.c} \qquad (15)$$

or

$$(a \vee b \vee c).\overline{a.b} \, . \, \overline{a.c} \, . \, \overline{b.c} \qquad (16)$$

or the combination

$$
\begin{aligned}
a &\equiv \bar{b}.\bar{c} \\
b &\equiv \bar{a}.\bar{c} \\
c &\equiv \bar{a}.\bar{b}
\end{aligned}
\qquad (17)
$$

which is meant to be a conjunction of the three lines. When a complete and exclusive disjunction of n terms is to be written, form (17) is the most suitable one:

$$
\begin{aligned}
a_1 &\equiv \overline{a_2.a_3} \ldots \overline{a_n} \\
a_2 &\equiv \overline{a_1.a_3} \ldots \overline{a_n} \\
a_n &\equiv \overline{a_1.a_2} \ldots \overline{a_{n-1}}
\end{aligned}
\qquad (18)
$$

In these relations, the implication from left to right expresses exclusiveness, whereas the implication from right to left expresses completeness. It can be shown that the last line in (18) and (17) can be omitted since it is derivable from the other lines. For (17) this is proved as follows. When both 'a' and 'b' are false, the first two lines show that both '$\bar{b}.\bar{c}$' and '$\bar{a}.\bar{c}$' must be false; therefore '$\bar{c}$' is false, and thus 'c' is true. This is the implication from right to left in the third line. Furthermore, when 'c' is true, the right-hand expressions in the first two lines are false; therefore the left-hand expressions must also be false. This is the implication from left to right in the third line. This proof can easily be extended to (18).

Examples of complete and exclusive disjunctions occur in the theory of probability. For instance, when the roulette is spinning we know that the ball will finally stop in one and only one of the 37 fields of the roulette table. This condition is expressed by (18) when the statement 'a_i' is interpreted in the form: 'the ball will stop in the ith field'.

Let us add that a repeated use of the equivalence sign, as in (1), does not, of course, have the meaning of a ternary operation. It is simply an abbreviation standing for a conjunction of equivalences in which one side occurs two times. Thus (1) means:

$$[a/b \equiv \overline{a.b}].[\overline{a.b} \equiv \bar{a} \vee \bar{b}] \tag{19}$$

Since the equivalence operation is transitive, this formula includes an equivalence between the first and the last term. The way of writing used in (1) is often employed for implications, too.

The question may be asked whether it is possible to define reversed operations for the operations of the propositional calculus. The term is meant as an analogue to the reversed operations of arithmetic, such as subtraction and division. For the 'or', for instance, we might attempt to define a reversed operation '$c{-}b$' by the equivalence

$$(a \vee b \equiv c) \equiv (a \equiv c{-}b) \tag{20}$$

This operation would correspond to subtraction, when the 'or' is regarded as the analogue of addition. It can be shown, however, that such a reversed operation cannot be defined.[1] Assume, for instance, that 'a', 'b', and 'c' are true; then '$c{-}b$' must be true in order to make (20) hold. Now let 'b' and 'c' be true, whereas 'a' is false; then both sides in (20) must be true, and, since 'a' is false, '$c{-}b$' must be false. It follows that, while 'b' and 'c' are true, the

---

[1] This fact was pointed out by C. S. Peirce; cf. his *Collected Papers*, Harvard University Press, Cambridge, 1933, Vol. III, p. 5.

expression '$c-b$' is sometimes true and sometimes false; therefore the truth-value of '$c-b$' is not determined by the truth-values of 'c' and 'b'. This means that the expression '$c-b$' does not have the nature of a propositional operation. A similar proof can be given for the 'and' and the implication. Only for the equivalence does a reversed operation exist; but this is also an equivalence, i.e., the equivalence is its own converse. This follows because the relation

$$[(a \equiv b) \equiv c] \equiv [a \equiv (c \equiv b)] \tag{21}$$

which is the analogue of (20), can be shown by case analysis to be a tautology. The impossibility of defining reversed operations for the sum and the product represents a remarkable difference between the algebra of logic and that of mathematics.

Corresponding results hold for the negations of the operations. When we replace the expression '$a \lor b$' in (20) by its negation, the expression '$c-b$' will assume the negations of the truth-values it had before; we therefore infer that the negation of disjunction does not have a reversed operation. The same holds for the negation of conjunction and implication. Only the negation of equivalence, i.e., the exclusive 'or', has a reversed operation, which is identical with it; that is, the exclusive 'or' is its own converse. This is shown by the tautology

$$[(a \land b) \equiv c] \equiv [a \equiv (c \land b)] \tag{22}$$

Furthermore, it is easily seen that the negation is its own converse, since we have the tautology

$$(\bar{a} \equiv c) \equiv (a \equiv \bar{c}) \tag{23}$$

which corresponds to (20).

§ 11. Derivations

One of the advantages of the symbolic technique is that it allows us to manipulate logical formulas like mathematical formulas. The laws of *commutativity*, *associativity*, and *distributivity* supply the reason that these manipulations closely resemble algebraic transformations. Thus formulas (4a and 4c, § 8) express a technique known as 'multiplying out' in algebra, when the 'or' is interpreted as addition and the 'and' as multiplication.

An important difference between the algebra of logic and the algebra of numbers, however, is that for logical algebra a second distributive law holds, stated in formula (4b, § 8), whose analogue in the algebra of numbers does not hold. In combination with the other rules this law leads to a certain duality of the 'or' and the 'and'. Certain formulas, such as (4a and 4b, 4c and 4d, 5c and 5d, § 8), are dual to each other; i.e., one results from the other if we substitute 'or' for 'and', and 'and' for 'or'.

This duality is also expressed in the *rules of de Morgan*, for nulated in (5a and 5b, § 8). These rules state that we can break a negation line over two propositions if we replace the 'and' by the 'or' and the 'or' by the 'and'. These rules may also be called *rules for the division of the scope of a negation*.

By means of the algebraic technique we can easily derive many formulas of the list on page 38 from others given there. Thus (1e, § 8) follows from (1f, § 8) if we apply (5a and 1d, § 8). Starting from (7a, § 8) we can derive (7b, § 8) as follows:

$$(a \equiv b) \equiv (a \supset b).(b \supset a) \equiv (\bar{a} \vee b).(\bar{b} \vee a) \equiv \bar{a}.\bar{b} \vee b.\bar{b} \vee \bar{a}.a \vee a.b$$
$$\equiv a.b \vee \bar{a}.\bar{b} \tag{1}$$

Here we have applied, successively, formulas (6a, 4c, 5a, § 8). We see that the last-mentioned formula plays an important role in that it gives us the means of shortening a formula. A similar role is played by (5c, § 8).

Another example is given by the derivation of (6c, § 8 from 6a, § 8):

$$a \supset b \equiv \bar{a} \vee b \equiv b \vee \bar{a} \equiv \bar{b} \supset \bar{a} \tag{2}$$

For the last step we have applied (6a, § 8) in the reverse sense, putting '*b*' for '*a*' and '*ā*' for '*b*'.

The procedure followed on the last step is called a *substitution*. We *substituted* '*b*' for '*a*' and '*ā*' for '*b*'. It is clear that such substitutions are permissible because, in our formulas, the letters '*a*', '*b*', etc., stand for any proposition; so we will be allowed to put, in particular, '*ā*' for '*a*', or for '*b*', etc. We can also use more complicated substitutions. A more elaborate treatment of substitutions is presented in § 12.

Each of the equivalences in (2) is a tautology, including the equivalence between the first and the last term, which holds because of the transitivity of the equivalence operation. The terms between which the equivalence sign is placed, however, are not tautologies, but synthetic expressions. We say that these expressions are transformed by the equivalences and call a derivation of the form (2) a *transformation*. This name is used to express the idea that the meaning of the synthetic expressions in (2) is the same, owing to the tautological nature of the equivalence.

Also the equivalences in (1) express a transformation. Here the first equivalence '$a \equiv b$' represents a synthetic expression; the other equivalences are tautological. The synthetic equivalence '$a \equiv b$' is here transformed into several other synthetic expressions, ultimately into the expression '$a.b \lor \bar{a}.\bar{b}$'. As a further example of a transformation let us start from the synthetic expression

$$\overline{a.(b \supset c)} \tag{3}$$

Applying (5a, § 8) we obtain

$$\bar{a} \lor \overline{b \supset c} \tag{4}$$

Using (6b, § 8), we transform (4) into

$$\bar{a} \lor b.\bar{c} \tag{5}$$

By means of (4b, § 8), this formula can be transformed into

$$(\bar{a} \lor b).(\bar{a} \lor \bar{c}) \tag{6}$$

All these expressions have the same meaning.

It should be noticed that in transformations as described some care must be taken with respect to parentheses. In the transition from (3) to (4) we have omitted the parentheses because the negation line in '$\overline{b \supset c}$' binds like parentheses. On the other hand, sometimes parentheses must be added. Thus in going from

$$a.\overline{b.c} \tag{7}$$

to

$$a.(\bar{b} \lor \bar{c}) \tag{8}$$

we must add parentheses in order to indicate the unit marked in (7) by the long negation line.

Often it is useful to transform a formula into its *simplest form*. We say that a formula has its *simplest form* if it contains *no parentheses* and only the operations '*and*', the *inclusive* '*or*', and *shortest negation lines*, i.e., negation lines extending only over elementary propositions. In our last transformation, expression (5) has this form; in (1), this form is given by the last expression.

The technical name of the simplest form is *disjunctive normal form.* This name expresses the fact that the formula in this form is written as a disjunction. Another important form is given by the *conjunctive normal form*, in which the major operations are conjunctions. Because of our rules about the binding force of symbols this form cannot be written without parentheses. As in the disjunctive normal form, the only operations used in the conjunctive normal form are conjunctions, disjunctions, and shortest negation lines; the terms of the disjunctions, i.e., the terms inside the parentheses, then must be so written that they consist of individual elementary propositions. This aim can be reached by further applications of the second distributive law (4b, § 8). An example of this form is (6). Every formula, synthetic or analytic, can be transformed into either of the two normal forms.

The normal forms are used when two different expressions are compared in order to determine whether they are tautologically equivalent. Thus the two expressions

$$\overline{(a \supset \bar{b}).(c \supset a)} \quad \text{and} \quad (b \supset \bar{a}) \supset \overline{c \supset a} \qquad (9)$$

can be shown to be tautologically equivalent when we transform them into the simplest form

$$a.b \lor c.\bar{a} \qquad (10)$$

This transformation is achieved by repeated use of formulas (5a, 6b, 6a, 1d, § 8). In order to compare the two formulas, however, we must take care that the normal form is constructed as short as possible. If it is not, it can be shortened by the use of the relations (5c and 5d, § 8). Thus the disjunctive normal form

$$a.b \lor a.\bar{b} \lor c \qquad (11)$$

can be shortened by the use of (4a, § 8), to the form

$$a.(b \lor \bar{b}) \lor c \qquad (12)$$

which can be replaced by

$$a \lor c \qquad (13)$$

Similarly, the conjunctive normal form

$$(a \lor b).(a \lor \bar{b}).c \qquad (14)$$

can be shortened to the form

$$a.c$$

when we use (4b, § 8) in the reverse direction for the two parentheses and then apply (5d, § 8). Formulas (5c and 5d, § 8), which play an important part in the shortening of expressions, hold also in a somewhat generalized form. Any tautology occurring as a factor can be canceled; thus '$a.(b.c \supset b)$' is the same as 'a'. Similarly, any contradiction occurring as a term of a disjunction can be canceled; thus '$a \lor b.\bar{b}.c$' is the same as 'a'.

The disjunctive normal form is used, furthermore, in order to find out whether a formula is a tautology. When two terms of this form together constitute a tautology '$a \lor \bar{a}$', the whole formula must be a tautology. This follows because the combination '$a \lor \bar{a}$' is sufficient to make the whole disjunction always true. Thus the formula

$$a \lor b.\bar{c} \lor d.a \lor \bar{a} \qquad (15)$$

is a tautology.

Similarly, a contradiction is recognizable from the fact that two of the factors of the conjunctive normal form together constitute a contradiction '$a.\bar{a}$'. Thus the formula

$$a.(b \lor \bar{c}).(d \lor \bar{a}).\bar{a}.c \qquad (16)$$

is a contradiction.

Both conditions are *sufficient* conditions, respectively, for tautological and contradictory character; i.e., when these conditions hold the formula certainly has the respective character. In order to make them *necessary conditions*, i.e., conditions which allow us to state that the formula does *not* have the character when the respective condition is *not* fulfilled, we must first reduce the normal form to a shorter form. Thus the formula

$$a.b \lor \bar{a}.b \lor a.\bar{b} \lor \bar{a}.\bar{b} \lor c \qquad (17)$$

is a tautology although it does not include the term '$a \lor \bar{a}$' directly. By the use of (4c, § 8) it can be shortened, however, to the form

$$(a \lor \bar{a}).(b \lor \bar{b}) \lor c \qquad (18)$$

Here we can cancel '$b \lor \bar{b}$' according to (5c, § 8). The formula then satisfies the criterion stated.

The example (17) shows that we can state our criterion also as follows: a disjunctive normal form is a tautology if some of its terms together constitute a tautology. In this version the criterion is sufficient and necessary. In (17) the tautology used is given by the first four terms.

On the other hand, in addition to canceling '$b \lor \bar{b}$' in (18) we can cancel even the term 'c'. This is possible because it represents a transition to an equivalent expression, namely, from an always-true formula to another always-true formula. The formula then assumes the form

$$a \lor \bar{a} \tag{19}$$

This is its shortest form. It is clear that we can do the same for every tautology. All tautologies have the same shortest disjunctive normal form, namely, (19). This result illustrates the fact that all tautologies mean the same; namely, they mean nothing. We said before that we distinguish this case, which we call *emptiness*, from the case of *meaninglessness*. To mean nothing is not the same as to have no meaning. Similarly, every contradiction is reducible to the form '$a.\bar{a}$'.

Though the shortest normal form has the advantages explained, another way of writing a normal form must also be particularly mentioned. When we analyze a formula by case analysis we find that for certain combinations of the truth-values of the elementary propositions the formula is true. We can therefore write the formula as a disjunction of its T-cases. Thus the formula

$$a \supset b \tag{20}$$

can be written in the form

$$a.b \lor \bar{a}.b \lor \bar{a}.\bar{b} \tag{21}$$

We shall call this an *expansion in T-cases*. This way of writing, too, represents a disjunctive normal form, though it need not be the shortest normal form; thus (21) can be shortened by the use of (4a, 5c, 5e, § 8) to '$\bar{a} \lor b$'. When the formula contains more than two variables and several operations we can apply the same procedure by going through all the possible truth-values of its elementary propositions. Thus the formula

$$a \equiv b \lor c \tag{22}$$

can be written in the form

$$a.b.c \lor a.b.\bar{c} \lor a.\bar{b}.c \lor \bar{a}.\bar{b}.\bar{c} \tag{23}$$

This is the expansion of (22) in its *elementary T-cases*. We can also expand the formula in its *major T-cases*, i.e., regarding the major terms as units. We then arrive at

$$a.(b \lor c) \lor \bar{a}.\overline{b \lor c} \tag{24}$$

This is the expansion of (22) in its major T-cases. This form, how-ever, is not a normal form; only the expansion in elementary T-cases is always a disjunctive normal form.

The transformations constitute only a special kind of derivation. Instead of equivalent expressions, we can also derive mere implica-tions. We then use one-sided implications within the derivations. For instance, when we start from the expression

$$\overline{a \vee b} \qquad (25)$$

we can first transform it by means of (5b, § 8) into the equivalent expression

$$\bar{a}.\bar{b} \qquad (26)$$

and then, by means of the one-sided implication (8b, § 8), derive the expression

$$\bar{a} \qquad (27)$$

Here we make use of the fact that, since (26) implies (27), (27) must be true if (26) is true. We say that we go from (26) to (27) by means of an *inference;* (27) is the *conclusion* of the inference. Since (26) is equivalent to (25), we thus have also shown that, if (25) is true, (27) must be true. We therefore say that we have derived (27) from (25). This does not mean that (27) will be equivalent to (25); on the contrary, (27) says less, and we have here only a one-sided implication

$$\overline{a \vee b} \supset \bar{a} \qquad (28)$$

On the other hand, the term 'derivable' does not require that the implication be one-sided. A formula 'q' is said to be *derivable* from another formula 'p' if we can show that, in case 'p' is true, 'q' must be true. Since this relation holds also between equivalent formulas, the term 'derivable from' includes both mere conse-quences and equivalent expressions. In transformations the deriva-tion is *reversible;* i.e., we then can derive not only 'q' from 'p' but also 'p' from 'q'. If the derived formula is a mere consequence, the derivation is *irreversible*. Similarly, the term 'inference' is not restricted to the derivation of a mere consequence; we speak of an inference also when we go from one formula to an equivalent formula, as in the transition from (25) to (26).

In our example, the implication (28) is a tautology. Derivability, therefore, is here translatable into the existence of a tautological implication. In the calculus of propositions, this holds for all practically important cases. We shall later inquire into the conditions of this relation (§ 14).

In order to give a more complicated example, let us show that from the expression

$$(a \supset b).(c \supset d) \tag{29}$$

we can derive the expression

$$a.c \supset b.d \tag{30}$$

This relation is stated in the one-sided implication (8j, § 8); but we shall give here a demonstration by deriving it from other formulas. We first transform (29) by means of (6a, § 8) into the form

$$(\bar{a} \vee b).(\bar{c} \vee d) \tag{31}$$

Applying twofold distribution, i.e., (4c, § 8), we write this

$$\bar{a}.\bar{c} \vee b.\bar{c} \vee \bar{a}.d \vee b.d \tag{32}$$

By means of (6a, § 8) this can be transformed into

$$\overline{\bar{a}.\bar{c} \vee b.\bar{c} \vee \bar{a}.d} \supset b.d \tag{33}$$

With the use of (5b, 5a, and 1d, § 8) the implicans can be transformed so that (33) assumes the form

$$(a \vee c).(\bar{b} \vee c).(a \vee \bar{d}) \supset b.d \tag{34}$$

By twice applying the rule (4c, § 8) of twofold distribution to the implicans we obtain

$$a.\bar{b} \vee a.c \vee a.c.\bar{b} \vee a.c \vee a.\bar{b}.\bar{d} \vee a.c.\bar{d} \vee c.\bar{b}.\bar{d} \vee c.\bar{d} \supset b.d \tag{35}$$

The long implicans, which may be abbreviated by 'p', contains the term '$a.c$' twice. Because of (1b, § 8) one of these terms can be canceled. Considering '$a.c$' as the 'a' of formula (8a, § 8) and the remainder of the implicans 'p' as the 'b' of this formula, we have

$$a.c \supset p \tag{36}$$

Since (35) can be written

$$p \supset b.d \tag{37}$$

we obtain, with the transitivity of implication formulated in (8l, § 8), the expression (30).

The practical use of derivations consists in the fact that they allow us to state the meaning of a propositional combination in a simpler and clearer form, and to state the consequences of a considered assumption. When we are told, for instance: 'it will rain tomorrow, and there will be an earthquake or it will rain', we do not easily realize that this sentence actually says nothing about the earthquake and is equivalent to the statement: 'it will rain tomorrow'. Formula (4e, § 8) shows us this equivalence. The most important use of derivations is made in the mathematical sciences. There the implications of a given assumption are constructed by means of a highly developed technique, and it would be impossible to realize the reach of the assumption if we were dependent on the power of unformalized thinking alone. [Ex.]

§ 12. The rule of substitution

We have so far treated the method of derivation in an informal way. That is, we have handled the method without clearly stating the rules we followed in doing so. We now turn to a formulation of these rules, thus setting up the *rules of derivation*.

Let us first consider a distinction concerning the symbols used in our formulas. When we compare a tautological expression, such as the formula

$$\overline{a \vee b} \equiv \bar{a}.\bar{b} \qquad (1)$$

with a synthetic expression like

$$a.(b \supset c) \qquad (2)$$

we notice a remarkable difference. The second expression is true only for certain special meanings of the propositions 'a', 'b', 'c'; a tautological expression like (1) is true for all meanings of these propositions. Therefore a tautological expression can be universally asserted; the propositional symbols in the formula then have the character of *free variables*. If the synthetic expression (2) is asserted, the propositional symbols in it have the character of *constants;* i.e., they stand as abbreviations for special propositions. An expression like (2), therefore, cannot be asserted before special values have been assigned to the variables contained in it. Variables subject to this restriction are called *uninterpreted constants*. We operate with them as though they were specialized and may therefore also call them *quasi-specialized variables*. Though we do not *assert* an expression containing such variables unless a suitable specialization of the variables has been given, we may at least *consider* the expression — for instance, with the intention of studying certain consequences depending on it.

It should be noticed that the difference between these two kinds of variables is not inherent in the nature of the variable itself but depends on the expression within which the variable is used. Thus, with respect to the expression '$a \vee \bar{a}$', the variable 'a' is a free variable; but with respect to the first part of the expression, namely, 'a', the same variable 'a' is an uninterpreted constant, or a quasi-specialized variable. In the list of formulas given on page 38, likewise, the variable 'a' is a free variable with respect to each individual formula, whereas it is an uninterpreted constant with respect to the first major term of the formula. When we speak of free or quasi-specialized variables, therefore, we should always add the phrase 'relative to . . .', indicating the expression relative to which the term is used.

There are also expressions of a mixed nature, containing both uninterpreted constants and free variables. When such formulas result from tautologies by a partial specialization, they are not essentially different from the original formulas. Thus when we put in (1) for 'a' the special sentence 'a_1', for instance the sentence 'the sun is hot', the sentence remains tautologous because *if* we regard 'a_1' as a variable the sentence '$\overline{a_1 \vee b} \equiv \bar{a}_1.\bar{b}$' holds for all values of this variable. Thus, if we put for 'a_1' the sentence 'the sun is cold', the formula will also be true. When, in addition, 'b' is specialized into 'b_1', the sentence will still be a tautology. Our term 'tautology', therefore, includes formulas in free variables, formulas in constants, and mixed formulas. On the other hand, these considerations show that it is always permissible to deal with a free variable as though it were an uninterpreted constant. The reverse relation, of course, does not hold.

There are other formulas of the mixed sort which do not hold for all values of the constants but only for all values of the free variables. Such expressions are synthetic. An example is given by the formula

$$a \vee \bar{a} \supset b_1 \tag{3}$$

which is meant to hold for all values of 'a', whereas 'b_1' is a constant. Obviously the formula will hold only if 'b_1' is true. Thus the formula will be true if we put for 'b_1' the sentence 'the sun is hot', while 'a' is left as a free variable. The formula will be false if we put for 'b_1' the sentence 'the sun is cold'.

We shall see later that such formulas as (3) change their meanings according to the way they are used, and we shall therefore call them *indefinite expressions*. We need not discuss formulas of this sort here because we shall show that they can be dispensed with (cf. § 42). Thus formula (3) can be replaced by the formula 'b_1' since it says no more than that 'b_1' is true. All the formulas used by us in derivations, and referred to in the following theory of derivation, will therefore be either tautologies or synthetic formulas containing only constants. Such formulas may be called *definite expressions*.

In (3) we have indicated by means of a subscript that 'b_1' is a quasi-specialized variable. Some such device is necessary because we have here a formula which contains both free and quasi-specialized variables (with respect to this formula). In a formula like (2) such an indication is not necessary; here it is clear from the synthetic character of the expression that all variables contained in it are quasi-specialized. Since we shall deal for the present only with definite expressions, we need not use subscripts. If a synthetic definite expression occurs, it is understood that we do not assert it unless a suitable specialization of the variables has been given.

Let us first consider a form of derivation which in its application to definite formulas refers only to tautologies since it concerns free variables. This procedure is based on the rule of substitution. It is in the nature of free variables that when they occur in an asserted expression we can substitute any other propositional expression for them. In § 11 we mentioned the procedure of substitution with reference to (2, § 11). Another kind of substitution is performed when we put a constant for a variable, for instance when we go from '$a \vee \bar{a}$' to the sentence 'it will rain tomorrow or it will not rain tomorrow'. We can also substitute other propositional variables, or compound propositional expressions, for free variables. The only condition for all substitutions is the *coupling condition* mentioned in § 3: the substitution must be made simultaneously in all places where the original symbol occurs. Thus in (6a, § 8) we can substitute '$c \vee d.e$' for 'a' and derive the formula

$$c \vee d.e \supset b \equiv \overline{c \vee d.e} \vee b \qquad (4)$$

We may even use in these substitutions symbols that occur in other places in the formula. Thus in (4c, § 8) we may put *'c'* for *'a'* and derive the formula

$$(c \vee b).(c \vee d) \equiv c.c \vee b.c \vee c.d \vee b.d \qquad (5)$$

which can be further simplified by putting *'c'* for *'c.c'*, according to (1c, § 8). It would be false, however, to substitute *'c'* for *'a'* only on the left-hand side and to leave the *'a'* on the right-hand side. Furthermore, it is not permissible to substitute one letter for a group of letters; only the reverse procedure is correct. Thus, if we put *'b'* for the combination *'a.ā'* in (1f, § 8), we obtain *'b̄'*; but this statement cannot be generally asserted, since it will be false for many propositions *'b'*.

We formulate this procedure by the following rule, using the term 'propositional expression' defined in § 6:

Rule of substitution. It is permissible to assert a formula resulting from a tautology when a propositional expression, which is arbitrarily chosen, is substituted in the place of any free elementary propositional variable, provided that this substitution is made in all places where the original variable occurs.

Like all rules, this rule is a *directive;* it states a permission as to the introduction of new formulas. It therefore requires a *justification*, i.e., an explanation of why it serves the aim of the manipulations to be performed on propositional expressions. Since this aim is to find *true* formulas, we must show that the rule leads to the construction of true propositional expressions.

That, in fact, the rule has the required property is demonstrable as follows. If a variable expression is substituted, the resulting formula must be true for all values of the substituted expression. This is clear because the substituted expression can have only the two truth-values T and F, like the elementary proposition which it replaces; since the original formula is true for both these cases, so must be the formula obtained by the substitution. If a special proposition is substituted, the resulting formula cannot be false, since it will be true whether the proposition substituted is true or false. If there are other free propositional variables left, the formula will be always true in these variables, and then further substitutions for these variables can be made. If no other propositional variables are left, the resulting formula will be a special proposition which is true.

It is evident, therefore, that, if the original formula was a tautology, the resulting formula will also be tautologous, since it must be true for all values of its elementary symbols. The substitution cannot have changed this property because it was made in a place for which both the values 'true' and 'false' will lead to a true-value of the formula. This result may be formulated as a *metatheorem*, i.e., a theorem to be stated in the metalanguage. A metatheorem is not a directive, but a cognitive statement which can be proved to be true. Synonymously with the word 'theorem' we shall also use the word 'law'.

Metatheorem 1. (*Law of substitution.*) When a substitution is made in a tautology, the resulting formula is a tautology.

This formulation makes clear the difference between the *rule* of substitution and the *law* of substitution.

§ 13. The rule of replacement

Whereas the procedure of substitution applies only to free elementary variables, there is a second method of introducing new expressions, by which all sorts of propositional expressions, including compound expressions and constants, can be replaced. This procedure of *replacement*, as we shall call it, is dependent, however, on the condition that the new expression is equivalent to the original expression. We used this procedure in the transition from (29, § 11) to (31, § 11). It has the advantage over the procedure of substitution that a replacement is not bound to a coupling condition; i.e., it need not be made in all places where the original term occurs. A further advantage is that a replacement can also be applied to constants and therefore can be used for definite synthetic formulas. We formulate this procedure as follows:

Rule of replacement. It is permissible to assert a formula resulting from any true formula when a propositional expression, which is equivalent to a particular propositional expression forming a part of the formula, is put in the place of the latter expression.

The justification of this rule is given by the following metatheorem:

Metatheorem 2. (*Law of replacement.*) When a replacement is made in a true formula, the resulting formula is true.

To prove this theorem we must first present some preliminary considerations concerning operations with equivalences. If we have an equivalence

$$p \equiv q \tag{1}$$

where 'p' and 'q' are abbreviations standing for different combinations of elementary propositional variables 'a', 'b', 'c' . . ., it is permissible to add identical terms on both sides of the equivalence by means of any of the propositional operations; the resulting formula then will also be an equivalence. If 'r' is a propositional variable, this rule can be expressed in the following formulas:

$$(p \equiv q) \supset (\bar{p} \equiv \bar{q}) \tag{2}$$
$$(p \equiv q) \supset (p \vee r \equiv q \vee r) \tag{3}$$
$$(p \equiv q) \supset (p.r \equiv q.r) \tag{4}$$
$$(p \equiv q) \supset (p \supset r \equiv q \supset r) \tag{5}$$
$$(p \equiv q) \supset [(p \equiv r) \equiv (q \equiv r)] \tag{6}$$

These formulas are easily verified as tautologies by case analysis; in this procedure the expressions 'p' and 'q' can be dealt with as elementary propositional variables.

Formulas (2)–(6) furnish a proof of metatheorem 2 on the basis of the following considerations. When a propositional expression 'p' occurring within a true formula is replaced by an equivalent expression 'q', we consider first the subunit consisting of 'p' and the term connected with 'p' by the propositional operation immediately following, for instance a subunit of the form '$p \vee r$'. According to one of the formulas (2)–(6) this subunit is equivalent to the corresponding subunit resulting from the replacement, for instance '$q \vee r$'. We now regard the original subunit as representing a new expression 'p', and similarly the new subunit as an expression 'q'. Since we know that the condition '$p \equiv q$' again holds, we can repeat the previous inference for the next higher subunit; and so on. Following the step structure of the formula we finally prove the equivalence to hold between the whole original formula and the new formula resulting from the replacement. This concludes our proof: since the new formula is equivalent to the original one, it must be true if the original formula is true.

Let us illustrate the rule of replacement by an example. Assume the synthetic expression

$$(a \supset b).(c \vee d) \vee (a \supset b).e \tag{7}$$

to be known as true; the letters 'a', 'b', etc., occurring in it, represent constants. Let a further relation introducing a new constant 'f'

$$a \supset b \equiv f \tag{8}$$

also be known as true. Then we can replace the first term '$a \supset b$' in (7) by 'f', leaving the other term '$a \supset b$' unchanged. The formula thus resulting

$$f.(c \vee d) \vee (a \supset b).e \tag{9}$$

will also be true.

The procedure of replacement acquires a particular interest when the equivalence used for the replacement is tautological, i.e., when formula (1) is a tautology. This requires, of course, that the expressions 'p' and 'q' be composed of elementary propositional variables, 'a', 'b', etc., or such propositions, so that (1) is a tautology in these variables. Since formulas (2)–(6) are tautologies, the implicates occurring in them must also be tautologies; that is, they must always be true in the elementary variables 'a', 'b', etc., of which 'p' and 'q' are composed, and in 'r'. This follows because, if it were not so, there would be a case where the implicans was true and the implicate false, a case excluded by the fact that relations (2)–(6) are tautologies. Following once more the step structure of the formula as was done in the preceding proof, we thus derive the theorem:

Metatheorem 3. (Law of tautological replacement.) When the rule of replacement is applied under the condition that the replaced term and the replacing term are tautologically equivalent, the resulting formula is tautologically equivalent to the original formula.

Continuing our example (7) we can illustrate this theorem by substituting for the first '$a \supset b$' in (7) the tautologically equivalent expression '$\bar{a} \vee b$'. We thus arrive at the formula

$$(\bar{a} \vee b).(c \vee d) \vee (a \supset b).e \tag{10}$$

It is easily seen that this synthetic expression is tautologically equivalent to (7). We have made replacements of this sort, and thus used metatheorem 3, in the transformations explained in § 11.

Since a formula that is tautologically equivalent to a tautology must also be a tautology, namely, must be always true in its elementary variables (otherwise it could not be equivalent to a tautology), we immediately derive from metatheorem 3 the theorem:

Metatheorem 4. When the rule of replacement is applied to a tautology under the condition that the replaced term and the replacing term are tautologically equivalent, the resulting formula is also a tautology.

The condition of tautological equivalence used in metatheorems 3 and 4 is always satisfied if the replaced and the replacing term contain no constants. When an equivalence between such terms is asserted it must hold in all free variables; therefore it will be tautological. This is the reason that we can make unqualified replacements as long as we deal exclusively with formulas containing no constants, i.e., in the derivation of tautologies containing only free variables from other such tautologies. The conditions of metatheorem 4 are then always satisfied, and this theorem, therefore, will apply to the results obtained.

We must now add some general remarks on the rule of replacement. In the proof given for metatheorem 2 we did not apply the rule of replacement; but we did apply the rule of substitution, since the use of tautologies (2)–(6) requires that the particular terms forming subunits be substituted in these tautologies. Furthermore, we used inferences; i.e., we applied the rule of inference which will be formulated later (cf. § 14). Since the proof describes a procedure that can be executed with the formulas under consideration, it is clear that the rule of replacement can be dispensed with; instead, we can always restrict ourselves to the use of the rule of substitution in combination with inferences. The rule of replacement, therefore, represents only a short cut by which a complicated procedure, feasible without it, is abbreviated. Rules of this sort may be called *secondary rules*. In contradistinction we shall use the name *fundamental rules* for indispensable rules, like the rule of substitution. The method of derivation is greatly simplified by the use of secondary rules, and we shall later introduce more rules of this sort.

Let us illustrate these considerations by example (7). The transition to (10) can be divided into the following steps, which represent substitutions and inferences, but not replacements. We start from the equivalence

$$a \supset b \equiv \bar{a} \vee b \tag{11}$$

§ 13. The Rule of Replacement 63

We now use (4), substituting there '$a \supset b$' for 'p', '$\bar{a} \vee b$' for 'q', and '$c \vee d$' for 'r'. We thus obtain

$$(a \supset b).(c \vee d) \equiv (\bar{a} \vee b).(c \vee d) \tag{12}$$

On our next step we use (3), considering the left-hand side of (12) as 'p', the right-hand side as 'q', and the expression '$(a \supset b).e$' as 'r'. We thus obtain

$$(a \supset b).(c \vee d) \vee (a \supset b).e \equiv (\bar{a} \vee b).(c \vee d) \vee (a \supset b).e \tag{13}$$

The right-hand side of this relation corresponds to (10). From the equivalence we immediately infer that, if the left-hand side is true, so must be the right-hand side.

This example makes clear that the rule of replacement can be dispensed with. It is not the rule of substitution alone, however, that takes its place; in addition, inferences are required. The nature of inference is explained in § 14.

An apparent paradox may be clarified in this connection. We saw that the rule of substitution cannot be applied to synthetic formulas containing only constants, whereas the rule of replacement can be applied to such formulas. How is this statement compatible with the eliminability of the rule of replacement? The answer is that, when the rule of replacement is eliminated, we do not make substitutions in the synthetic formulas; we rather substitute such formulas, or parts of them, in tautologies. Thus (12) is derived from (11) by substituting the synthetic parts of (11) in the tautology (4).

Let us now regard a general relation holding between a given formula 'p' and another formula 'q' derived from it by a substitution or a replacement. A substitution, as far as definite formulas are concerned, can be made only in a tautology; since, according to metatheorem 1, § 12, we thus arrive at a tautology, the implication '$p \supset q$' must also be tautologous. The same result must hold when a replacement is made by the use of a tautological equivalence, according to metatheorem 3. Finally, we can show that this result can be extended to a nontautological replacement when we include in the premise 'p' the equivalence used. Thus, when we regard both (7) and (8) as premises, the formula resulting from the replacement, in our case formula (9), will then be tautologically implied by the premises. This follows because relations (2)–(6) used for the proof of metatheorem 2 are tautologies. To see this more clearly, let us replace the second equivalence in these relations by an implication; then (3), for instance, assumes the form

$$(p \equiv q) \supset (p \vee r \supset q \vee r) \tag{14}$$

According to (6d, § 8) this can be written

$$(p \equiv q).(p \vee r) \supset (q \vee r) \tag{15}$$

This, of course, is also a tautology. It shows that a derivation made by a replacement alone will always lead to a formula which is tautologically implied when the equivalence used for the replacement is included in the premises. We therefore can now state the following theorem:

Metatheorem 5. When a formula 'q' is derivable by the rules of substitution or replacement from a set 'p' of definite formulas, the formula '$p \supset q$' is a tautology.[1]

§ 14. The rule of inference

We now turn to another rule, which is also a rule of derivation, and which has frequently been applied in the preceding considerations. This is the rule of inference. In contradistinction to some other rules of inference it is also called the *fundamental rule of inference*. We formulate it as follows:

Rule of inference. (*Modus ponens.*[2]) If '$a \supset b$' is true, and 'a' is true, 'b' can be asserted.

This rule is usually indicated by the schema

$$\frac{\begin{array}{c} a \supset b \\ a \end{array}}{b} \tag{1}$$

The first two lines of (1) are called *premises;* the third line is called *conclusion.* An example is given by the propositions:

> If an electric current is passed through this coil of wire, there is a magnetic field in the environment of the coil.
>
> An electric current is passed through this coil of wire.
>
> ――――――――――――――――――――――――――
>
> There is a magnetic field in the environment of this coil. $\tag{2}$

――――――――――――――――

[1] That this theorem does not hold for indefinite formulas is shown in (9, § 42).

[2] The name 'modus ponens' was introduced in medieval logic; cf. p. 71 and p. 208.

We applied this rule, for instance, in the derivation of (27, § 11) and also in the transition from (36 and 37, § 11) to (30, § 11). If the latter inference is to be completely stated, it must be written in the form:

$$\frac{(a.c \supset p).(p \supset b.d) \supset (a.c \supset b.d)}{\underline{(a.c \supset p).(p \supset b.d)}} \qquad (3)$$
$$a.c \supset b.d$$

We see here the way in which the rule of inference works. The first line of (3) is obtained from (81, § 8) by means of the rule of substitution. The second line of (3) is derived in a more complicated way (in which the rule of inference has been used repeatedly). The third line of (3) cannot be derived from the other two lines by a substitution. It is the rule of inference which permits us to separate the implicate from the first line of (3), and to assert it independently.

This consideration makes clear the distinction between inference and implication. An implication is a statement; it is used for inferences, but it cannot take the place of an inference because an inference is a *procedure* and not a *statement*. This procedure can be described only in a rule, formulated in the metalanguage, and symbolically expressed by a schema. The inferential implication (8e, § 8), on the other hand, belongs to the object language and therefore cannot replace this schema.

It is important to realize that the rule of inference holds for every sort of implication. In (3) the implication is a tautological implication; in the example (2) we have used a connective implication which is not tautological. The rule holds, however, even if we use a merely adjunctive implication, as for instance in the example:

$$\frac{\text{If snow is white, then sugar is sweet.}}{\text{Snow is white.}} \qquad (4)$$
$$\text{Sugar is sweet.}$$

The conclusion is true, although there is no connection in the implication of the first premise, which holds only in the adjunctive sense. It can easily be shown that the so-called paradoxes (cf. § 7) of adjunctive implication cannot lead to false conclusions. If we have an adjunctive implication with a false implicans, the first line of the schema will be true; but, since the implicans cannot be asserted, the second line is not true and the conclusion cannot be derived.

If we have an adjunctive implication with a true implicate, the first line will also be true; if in addition the implicans is true, as in (4), the conclusion can be derived; but then it is true, and the inference thus furnishes a true conclusion. The rule of inference, therefore, is not subject to any restriction as to the nature of the implication used.

It is obvious, however, that the practical value of the rule of inference originates from the use of connective implications. If an implication is established merely adjunctively, by showing that the truth-values of 'a' and 'b' furnish a 'T' in the truth tables of implication, and if 'a' is true so that an inference can be drawn, this inference is no longer necessary, since the truth of 'b' was known before. Thus in order to know that the first premise of (4) is true, we must know that sugar is sweet; hence the conclusion does not tell us anything which could not be asserted before. The knowledge of the truth of the implicate is a necessary condition for the case that an inference based on a merely adjunctive implication can be made; therefore the inference is of no use. It is only in its application to connective implications that the inference has practical value. The truth of the implication is then established by means other than an examination of the truth of the implicate; therefore the assertion of the conclusion represents a statement which is new, psychologically speaking. The connective implications to be used may be either of the tautological kind or of the synthetic kind (cf. § 9 and chapter VIII).

Like all other rules, the rule of inference is a directive; it expresses a permission as to the assertion of a formula. It therefore needs a justification. Regarding, as before, the establishment of true formulas as the aim of logical manipulations, we can give this justification by summarizing the preceding considerations in the following short analysis. From the definition of implication in the truth tables I b it follows that the only case in which a 'T' occurs both in the column headed by 'a ⊃ b' and the column headed by 'a' is the first horizontal line of the table; but in this case 'b' is also true. This holds equally for adjunctive and connective implications. Therefore the rule of inference is justified because, if true premises are used, the conclusion must be true. We express this result in the theorem:

Metatheorem 6. (*Law of inference.*) If a formula is derived by the rule of inference from true premises, it is true.

The justification of the rule of inference given by this metatheorem may seem trivial because the theorem appears so obvious in view of the definition of implication. Nonetheless it is important to realize that a justification must be required for a rule. Rules are not arbitrary conventions about the use of language; they are established for the purpose of finding true statements. If it could not be proved that an inference from true premises will lead to a true conclusion, nobody would use the rule of inference in deductive logic. It is different with inductive logic. Here we have a rule of inference expressed by the rule of induction by enumeration, which states in its simplest form that if an event has repeatedly occurred, without exception, in combination with certain circumstances, we may assume that under the same circumstances it will occur again. This rule is also a directive, since it states a permission for asserting a statement. Now we cannot prove that the conclusion of an inductive inference must be true; therefore the problem of justification is here of an intricate nature and requires for its solution a reinterpretation of the objectives of knowledge.[1] On the other hand, the fact that for deductive rules the justification is so easily given has misled some logicians to deny the existence of a problem of justification. A simple consideration of the nature of rules as directives given with the intention of finding true statements shows that this attitude is not admissible.

The problem of justification cannot be evaded by the elimination of the rule and its replacement by the metatheorem. The rule is indispensable because it includes a reference to the aim of derivation; it therefore goes beyond the results of the metatheorem. Thus metatheorem 6 taken alone does not justify the rule of inference; it does so only when the aim of derivation is stated as the construction of true formulas. If we set up the aim of deriving contradictions, the rule of inference would not be a suitable instrument and therefore would not be justifiable.

All that is required of the rule of inference is that it leads from

[1] Cf. the author's *Experience and Prediction*, University of Chicago Press, Chicago, 1938, § 39.

true statements to true statements. If false premises are used, the conclusion need not be false; it will be false in some cases, true in others. Consider for instance the inference:

> If this piece of wood is heavier than an equal
> volume of water, it will float.   (5)
>
> This piece of wood is heavier than an equal vol-
> ume of water.
> _____
> This piece of wood will float.

Here the second premise is false; the first is also false when the implication is interpreted in the connective sense. But the conclusion is true.

Whereas the propositions occurring in an inference are either *true* or *false*, the inference itself is not called true or false. Instead, we use the terms *valid* and *invalid*. An inference is valid if it furnishes a true conclusion for any set of true premises. Thus the inference expressed in schema (1) is valid; so is the inference given in the examples (2)–(5). The latter example shows that an inference can be valid although the premises are false. We shall call an inference valid even when both premises and conclusion are false, if only it is so constructed that, if the premises were true, it would yield a true conclusion. If an inference is valid and the premises are true, the inferenc is called *conclusive*.

In order to construct an invalid inference let us consider the schema

$$\begin{array}{c} a \supset b \\ \dfrac{b}{a} \end{array} \qquad (6)$$

This is invalid because 'a' can be false when both premises are true. Russell has given a witty illustration of this sort of inference:

> If pigs have wings, some winged animals are good
> to eat.   (7)
> _____
> Some winged animals are good to eat.
> Pigs have wings.

We can, of course, present other examples of this inference in which both premises and conclusion are true; an illustration is offered when we replace the word 'pigs' in (7) by the word 'turkeys'. The inference, however, remains invalid, because its form is no guarantee for the truth of a conclusion derived from true premises.[1]

The invalid inference (6) may be called a *fallacy*. The word 'fallacy' does not have an unambiguous meaning, since not every incorrect derivation is called a fallacy. We speak of a fallacy only if the invalid derivation *resembles* a valid derivation to a certain extent. The word 'resembles' as used here has, of course, a rather vague meaning. But it appears permissible to say that the schema (6) resembles the schema (1). And it is a matter of fact that the fallacy (6) is often committed. It is made, for instance, by the politician who argues: a communist will advocate price control; congressman X advocates price control, therefore congressman X is a communist.

Let us now consider some further properties of the rule of inference. First, it is clear that the schema (1) can also be used in combination with the rule of substitution; i.e., it is permissible to substitute for 'a' or 'b' any propositional expression. We have given an example in (3). The first premise here is a tautology; the second is synthetic. The conclusion is also synthetic. It will always be so when synthetic premises are *essentially used*, i.e., when they cannot be eliminated in the inference. On the other hand, it is always permissible to add tautologies to synthetic premises because tautologies are always true. The addition of tautologies supplies the source from which springs the far-reaching power of deductive methods — a power that manifests itself in the widespread application of such methods in the empirical sciences.

Turning to premises all of which are tautologies we can derive a further result for schema (1). By definition, such premises are true for all values of the elementary variables contained in 'a' and 'b'. But then 'b', taken separately, must also be true for all values of its elementary variables; otherwise a combination of these variables

---

[1] In the theory of inductive knowledge the opinion has been stated that the schema (6), although it does not necessarily lead to a *true* conclusion, can at least make its conclusion *probable*, or confer to it a certain *degree of confirmation*. This is a misinterpretation of the procedure of indirect evidence. When inferences apparently similar to (6) are made, such as in the evidence derived for a scientific hypothesis from a verification of its implications, the actual schema of the inference is of a much more complicated structure. This structure can be made clear, and the inference justified, only when certain theorems of the calculus of probabilities are adduced. Cf. the author's *Experience and Prediction*, University of Chicago Press, Chicago, 1938, § 41.

would exist such that 'a' is true and 'b' is false, which is not compatible with the assumption that '$a \supset b$' is always true. We thus derive the theorem:

Metatheorem 7. When the premises of an inference are tautologies, the conclusion will be tautologous.

When the premises are synthetic, this result, of course, cannot be derived. But it is easy to see that then, at least, the implication between premises and conclusion is tautological. This follows because the inferential implication

$$a.(a \supset b) \supset b \qquad\qquad (8)$$

as stated in (8e, § 8) is a tautology. Since this result holds equally when the premises are tautological, we have the theorem:

Metatheorem 8. The conclusion of an inference is tautologically implied by its premises.

It is important that this theorem can be stated without any restriction as to the form of the premise, in contradistinction to metatheorem 5, § 13.

Since we have shown that all derivations, in the calculus of propositions, are reducible to the rules of substitution and inference, we can now establish the two following general theorems. The first represents a combination of metatheorem 7 with metatheorem 1, § 12; the second combines metatheorem 8 with metatheorem 5, § 13.

Metatheorem 9. (*Law of the transfer of tautological character.*) When a formula, in the calculus of propositions, is derived from tautologies alone, it is a tautology.

Metatheorem 10. (*Law of tautological derivability.*) When a formula 'q', in the calculus of propositions, is derivable from a set 'p' of definite formulas, the formula '$p \supset q$' is a tautology.

As to the definition of the term 'definite' we refer to p. 57. The significance of this theorem is discussed later.[1]

§ 15. Secondary rules of inference

Like the rule of substitution, the fundamental rule of inference can be supplemented by similar rules which, in principle, can be

---

[1] Cf. § 27 and § 32, and some further remarks on this theorem in § 16.

dispensed with, but which represent short cuts simplifying derivations. Following the notation introduced in § 13 we shall speak here of *secondary rules of inference*.[1]

Secondary rules of inference were developed in traditional logic. The fundamental rule of inference was called there *modus ponens*, i.e., mood of affirming. A similar rule was called *modus tollens*, i.e., mood of denying. These rules were also called *hypothetical syllogisms;* but this notation is misleading since the inference used here is not a syllogism. We shall deal with the syllogism, which is a rather special kind of inference, in § 36. The modus tollens is given by the following rule.

Rule of negative inference. (Modus tollens.) If '$a \supset b$' is true, and 'b' is false, then '$\bar{a}$' can be asserted.

The corresponding schema is:

$$\frac{\begin{array}{c} a \supset b \\ \bar{b} \end{array}}{\bar{a}} \tag{1}$$

Using example (2, § 14) we can illustrate this as follows:

$$\frac{\begin{array}{l} \text{If an electric current is passed through this coil} \\ \text{of wire, there is a magnetic field in the environ-} \\ \text{ment of the coil.} \\ \text{There is no magnetic field in the environment of} \\ \text{the coil.} \end{array}}{\begin{array}{l} \text{There is no electric current passing through this} \\ \text{coil of wire.} \end{array}} \tag{2}$$

It is easy to show that the rule of negative inference is eliminable and can be replaced by the fundamental rule of inference. From (6c, § 8), we have the tautological equivalence

$$a \supset b \equiv \bar{b} \supset \bar{a} \tag{3}$$

Therefore (1) can be written

$$\frac{\begin{array}{c} \bar{b} \supset \bar{a} \\ \bar{b} \end{array}}{\bar{a}} \tag{4}$$

---

[1] The selection of one of these rules as fundamental is arbitrary; either of the rules (1) or (11) might be used equally well for this purpose. The other rules then are derivable from it. The rule given in § 14 seems to be the most convenient one for use as the fundamental rule.

II. The Calculus of Propositions

This is the same form as $(1, \S\ 14)$, since we can substitute there any sort of proposition for 'a' and 'b'.

There are two further forms of secondary rules of inference, in which a disjunction is used instead of an implication.

First rule of disjunctive inference. (*Modus ponendo tollens.*) If '$a \wedge b$' is true, and 'a' is true, then '$\bar{b}$' can be asserted.

The schema is

$$\frac{\begin{array}{c} a \wedge b \\ a \end{array}}{\bar{b}} \qquad (5)$$

An example is given by the inference:

$$\frac{\begin{array}{l} \text{This metal is either sodium or potassium.} \\ \text{This metal is sodium.} \end{array}}{\text{This metal is not potassium.}} \qquad (6)$$

It is clear that this inference is restricted to the use of the exclusive 'or'. We can show, as before, that this rule of inference can be reduced to the fundamental rule of inference. From $(2, \S\ 10)$ we have with $(6a, \S\ 8)$

$$a \wedge b \equiv (a \vee b).(\bar{a} \vee \bar{b}) \qquad (7)$$
$$\equiv (\bar{a} \supset b).(a \supset \bar{b})$$

Using $(8b, \S\ 8)$ we derive

$$(a \wedge b) \supset (a \supset \bar{b}) \qquad (8)$$

We now perform the inference (5) in two steps, each of which is made by the fundamental rule of inference:

$$(a \wedge b) \supset (a \supset \bar{b}) \qquad (9)$$
$$\frac{a \wedge b}{a \supset \bar{b}}$$
$$\frac{\begin{array}{c} a \supset \bar{b} \\ a \end{array}}{\bar{b}} \qquad (10)$$

The schema (9) can therefore be considered as the contraction of two inferences, in terms of the fundamental rule, into one inference.

Second rule of disjunctive inference. (*Modus tollendo ponens.*) If '$a \vee b$' is true, and 'a' is false, then 'b' can be asserted.

The schema is

$$\frac{\begin{array}{c} a \vee b \\ \bar{a} \end{array}}{b} \qquad (11)$$

Here it is not necessary, though permissible, to use the exclusive 'or'. An example is given by the inference:

> The pains in the arm of the patient are caused by a heart disease or by rheumatism. (12)
>
> The cardiogram shows that there is no heart disease.
> _____
> The pains are caused by rheumatism.

This inference is valid although both causes may exist simultaneously. The reduction of this rule to the fundamental rule of inference is given when we replace the first premise by its equivalent '$\bar{a} \supset b$', as stated in (6a, § 8); we then have

$$\frac{\begin{array}{c} \bar{a} \supset b \\ \bar{a} \end{array}}{b} \qquad (13)$$

Another secondary rule of inference, which, however, was not explicitly stated in traditional logic, is given by the use of an equivalence instead of an implication. Since an equivalence means the same as two implications running in opposite directions, we can immediately formulate the rule:

Rule of equivalent inference. If '$a \equiv b$' is true, and 'a' is true, we can assert 'b'.

The schema of this rule is:

$$\frac{\begin{array}{c} a \equiv b \\ a \end{array}}{b} \qquad (14)$$

We have applied this rule frequently in the transformations presented in § 11, thus in the transition from (3, § 11) to (4, 5, and 6, § 11).

Let us add that the schemas (5), (11), and (14) are symmetrical in 'a' and 'b'; therefore the roles of 'a' and 'b' in the second premise and in the conclusion can be exchanged. This is not true for (1); nor is it true for the fundamental rule (1, § 14).

Another frequently applied rule is given by the construction of a chain of inferences. It can be stated as follows:

Chain rule of inference. If a set of implications '$a \supset b$', '$b \supset c$', ... '$p \supset q$', '$q \supset r$', is true, the implication '$a \supset r$' can be asserted; and if, in addition, 'a' is true, 'r' can be asserted.

The schemas for the two forms are as follows:

$$
\begin{array}{cc}
 & a \\
a \supset b & a \supset b \\
b \supset c & b \supset c \\
\cdots & \cdots \\
\cdots & \\
p \supset q & p \supset q \\
q \supset r & q \supset r \\
\overline{a \supset r} & \overline{r}
\end{array}
$$

(15)   (16)

This rule is easily proved to furnish true conclusions by the transitivity of implication.

An important schema is the *dilemma*, well known in traditional logic. It has the form

$$
\begin{array}{c}
a \lor b \\
a \supset c \\
\underline{b \supset d} \\
c \lor d
\end{array}
$$

(17)

This schema is derivable from (15) when we write it in the form

$$
\begin{array}{c}
\bar{c} \supset \bar{a} \\
\bar{a} \supset b \\
\underline{b \supset d} \\
\bar{c} \supset d
\end{array}
$$

(18)

The conclusion is the same as '$c \lor d$'. An example for (17) is given by the inference:

> A democracy can either grant its adversaries all democratic liberties or exclude them from the freedom of speech and of political organization.
>
> If it grants all democratic liberties to its adversaries, it will be in danger of being overthrown.   (19)
>
> If it excludes its adversaries from the freedom of speech and of political organization, it violates some of the principles of democracy.
>
> A democracy either will be in danger of being overthrown by its adversaries, or will violate some of its principles.

In the medieval art of disputation the dilemma was often used for the refutation of an opponent: it was demonstrated that the argument of the opponent was reducible to two alternatives, called the *horns of the dilemma*, each of which was equally detrimental to his position. From this application the word 'dilemma' has acquired a second meaning and is used to designate a predicament which leaves a man only a choice between two evils.

When, in particular, the propositions 'c' and 'd' in (17) are identical, we have the simplified schema:

$$\begin{array}{l} a \vee b \\ a \supset c \\ \underline{b \supset c} \\ c \end{array} \qquad (20)$$

This *simple dilemma* is often used for the proof of an assertion. Thus it can be used in the following example:

> In 1940, either a Republican or a Democrat was to be elected President of the United States.
>
> If a Democrat was elected, he would be compelled by the circumstances to lead the United States into the European war.   (21)
>
> If a Republican was elected, he too would be compelled by the circumstances to lead the United States into the European war.
>
> The United States was to enter the European war.

The dilemma acquires particular interest when the disjunction used as a premise is analytic, i.e., has the form '$a \vee \bar{a}$'. Because of its tautological character this premise can be omitted, and the schema assumes the shorter form:

$$\begin{array}{l} a \supset c \\ \underline{\bar{a} \supset d} \\ c \vee d \end{array} \qquad (22)$$

When, as in (20), the two propositions 'c' and 'd' are identical, i.e., when we have a simple dilemma, we arrive at the schema:

$$\begin{array}{l} a \supset c \\ \underline{\bar{a} \supset c} \\ c \end{array} \qquad (23)$$

This sort of inference, which may be called an *analytic dilemma*, is often used in mathematics. It may be illustrated by the example:

> If a given integer is not a prime number, it can be written as a product of prime numbers.
>
> If the integer is a prime number, it can be written (24) as a product of itself and the number 1, and therefore also as a product of prime numbers. _____
>
> Any given integer can be written as a product of prime numbers.

Another schema, which may be called *redundance of an alternative*, is as follows:

$$\frac{\begin{array}{c} a \lor b \\ b \supset a \end{array}}{a} \qquad (25)$$

When we write '$a \lor \bar{b}$' for the second premise, this schema follows from the tautology '$(a \lor b).(a \lor \bar{b}) \equiv a$'. An example is given by the inference

> Man descends from animals or from the apes.
>
> If man descends from the apes, he descends from animals. _____ (26)
>
> Man descends from animals.

Since the secondary rules of inference represent only short cuts and can be replaced by a longer derivation containing only the fundamental rule of inference and the rule of substitution, metatheorems 6–8, § 14, must hold also for these other rules of inference. Therefore the justification of the secondary rules is included in that of the fundamental rule. The general properties of the theory of derivation are not changed by the use of abbreviated methods. [Ex.]

§ 16. Remarks on the method of derivation

In the rules of substitution and inference, the secondary rules introduced in § 13 and § 15, and metatheorems 1–10, we have described the procedure of derivation and stated its essential features. It is clear that this procedure can be applied only when a set of formulas is given, since it represents a method that leads from a given formula to a new formula. Let us therefore add a word about the sets of formulas from which we depart.

When we are concerned with derivations from synthetic formulas, the initial formulas must be verified by methods other than logical. They may be verified by direct observations; or they may be established by more comprehensive methods including inductive inferences, such as are used in the construction of scientific knowledge. The verification, however achieved, will be of an *empirical* nature.

When we turn to derivations of *logical formulas* we must not use empirical premises. The only admissible initial formulas are analytic formulas, or tautologies. In order to find initial tautologies, i.e., tautologies known to be such without the use of derivations, we apply the method of *case analysis*. This method does not represent a derivation; rather it constitutes a *criterion* of tautological character. It allows us to establish by simple methods a fairly large stock of initial formulas. Since, within the calculus of propositions, case analysis can be applied to every formula, the method of derivation can be dispensed with in principle; it is used, however, because on the whole it is a much simpler method than case analysis, which in application to complicated formulas is a rather clumsy instrument.

Formulas used as initial formulas for derivations are also called *axioms*. This term applies both to synthetic and to analytic initial formulas. Derived formulas are also called *theorems*. In general, however, these two terms are used only when the context of derivation is rather comprehensive. We speak of the axioms and theorems of mechanics, of arithmetic, of logic. With reference to a short, or less important, context of derivation we speak of *premises* and *conclusions*, or *consequences*. It should be noticed that both 'axiom' and 'theorem' are relative terms, i.e., a formula is an axiom, or a theorem, for a certain derivation. Within another context, a formula that previously was an axiom may become a theorem, and vice versa.

This may be made clear by the use of derivations of tautologies. In (1, § 11) we use the formula (7a, § 8) as an axiom, and then derive (7b, § 8). This procedure can be reversed; then (7b, § 8) will be an axiom and (7a, § 8) will be a theorem. In fact, most of

the tautologies given on p. 38 can be regarded as theorems; they
can be derived from a much smaller set of formulas. We even can
choose the initial set in various ways and thus introduce various
sets of logical axioms. One of the aims of the axiomatic method is
to reduce the axioms to a *minimum set*. It has been shown that such
a set is given by the formulas:

$$a \vee a \supset a \tag{1}$$
$$a \supset a \vee b \tag{2}$$
$$a \vee b \supset b \vee a \tag{3}$$
$$(a \supset b) \supset (c \vee a \supset c \vee b) \tag{4}$$

Applying to these formulas the procedures of substitution and inference we can derive all tautologies of the calculus of propositions.
For the proof we refer the reader to the literature.[1] The tautological
character of formulas (1)–(4), of course, is easily verified by case
analysis.

As long as we are concerned with derivations from a minimum
set, the only formulas admissible as premises are the given set and,
in addition, those formulas that we have already derived from the
minimum set. It is different when we make derivations in general.
Here we are concerned only with the establishment of true formulas;
and it is therefore permissible to use for the introduction of initial
formulas every source of knowledge. In a derivation of logical
formulas we therefore may use any formula whose tautological
character has been established by case analysis. When we make
derivations from synthetic formulas, we shall be allowed to add
any tautology, since a tautology is an always-true formula. For
this reason, when we say that a synthetic formula 'q' is derivable
from a set 'p' of synthetic formulas, it is understood that this means:
'q' is derivable from 'p' in combination with a set of tautologies.
The term 'derivable', in general, is used in this wider sense.

This notation can also be used for the interpretation of metatheorem 10, § 14. Assume that a synthetic formula 'q' is derived
from a synthetic formula 'p' in combination with a set 't' of tautologies. According to metatheorem 10, the relation '$p.t \supset q$' then
is a tautology. It follows then that the relation '$p \supset q$' must also

---

[1] Cf. D. Hilbert and W. Ackermann, *Theoretische Logik*, J. Springer, Berlin, 1928,
§ 10.

be a tautology. If it were not, the formulas 'p' and 'q' must contain an elementary variable 'a' such that, for a certain truth-value of 'a', 'p' would be true and 'q' would be false. But then the same must hold for the relation '$p.t \supset q$' since the expression 't' cannot be made false by any choice of the elementary variables. It follows that metatheorem 10 holds also when the term 'derivable' is used in its wider sense. Only the synthetic axioms used for the derivation of the synthetic formula 'q' must be completely included in 'p' when we say that 'q' is derivable from 'p'.

With reference to the derivation of tautologies, the term 'derivable' in the wider sense becomes empty. If it is permissible to add tautologies to given premises, we can always add the tautology to be derived to the premises, and can therefore say that every tautology is derivable from every tautology. When we speak of the derivation of tautologies we therefore shall use the term 'derivable' in the narrower sense.

III. The Simple Calculus of Functions

§ 17. Propositional functions

In the preceding chapter, on the calculus of propositions, we regarded propositions as wholes, and the only structures considered were those of relations between different propositions. We now turn to an analysis of the inner structure of propositions, and with this to the calculus of functions. The name of this part of logic is derived from the fundamental concept of propositional function.

Consider the proposition 'Aristotle was a Greek'. It is an atomic proposition since it includes no propositional operations, and it would be symbolized, in the calculus of propositions, by one letter. If we regard its inner structure, however, we see that the proposition consists of two parts. Grammatically speaking it contains a *subject*, the word 'Aristotle', which refers to a man, and a *predicate*, the word 'Greek', which refers to a property of that man. These two parts are not of a similar logical nature but represent the general distinction between a *thing* and a *property*. The proposition tells us that the thing has this property; in order to do so it contains the phrase 'was a', which indicates that the thing-property relation holds between the objects denoted by the words 'Aristotle' and 'Greek'.

Both the thing and the property belong to the sphere of objects. The subject and the predicate constitute their correlates in the sphere of signs. The plurality of things and properties is portrayed by the plurality of subjects and predicates. In the sphere of signs, however, we can now introduce variables, i.e., signs which are not restricted to denoting one individual thing, or property, but which can be used to denote any thing or any property. As variables denoting things we shall use letters 'x', 'y', 'z', etc.; as predicate variables, letters 'f', 'g', etc.

The introduction of variables suggests the analogy to a mathematical function. For the form of the sentence cited, therefore, the symbolization

$$f(x) \qquad (1)$$

has now been generally accepted. The property is here conceived as a function of the thing, which in turn is regarded as the argument of the function. The same function can have various arguments; thus Plato is another argument of the function Greek. On the other hand, not all things are arguments of the function Greek. Thus David Hume is not an argument of this function; i.e., David Hume was not a Greek.

Argument and property, in combination, determine a situation. Thus it is a specific situation that Aristotle is a Greek. When the argument varies while the function remains constant, we obtain a set of different situations. It is this variation of the situation with the argument which constitutes the root of the name 'function', in analogy to mathematical functions. Whereas mathematical functions, however, usually vary continuously with the argument, we have no such continuity for logical functions. The term 'function', therefore, is used here in a somewhat wider sense. Incidentally, such a wider sense is frequently also used in mathematics.[1]

The functional conception of properties makes necessary a revision of the notation referring to the sphere of signs. The signs which denote function and argument will be called *function-name* and *argument-name* or *function-sign* and *argument-sign*. Synonymously with the term 'function-name' we use the traditional term 'predicate'. The term 'subject', which in traditional grammar is used synonymously with our term 'argument-name', has been abandoned in symbolic logic, for reasons to be explained later (§ 45). The expression '$f(x)$', including both predicate and argument-sign, will be called a *functional*.[2]

In the proposition, the function-name has a certain syntactical relation to the argument-name. In conversational language, this relation is given by word order: the function-name usually follows the copula 'is', or the phrase 'is a', or 'was a', whereas the argument-name precedes these terms. In symbolic logic, this syntacti-

---

[1] As to the logical nature of mathematical functions, cf. § 54.
[2] Instead of our term 'functional', Russell uses the term 'ambiguous value of a propositional function'; *Principia Mathematica*, Vol. I, pp. 15, 38–41.

cal relation is created by the parentheses; the function-name is outside the parentheses, and the argument-name is inside. With respect to this syntactical relation, the function-name is also called a *propositional function*. A propositional function, therefore, can be defined as a function-name within a certain syntactical usage.

This terminology requires some clarification. Let us consider the expression ' . . . is tall'. When we fill out the blank by a proper name, the expression will be made a proposition, which is true or false depending on the argument-name chosen. Thus 'Peter is tall' may be true, whereas 'Paul is tall' may be false. The phrase ' . . . is tall', therefore, can be used as a correlator; it correlates a proposition to argument-names like 'Peter' and 'Paul'. It is this usage as a correlator which we refer to when we say that the phrase ' . . . is tall' is a propositional function. This phrase determines, not one proposition, but a set of propositions. It resembles thus far a mathematical function, which also can be regarded as a correlator. Thus the mathematical function '$\sqrt{}$' correlates to every number another number which is its square root.

In order to indicate this usage of the function-name we shall introduce, with Russell, a notation by means of a circumflex. We shall say that '$f(\hat{x})$' is a propositional function. This means: the sign 'f', when used with an argument in the place of the sign 'x', is a propositional function. The sign 'x' covered by the circumflex has here the same task as the blank indicated by the periods in the expression ' . . . is tall'. Therefore '$f(\hat{x})$' does not mean the same as '$f(x)$'. In '$f(x)$', the functional, the variable 'x' has been inserted, whereas in '$f(\hat{x})$' the insertion, so to speak, has not yet been made. We therefore could also say that '$f(\ )$' is a propositional function. But the circumflex notation has some advantages which we shall explain presently.

Whereas the propositional function is in the linguistic sphere, the function denoted by the function-name is in the object sphere. We shall therefore call the latter function a *situational function*. Thus, being tall, or tallness, is a situational function; the word 'tall', on the other hand, is used as a propositional function. We see that the term 'function' is ambiguous as to level of language, like the term 'proposition', the term 'sign', and other terms (cf. § 3). By the addition of an adjective, such as 'propositional', or 'situational', we indicate the level on which the term 'function' is used. We shall later introduce other functions that also may occur on different levels of language (cf. § 54). In order to have a general notation, we call functions belonging to the level of objects *objective functions;* those belonging to the level of signs we call *linguistic functions.* Propositional functions are linguistic functions; situational functions are objective functions. These modifiers may be dropped when it is clear from the context which sort of function, or argument, is meant.

There are also functions of two variables. For instance, the sentence 'Peter is taller than Paul' refers to the situational function *being taller* and thus contains the propositional function 'taller'. Both functions are two-place functions; i.e., they possess two arguments. Accordingly, we shall write the sentence in the form

$$f(x, y) \qquad\qquad (2)$$

When the symbols '*x*' and '*y*' are regarded as variables, (2) represents a two-place functional. For the function-name we have chosen here the same symbol as in (1), the two-place character being expressed by the two variables of the functional. When the sign '*f*' is used alone, the place-number of the function is not expressed. We shall therefore resort to the circumflex symbol whenever we wish to indicate this number in a function not written within a functional; thus we say that the functional (2) contains the function '$f(\hat{x}, \hat{y})$'.

The two-place situational functions are also called *relations*. This term, therefore, corresponds to the term 'property', which is usually applied only to one-place situational functions. By predicates we shall understand the names both of properties and of relations. Further examples of two-place functions are given in sentences like 'Peter is a brother of Paul', or 'John loves Mary'. We see that functions are expressed, in conversational language, by various word categories; they may be nouns, adjectives, or verbs. In the last case, the copula 'is' is not used. We shall speak of these grammatical questions later (cf. § 45). There are also functions with more than two variables. A three-place function is given by the verb 'gives' in the sentence 'Peter gives Paul a book'. The corresponding situational function is called a three-place relation.

In this group belongs also the relation *between* since it is used in sentences of the form 'x is between y and z'. Functions with more than three variables occur in more complicated expressions, for instance, when space-time indications are regarded as arguments (cf. § 47).

A two-place function can also be given by way of propositional operations. Consider the sentence, uttered with reference to a certain couple: 'he is a composer or she is a writer'. It is symbolized in the form

$$f(x) \vee g(y) \qquad\qquad (3)$$

Compare this with the sentence 'he is a composer or a writer', which we symbolize as

$$f(x) \vee g(x) \qquad\qquad (4)$$

In (3) the function-name '$f \vee g$' is used as a two-place function; in (4) the same name is used as a one-place function. We see that the function-name, taken alone, does not say here in which way it is used. This is a case where the circumflex notation cannot be dispensed with. We say that (3) contains the function '$f(\hat{x}) \vee g(\hat{y})$', while (4) contains the function '$f(\hat{x}) \vee g(\hat{x})$'. Were we to use blanks instead of the circumflex notation, we could not express this difference.

In conversational language the function-name is usually so chosen that it immediately indicates the number of variables of the function. Thus 'brother' and 'taller' are two-place functions; 'tall' and 'house' are one-place functions. But there are also words which are used both ways. We say 'Peter is married' and 'Peter is married to Mary'. Similarly, the function 'composer or writer' can be used as a one-place or a two-place function. We shall therefore use the circumflex symbol occasionally also in combination with conversational language, and distinguish between the functions '$\hat{x}$ is a composer or $\hat{y}$ is a writer' and '$\hat{x}$ is a composer or $\hat{x}$ is a writer'.

While these considerations make clear the use of the circumflex symbol, we must add some remarks about its logical nature. We can regard the sign '$f(\hat{x})$' as a name of the corresponding situational function, synonymous with 'f'; the first name carries the advantage of expressing the inner structure of the function. In the functional '$f(x)$', however, we do not use the name '$f(\hat{x})$', but the name 'f';

that is, we do not write the form '$f(\hat{x})(x)$', but rather write '$f(x)$'. The shorter name is here equally suitable because the structure of the function is expressed by the form of the functional. We say, however, that the functional '$f(x)$' contains the propositional function '$f(\hat{x})$'. This means: it contains the function-name 'f' used with one variable. Generally speaking, a sentence containing the phrase 'the propositional function '$f(\hat{x})$'' is translatable into a sentence which contains in the same place the phrase 'the function-name 'f', used with one argument variable'. A corresponding translation holds when functions of several variables are used. It follows that in the expression 'the propositional function '$f(\hat{x})$'' the circumflex must be regarded, not as inside the quotes, but as outside and belonging to the metalanguage. The phrase stands as an abbreviation for 'the propositional function resulting from the functional '$f(x)$' when the 'x' is canceled'. We indicate this unusual meaning of the combination of quotes and circumflex by the prefixed phrase 'propositional function'.

This discussion shows that a propositional function is not precisely the same as a function-name; it is a function-name in combination with the rules of its usage. This somewhat peculiar meaning of the term has its analogue in other terms of conversational language. When a girl winds a scarf around her head, we say that she wears a turban. She could use the same scarf to make a sling for somebody's broken arm. Is the turban the same thing as the sling? We would not say that. Although both are made of the same scarf, we call them different things because in the two cases the scarf is used in different ways. A turban is a scarf in a certain usage; a sling is a scarf in another usage. Words like 'turban' and 'sling', therefore, may be called *usage-predicates;* they refer to a thing in combination with its usage.

When we wish to avoid this mode of speech we must say that a turban is, not a thing, but a property of a thing, namely, of a scarf; similarly, a sling would be another property of a scarf. Then it is clear that a turban is not a sling. But this would appear an unusual mode of speaking; we would not like to say that a man carries his broken arm in a property of a scarf. Similarly, we might say that a propositional function is, not a function-name, but the property of the sentence consisting in the occurrence of the function-name at a certain place of the sentence.[1] But this mode of speaking, although precise, deviates so much from the usage now generally accepted in symbolic logic that we shall not follow it.

The difference between the propositional function and the functional may be made clear as follows. The functional is a propositional form, differing from a proposition only in so far as it is a variable. Functionals are, therefore, propositional variables of the same sort as our symbols 'a', 'b', etc., with the difference that the inner structure is indicated. Two different functionals '$f(x)$' and '$f(y)$' agree in so far as they include the same function. Thus we can express a relation between two propositional variables, namely, a partial identity, without specifying them. A functional is true or false, depending on the meaning of 'f' and 'x' to be used. The function '$f(\hat{x})$', on the other hand, is a part of the functional. It therefore is not true or false; these concepts can be applied only to a proposition, not to a part of it. The propositional function is, so to speak, a mold in which we cast a content by applying it to a special argument. This operation is called *specialization*. By the operation of specialization, i.e., by putting a special value 'x_1' in the place of the argument, we construct a proposition '$f(x_1)$' from a propositional function '$f(\hat{x})$'. We use the subscript in 'x_1' to indicate that the variable is specialized. A specialized variable is also called a *constant*, or an *individual-sign*. The proper names of conversational language are individual-signs.

To every one-place propositional function '$f(\hat{x})$' there belongs a set of argument-signs 'x' which make the resulting propositions true; we call the set of all these signs the *class of satisfying argument-signs* of the propositional function. As each such argument-sign corresponds to a thing, we also have an object class coordinated to the propositional function. Using a well-known term of traditional logic we call this class the *extension* of the propositional function,

---

[1] We should like to add here a note, even though it will not be understandable to the beginner in logistics before he has followed this text up to § 37. Combining the above interpretation with Russell's principle of abstraction, we can regard the propositional function, in this conception, as the class of sentences resulting when the argument variable (or variables) runs through all its values while the function-name remains constant. The circumflex appears here as a symbol of the metalanguage, indicating transition to a class and thus binding the variable. Only in this interpretation is the propositional function strictly analogous to the situational function, which is the class of objects x for which '$f(x)$' is true.

or of the situational function. The extension is therefore the class of things for which the predicate holds. If no argument satisfies the function, as is the case for the function '$\hat{x}$ is a unicorn', we say that the corresponding class, or the extension, is *empty*. Object classes will be denoted by capital letters which we choose, if possible, in correspondence to the letters used for functions. Thus F is the extension of the propositional function '$f(\hat{x})$', or of the property f. Instead of saying that a thing x has the property f we therefore can equally well say that x belongs to the class F, or to the extension of f. For the extensions of more complicated functions we use other capital letters.

The extension of a *many-place function* is defined similarly. In order to satisfy a two-place function we must give an argument-couple, consisting of one argument for the place of 'x' and another for the place of 'y'. The class of satisfying couples is the extension of a two-place function. Similarly, the class of satisfying triplets is the extension of a three-place function, and so forth.

The functions we consider here have arguments of the thing type; such functions are called *simple functions*. There are other functions which have functions as arguments. Thus in the sentence 'hatred is destructive' the argument of the function 'destructive' is the function 'hatred'. Such functions are called *higher functions*. For the present we shall deal only with simple functions. The rules and operations holding for simple functions are comprised in the *simple calculus of functions*. It is only after the exposition of this calculus that we shall turn to the *higher calculus of functions* (cf. chapter VI). [Ex.]

§ 18. Binding of variables

The operation of specialization is not the only way of constructing a proposition from a propositional function. Another way is given by operations which we call *binding of variables*.[1]

---

[1] *Terminological note.* For the operations called by us 'binding of variables' the term 'quantification' is frequently used; operators then are called 'quantifiers'. We do not employ these terms because neither the all-statement nor the existential statement is a quantitative statement; they are qualitative statements. Sometimes the term 'generalization' is used to include both operations; it seems advisable, however, to reserve this term for the all-operation. We therefore suggest our term 'binding of variables' which has the advantage that it is derived from the now widely accepted term 'bound variable'. Russell's terms 'real variable' and 'apparent variable' (carried over from Peano) have now been generally replaced by 'free variable' and 'bound variable'.

The first of these operations is *generalization*. In this operation
we proceed from the propositional function '$f(\hat{x})$' to the proposition
'for all $x, f(x)$', which we write

$$(x)f(x) \tag{1}$$

For a given meaning of 'f', (1) is not a propositional variable, but a
proposition, although it contains the variable 'x'. This is clear
because the expression (1) then has a determinate truth-value.
Thus if '$f(x)$' is to mean 'x is blue', the proposition (1) is false; if
'$f(x)$' is to mean 'x is denotable', i.e., 'it is possible to introduce a
name for x', (1) is true. We find better examples of a true all-
statement if the function '$f(\hat{x})$' is not simple, but a combination of
other functions. Thus the formula

$$(x)[g(x) \supset h(x)] \tag{2}$$

has many true interpretations. Let 'g' be 'man', and 'h' be
'mortal'; then (2) is the proposition 'for all x, x is a man implies x
is mortal', or 'all men are mortal'. If '$g(x)$' means 'x is a magnet',
and '$h(x)$' means 'x attracts a piece of iron', (2) symbolizes the
proposition 'all magnets attract a piece of iron'.

We call the symbol '(x)' the *all-operator;* the content of the
brackets in (2) is called the *operand* of the formula. The whole
formula is called an *all-statement*, or a *general statement*. When the
operand, in particular, has the form of an implication as in (2),
the formula is called a *general implication*.

The variable 'x' in (1) or (2) is called a *bound variable;* it is bound
by an all-operator. On the other hand, the 'x' occurring in the
functional '$f(x)$' is called a *free variable*. What distinguishes the two
kinds of variables is that the expression containing the free variable
has no definite truth-value as a whole; it has a truth-value only
for each value of 'x'. The expression containing the bound variable,
on the other hand, has a definite truth-value, while the variable 'x'
assumes all values, one after another. A general statement always
refers to the totality of things; this is the reason that it must include
a variable running through this totality. If we call such variables
bound variables this name may be conceived as meaning 'bound to
run'. We may compare such a statement to a watch, which works
only as long as it is running; once the spring is wound it is bound to

run through all its states, and in doing so it tells the time. Similarly, a bound variable is a 'nonstop variable' which tells us something by its running. A free variable, on the other hand, can be 'stopped' by the insertion of a special value of 'x'; it is only then that the expression tells us something, i.e., becomes a statement.

When an all-statement like (2) is true, it is permissible to proceed to the specialized statement

$$g(x_1) \supset h(x_1) \tag{3}$$

This transition, however, does not constitute the operation of specialization leading from a function to a proposition. The transition from (2) to (3) is a *derivation;* that is, if (2) is true, (3) must be true.

We now turn to a second form of binding of variables. This operation obtains when we go from the propositional function '$f(\hat{x})$' to the proposition 'there is an x, such that $f(x)$'. We symbolize this proposition by

$$(\exists x) f(x) \tag{4}$$

If '$f(x)$' means, for instance, 'x is blue', (4) means 'there is a thing which is blue', which is a true sentence. If '$f(x)$' means 'x is a sea serpent', (4) is false.

We call the symbol '$(\exists x)$' the *existential operator;* the term following the operator is called, as before, the *operand*. Expression (4) is called an *existential statement*. It is important to realize that an existential statement contains a bound variable, like an all-statement, as it likewise refers to a totality; it says 'among the totality of things there is at least one which has the property f'.

With the expression of existence by means of an operator and a bound variable, symbolic logic settles the historic controversy about existence in a way that definitely excludes misuses and pseudo-demonstrations manipulating the term. The use of the operator compels us to give to an existential statement a form in which it asserts the existence of a thing having a certain property; thus the existential assertion is always qualified with respect to this property. It is only in this form that an existential statement tells us something new. When we say of a given physical object that it exists, we have made a trivial statement, since all physical objects exist.

What can reasonably be asked is whether an object of a certain kind exists. This question is to be answered by a synthetic statement of the form (4). Thus we can ask whether there exists a man who is seven feet tall. Let us assume that such a man has been found, and that his name is 'Peter'. We then know that there exists a man of a height of seven feet. But to say 'Peter exists' does not add anything to our knowledge. We shall later show [cf. (11, § 56)] that the empty statement 'Peter exists' also can be formulated; but we shall never need this statement.

This treatment of the existence concept is necessary because otherwise we are led to a fallacy in the usage of definitions. Definitions are arbitrary, and we may include whatever predicates we wish in a defined term; but after a definition is given there always remains the question whether there is a corresponding thing. The answer is to be given in the form (4) and requires verification by empirical methods [cf. also (13, § 56)].

Existence is formulated by an operation in the object language; its corresponding expression in the metalanguage is *truth*. Whenever a sentence is true, it can be translated into a statement that something exists in the sphere of objects. This relation between truth and existence can be given a correlate within the object language when we use, instead of the statement that the sentence is true, the asserted sentence itself. The resulting statement is expressed by the general formula:[1]

$$f(x_1) \equiv (\exists x)f(x).(x = x_1) \tag{5}$$

The equality sign used here represents *identity*, a relation holding between arguments, whereas the sign of equivalence, expressed by three horizontal bars, stands between propositions. For the present, we shall regard identity as a primitive term; a reduction to other logical terms is given later (§ 43). For the negation of identity we shall write '$x \neq y$', as an abbreviation for '$\overline{x = y}$'.

The existential operator is used not only in the symbolization of propositions which explicitly contain an expression like 'there is', or 'there exists'; we apply it also for the symbolization of the indefinite

---

[1] Cf. the discussion of this formula in § 46 and at the end of § 49, in particular, with respect to (18, § 49); furthermore, our remarks concerning (19, § 43) and (11, § 56).

article.[1] Thus 'Jeanne eats an apple' will be symbolized in the form

$$(\exists y)e(x_1, y).a(y) \tag{6}$$

where 'e' stands for 'eats', 'a' for 'apple'. The existential operator is necessary here because the individual apple that is eaten is not named; what is said is only that there is an apple which is eaten. [Ex.]

§ 19. Negation of operators

The two operators that we have introduced are connected in such a way that the negation of the one is expressible in terms of the other. This relation is stated in the following two formulas:

$$\overline{(x)f(x)} \equiv (\exists x)\overline{f(x)} \tag{1}$$
$$\overline{(\exists x)f(x)} \equiv (x)\overline{f(x)} \tag{2}$$

These two formulas make use of the difference between denying the whole statement and denying the operand; a shortening of the negation line, i.e., of the scope of the negation, is accompanied by a transition to the other operator. Thus the left-hand side of (1) may mean 'not all things x are animals'; the right-hand side states the same in the form 'there is a thing x which is not an animal'. As to all-statements, the English language expresses the difference between the long and the short negation line by means of the phrases 'not all . . . are' and 'all . . . are not'. Thus the right-hand side of (2) may be interpreted as meaning 'all things x are not sea serpents'. The same is stated in the left-hand side of (2) in the form 'there is no thing x which is a sea serpent'; or expressed more shortly: 'No thing is a sea serpent'. The rule stating that the word 'no' has the meaning of the phrase 'all . . . are not' is therefore expressed in the equivalence (2). Conversational language does not always keep to this rule. Thus the adage 'all that glitters is not gold' should be correctly formulated as 'not all that glitters is gold'.

If we add a long negation line on each side of (1) and (2), the two formulas are transformed into

$$(x)f(x) \equiv \overline{(\exists x)\overline{f(x)}} \tag{3}$$
$$(\exists x)f(x) \equiv \overline{(x)\overline{f(x)}} \tag{4}$$

[1] For another meaning of the indefinite article, cf. p. 111.

These equivalences enable us to define one of the two operators in terms of the other. It follows that only one of the operators is to be considered a primitive term.

Rules (1) and (2) for the shortening of the negation line show a certain duality between the two operators, which corresponds to the duality between the 'and' and the 'or'. This analogy has a deeper root; it originates from the fact that for a finite number of arguments 'x' the all-statement is identical with a conjunction, the existential statement with a disjunction. Thus we have in such a case

$$(x)f(x) \equiv f(x_1).f(x_2) \ldots f(x_n) \tag{5}$$
$$(\exists x)f(x) \equiv f(x_1) \lor f(x_2) \lor \ldots \lor f(x_n) \tag{6}$$

The all-statement can therefore be considered as a conjunction of an infinite number of terms; the existential statement, as a disjunction of an infinite number of terms. However, it would be incorrect to say that (5) and (6) are definitions of the operators. Conjunction and disjunction are operations defined for only a finite number of terms. To extend these operations to an infinite number of terms requires new primitive terms. The correct form of statement is therefore that a conjunction and a disjunction of an infinite number of terms are defined by the operators.

In addition to relations (1) and (2) between the two operators we have the relation

$$(x)f(x) \supset (\exists x)f(x) \tag{7}$$

Employing a term from traditional logic we call the existential statement the *subaltern* of the all-statement, using this word in the sense of 'being implied by'. It may seem strange that we infer here an existential statement from an all-statement although the word 'all' does not include any reference to existence. However, all we assume in (7) is that there are things x in this world, i.e., that the domain of things considered by logic is not empty. The acceptance of (7) in the list of logical formulas may therefore be considered as the formulation of this assumption. For an empty world, logic would degenerate into an idling system of formulas including contradictions; thus both the statements '$(x)f(x)$' and '$(x)\overline{f(x)}$' would be true.

The traditional *square of opposition* can be written in the form given in Fig. 1.

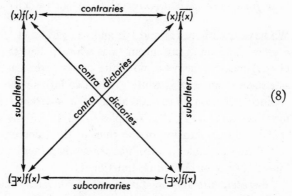

FIG. 1.   Square of opposition.

The term 'contradictory' expresses the fact that one statement is the negation of the other; the term 'contrary' expresses exclusion. The term 'subaltern' refers to the existential statements; each existential statement is the subaltern of the all-statement written above it. The subcontraries do not exclude each other; but they cannot be both false.

When an operand includes a combination of functionals, it is permissible to deal with the functionals according to the rules established for propositional variables. We therefore can apply here the method of transformation introduced in the calculus of propositions (cf. § 11). Combining this method with the rules holding for the negation of operators, we have, for instance:

$$\overline{(x)[f(x) \supset g(x)]} \equiv (\exists x)\overline{[f(x) \supset g(x)]} \equiv (\exists x)\overline{[\overline{f(x)} \vee g(x)]} \qquad (9)$$
$$\equiv (\exists x)f(x).\overline{g(x)}$$

We have applied here (1) and (6a and 5b, § 8). We see that the negation of a general implication is given by an existential conjunction. When we interpret '$f(x)$' as 'x glitters', '$g(x)$' as 'x is gold', the left-hand side means 'not all that glitters is gold'; this sentence is translated on the right-hand side into 'there is something which glitters and which is not gold'. The two forms are obviously equivalent.

Similarly we have

$$(\overline{\exists x})\overline{f(x).g(x)} \equiv (x)\overline{f(x).g(x)} \equiv (x)\overline{[\overline{f(x)} \vee \overline{g(x)}]} \equiv (x)[f(x) \supset \overline{g(x)}] \quad (\text{10})$$

We have used here (2) and (5a and 6a, § 8). Let '$f(x)$' mean 'x is a crow', and let '$g(x)$' mean 'x is white'; then the left-hand side of (10) means 'there are no white crows', whereas the right-hand side means 'no crow is white'. The two forms mean the same.

Since negation lines can always be shortened by the use of formulas (1) and (2), the examples (9) and (10) show, furthermore, that by the application of the rules of the calculus of propositions it is always possible to introduce *shortest negation lines,* i.e., negation lines drawn over individual functionals.

All-statements and no-statements of conversational language are mostly implications; existential statements are mostly conjunctions. This appears to be the reason that traditional logic has introduced a special notation for statements of this form, while it possesses no notation for other forms. Since in the scholastic notation properties are conceived as classes rather than as functions, we shall use capital letters 'S' and 'P' in this notation, and corresponding small letters 's' and 'p' in the expressions using the modern functional notation. The scholastic notation can thus be represented as follows:

$$S \, A \, P =_{Df} \text{all } S \text{ are } P$$
$$S \, E \, P =_{Df} \text{no } S \text{ is } P \qquad\qquad (\text{11})$$
$$S \, I \, P =_{Df} \text{some } S \text{ are } P$$
$$S \, O \, P =_{Df} \text{some } S \text{ are not } P$$

In logistic symbolism we have:

$$S \, A \, P =_{Df} (x)[s(x) \supset p(x)]$$
$$S \, E \, P =_{Df} (x)[s(x) \supset \overline{p(x)}] \qquad\qquad (\text{12})$$
$$S \, I \, P =_{Df} (\exists x)s(x).p(x)$$
$$S \, O \, P =_{Df} (\exists x)s(x).\overline{p(x)}$$

We can now write the square of opposition in the scholastic form:

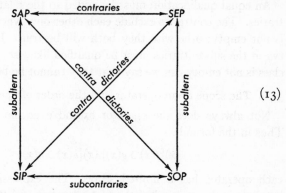

FIG. 2. Square of opposition, scholastic form.

We can easily prove, using relations (9) and (10), that as before the contradictories are negations of each other. There is to be added, however, a qualification concerning the relation 'subaltern'. The statements '$S I P$' and '$S O P$' can be derived from the corresponding statements '$S A P$' and '$S E P$' only if we assume that the class S is not empty, i.e., that the statement

$$(\exists x)s(x) \tag{14}$$

holds. The inference from 'all' to 'some', these concepts being taken in the scholastic form (11), is therefore bound to the assumption (14). This assumption must be carefully distinguished from the general assumption that the domain of things is not empty; (14) requires much more than the general assumption, and there are many functions '$s(x)$' for which (14) does not hold. For such functions the inference from '$S A P$' to '$S I P$' is not permissible. Thus although it may be true to say that all sea serpents live in the sea, it would be false to infer that some sea serpents live in the sea, the latter statement being equivalent to saying 'there are some sea serpents living in the sea'. The square of opposition in its scholastic form is therefore correct only when we interpret the relation 'subaltern' as meaning: 'being implied by . . . in case the first class is non-empty'.[1]

---

[1] We follow here the now generally accepted interpretation of the words 'all' and 'some' of conversational language. It should be kept in mind, however, that usually the word 'all' of conversational language is also stated with an existential implication,

An equal qualification must be added to the relation of the contraries. The contraries exclude each other only when the first class is not empty; otherwise they both will be true. The relation between the subcontraries must be qualified likewise; only if the first class is not empty can we say that they cannot be both false. (Ex.)

§ 20. The scope of an operator and the order of operators

Not always does the operator extend over the whole formula. Thus in the formula

$$(x)[f(x) \supset g(x)].(x)[h(x) \supset k(x)] \tag{1}$$

each operator has a limited scope. The expression in the first brackets is the operand of the first operator, and the expression in the second brackets is the operand of the second operator. By the *scope of an operator* we shall understand the combination of an operator and its operand. When the operands are separated, as in (1), we can denote the variables of the different scopes by different letters; thus (1) is the same as

$$(x)[f(x) \supset g(x)].(y)[h(y) \supset k(y)] \tag{2}$$

It is not necessary, however, to use two operators in our example; the two operators can be merged into one, and we can write instead of (1)

$$(x)[f(x) \supset g(x)].[h(x) \supset k(x)] \tag{3}$$

We call the step from (1) to (3) a *fusion of operands;* the inverse step, proceeding from (3) to (1), is called a *division of the operand.* For all-operators referring to expressions connected by an 'and' both a fusion and division is permissible; this fact is expressed by the equivalence of (1) and (3). For other operators or other propositional operations the step can frequently be made only in one

---

i.e., with the inclusion of the assumption (14). In this interpretation of the words 'all' and 'no' the statement (14) must be added to the right-hand side of the first two statements of (12). The inference from '*S A P*' to '*S I P*', and from '*S E P*' to '*S O P*', is then correct. On the other hand, it is also possible to construe the some-statement, by analogy with the all-statement, as not including the existential assumption (14); the statement '*S I P*' then would be interpreted as meaning '$(\exists x)s(x) \supset (\exists x)s(x).p(x)$'. With this interpretation the implication from 'all' to 'some' is also secured. It is the disadvantage of both these interpretations, however, that the statements combined by a diagonal line are not the negations of each other. These considerations show that changes in the interpretation of all-statements or some-statements do not lead to a correct form of the square of opposition.

direction because the two formulas under consideration are not equivalent; thus from

$$(x)f(x) \vee (x)g(x) \tag{4}$$

we can go to

$$(x)[f(x) \vee g(x)] \tag{5}$$

but not vice versa. We therefore can perform here only a fusion of operands, not a division. The reason is that we have only an implication leading from (4) to (5), but no implication in the other direction. That no such implication exists is clear if we put ‘$\overline{f(x)}$’ for ‘$g(x)$’; then (5) is true, whereas (4) may be false.

To simplify notation we set up the rule that, if no parentheses are given, the scope of the operator extends as far as the last term in which the variable of the operator occurs but stops if an equal operator with the same variable intervenes. This rule permits us to omit the brackets in (5).

An example in which only a division of the operand can be performed is given by the implication

$$(x)[f(x) \supset g(x)] \tag{6}$$

From formula (6) we can go to

$$(x)f(x) \supset (x)g(x) \tag{7}$$

but the reversed step cannot be made. Expression (7) says nothing if the function ‘$f(x)$’ does not hold for all ‘x’ since the implicans then is false. Expression (6), however, states that, in such a case, when ‘$f(x)$’ is true for a certain ‘x’, ‘$g(x)$’ must also be true for this ‘x’. Now it is clear that, if the property f should hold for all things, the property g then must also hold for all things. This is what (7) says. Thus, if (6) means ‘all men are mortal’, it follows that, if all things were men, all things would be mortal. But this is only a consequence, which says less than the first statement.

There are special cases in which one of the functionals within the scope of an operator is replaced by a term a which does not depend on the bound variable. We then call a a constant of the formula. Of this kind is the formula

$$(x)[f(x) \supset a] \tag{8}$$

Formula (8) does not state the same thing as

$$(x)f(x) \supset a \tag{9}$$

Formula (9), in which the operand consists only in the functional '$f(x)$', requires that 'a' be true if '$(x)f(x)$' is true, whereas 'a' may be true or false if '$(x)f(x)$' is false. In (8), however, 'a' must be true if there is at least one 'x' such that '$f(x)$' is true, as in this case the implicans is true for that 'x'; since the truth-value of 'a' does not vary with 'x', 'a' must therefore be true independently of 'x'. Formula (8) is therefore equivalent to

$$(\exists x)f(x) \supset a \tag{10}$$

In this formula the operand is given only by the functional '$f(x)$'. An example for (10) is the statement: 'if there is one good deed in this world, life is worth living'. This statement can obviously be pronounced in the form (8): 'for all things, if this thing is a good deed this implies that life is worth living'.

A similar transformation can be given for formula (9). Formula (9) means the same as

$$(\exists x)[f(x) \supset a] \tag{11}$$

Formula (11) will be false only when '$f(x)$' holds for all 'x' while 'a' is false; namely, if '$f(x)$' does not hold for all 'x', we can choose an 'x' for which '$f(x)$' is false and thus make the implication true. But these truth conditions are the same as explained for (9). The equivalence of (11) and (9) may be illustrated by the following example. The form (9) is given by the statement: 'if all things are dust, life is worth nothing'. In the form (11) the statement reads: 'there is a thing such that, if this thing is dust, life is worth nothing'. As long as this thing is not specified, we cannot know from the statement whether the consequence that life is worth nothing must be drawn; maybe the particular thing whose being dust would imply this consequence is not dust. Only when we have ascertained that all things are dust do we know that the unknown particular thing having the grave consequence must also be dust; therefore in this case alone can we draw the consequence. We see that the two statements amount to the same because (11) states an unspecified existence.

Since these examples illustrate that the meaning of a sentence will depend on whether a constant is included in the scope of an operator, or excluded from it, it will be logical even to admit an operand which consists of a constant alone. This admission is also made necessary in order to make the technique of the calculus consistent, as will be seen later. We shall therefore regard an expression of the form '$(x)a$' as meaningful, and as meaning the same as 'a'. Similarly, the expression '$(\exists x)a$' is also admitted as meaningful; this expression, too, means the same as 'a'. This terminology allows us also to speak of fusion or division of operands in such cases as the transition from (8) to (10).

When the constant is in the implicans a fusion or division of operands leaves the operator unchanged. We have here the relations

$$(x)[a \supset f(x)] \equiv a \supset (x)f(x) \tag{12}$$
$$(\exists x)[a \supset f(x)] \equiv a \supset (\exists x)f(x) \tag{13}$$

Formula (12) is clear because, if 'a' is true, the left-hand side requires that '$f(x)$' be true for all 'x'; otherwise the implication would be false for some 'x'. This is stated on the right-hand side. The left-hand side of (13) would be false if 'a' is true, while '$f(x)$' is false for all 'x'; this is expressed by the right-hand side.

The equivalences between (8) and (10) and between (9) and (11), and, in addition, the formulas (12)–(13), show that for implications both a fusion and a division of operands can be obtained when one of the terms is a constant. When the constant stands as implicate this process is accompanied by a change in the operator. Only when both terms are functionals depending on the same variable, as in (6), are we restricted to a division.

A collection of formulas governing the fusion or division of operands is given in formulas (9a–12f, § 25). The proof of all these formulas is given later (§ 25) by means of a general method which is more convenient than the inferences we have used in the preceding examples.

The binding of variables can be extended to functions of more than one variable. The variables can be bound, either all by the same operator, or by different operators. Thus the statement 'all mothers love their children' will be expressed by means of two all-operators, namely, in the form

$$(x)(y)[m(x, y) \supset l(x, y)] \tag{14}$$

Here 'm' stands for 'mother', 'l' stands for 'loves'. The statement 'there are happy marriages', interpreted in the form 'there are married persons who are happy', includes two existential operators:

$$(\exists x)(\exists y)m(x, y).h(x).h(y) \tag{15}$$

where 'm' stands for 'married to', and 'h' for 'happy'.

An example containing both kinds of operators is given by the statement 'every man has a father', which we write in the form:

$$(x)(\exists y)[m(x) \supset f(y, x)] \tag{16}$$

Here 'm' stands for 'man', 'f' for 'father'. This expression must be distinguished from the formula resulting by reversion of the operators:

$$(\exists y)(x)[m(x) \supset f(y, x)] \tag{17}$$

Using the same meaning for the letters we can read (17) as 'all men have a father'. This statement is not true for human fathers since it would mean that all men have the same person as their father. We must leave the proof of such a statement to the adherents of various religions, who maintain that they are able to prove this assertion in a symbolic meaning of the word 'father'.

By the distinction of (17) from (16), logistic thus expresses the difference between 'all' and 'every', which was not always clearly seen by the older logicians. Sometimes the word 'every' is also expressed by means of the word 'all' in combination with a plural in the second propositional function; thus (16) may be read as 'all men have fathers'. For one-place functions, the difference under consideration disappears and the terms 'all' and 'every' may be identified.

A further example, taken from arithmetic, may illustrate the difference between 'all' and 'every'. Let us consider the two statements

$$(x)(\exists y)[g(x) \supset g(y).h(x, y)] \tag{18}$$
$$(\exists y)(x)[g(x) \supset g(y).h(x, y)] \tag{19}$$

each of which differs from the corresponding statement (16) or (17) only by the more complicated form of the operand. Let '$g(x)$' mean: 'x is a natural number' (i.e., a positive whole number, zero included), and '$h(x, y)$': 'x is smaller than y'. Then (18) is true, as it means: 'for every natural number there is a greater natural number'. However, (19) is not true since it would mean that there is one natural number greater than all others. We obtain a true interpretation of (19) by putting for '$h(x, y)$' the expression 'x is not smaller than y'. The statement then means that there is a number y such that all numbers are not smaller than y; or, in other words, there is one number y such that y is smaller than all other numbers. Such a number exists indeed; it is the number zero.

These considerations lead to the following rules for the order of operators. When two operators of the same sort occur, such as in (14) or (15), their order can be reversed. However, when two different operators occur, such as in (16) or (17), a reversal of their order will change the meaning of the expression. It can easily be seen that for different operators at least a relation of derivation holds between the two expressions: if (17) is true, (16) is also true; but not vice versa. These rules are expressed in the following formulas:

$$(x)(y)f(x, y) \equiv (y)(x)f(x, y) \qquad (20)$$
$$(\exists x)(\exists y)f(x, y) \equiv (\exists y)(\exists x)f(x, y) \qquad (21)$$
$$(\exists y)(x)f(x, y) \supset (x)(\exists y)f(x, y) \qquad (22)$$

There are special cases in which all-operators and existential operators are commutative. This occurs when the two-place function '$f(x, y)$' is replaced by two one-place functions '$f(x)$' and '$g(y)$'. Thus we have

$$(\exists x)(y)[f(x) \supset g(y)] \equiv (y)(\exists x)[f(x) \supset g(y)] \qquad (23)$$

In the proof of this formula we can regard a functional, containing a variable different from that of an operator, as a constant with respect to this operator. Applying (12) on the left-hand side, we transform this side into

$$(\exists x)[f(x) \supset (y)f(y)] \qquad (24)$$

We now use the equivalence between (11) and (9), and transform (24) into

$$(x)f(x) \supset (y)f(y) \qquad (25)$$

Applying the same transformations in the reversed order to the right-hand side, we arrive at the same result. A collection of corresponding formulas is given in (16b–16d, § 25).

Let us add now some notational remarks. When a formula has

several operators, the operand of one of the operators will include all succeeding operators. Thus in the formula

$$(x)(y)f(x, y) \qquad (26)$$

the operand of '(x)' is the expression '$(y)f(x, y)$'; and the scope of '(x)', which includes this operator, is the whole formula. The operator '(y)' has only the expression '$f(x, y)$' as its operand, and its scope is the expression '$(y)f(x, y)$'. This distinction is necessary because the meaning of a formula containing several operators is definable only by reference to the order of these operators. Thus (26) means: first, keep 'x' constant and consider the statement '$(y)f(x, y)$'; then, generalize in 'x'. Now (20) shows that two all-operators are commutative; therefore, using the right-hand side of (20), we can also conceive '$f(x, y)$' as the operand of '(x)', and '$(x)f(x, y)$' as the operand of 'y'. For this reason, the distinction between the operands is not very important when only all-operators are concerned. Similar considerations hold for formulas containing several existential operators, such as (21). It is different when both existential operators and all-operators occur, as in (22). On the right-hand side of (22), the operand of '(x)' is the expression '$(\exists y)f(x, y)$'. This means: first, keep 'x' constant; then there is a 'y' such that '$f(x, y)$' holds; then, proceed likewise for all other values of 'x'. We see that this interpretation corresponds to the meaning of 'every'. On the left-hand side of (22), the operand of '$(\exists y)$' is the expression '$(x)f(x, y)$'. This means: consider the functional '$(x)f(x, y)$', which refers to all values of 'x', and therefore represents a functional of 'y' alone; there is a value of 'y' which makes this expression true. The interpretation constitutes the meaning of 'all'. The two terms 'all' and 'every' are therefore distinguished by the rule concerning the operands of consecutive operators.

On the other hand, it is often useful to regard operators of this sort, i.e., consecutive operators whose scopes terminate at the same place, as one *set of operators*. By the *operand of the set* we shall understand the common part of all the individual operands. Thus in (26) the expression '$f(x, y)$' is the operand of the set of operators '$(x)(y)$'; on the left-hand side of (22), '$f(x, y)$' is the operand of the set '$(\exists y)(x)$'. Furthermore, we define the scope of the set so that it includes the whole set. When the scope of a set of operators is the whole formula, we say that the operators of the set are *major operators*. The operand of the set will be called the *major operand* when the set is complete, i.e., when it includes all consecutive operators at the beginning of the formula.

A major operand may contain other operators. Thus in

$$(x)[g(x) \supset (\exists y)g(y).h(x, y)] \qquad (27)$$

the major operand is the expression in the brackets, which contains an operator '$(\exists y)$' whose operand is only the expression '$g(y).h(x, y)$'. Here the two operators '(x)' and '$(\exists y)$' do not constitute a set because they are separated by a term '$g(x)$' which is not an operator. There are other formulas which do not possess major operators, such as formulas (20)–(22). A formula possessing a major operand that does not contain any operators will be called a *one-scope formula*. The whole formula then is one, and only one, scope. Of this sort are formulas (14)–(19). If a negation line is drawn over a one-scope formula, the resulting expression is not a one-scope formula and the operators are no longer major operators, because the scope of the operators does not include the negation line. Examples are given by the left-hand sides of (1 and 2, § 19).

A formula that contains no operators at all may also be called a one-scope formula. We thus use the term 'scope' in a somewhat wider sense than the term 'operand', the latter being applicable only when the formula contains operators.

It can be easily seen that every formula in the calculus of functions can be transformed into a one-scope formula. Sometimes it is necessary first to introduce different letters for bound variables, as explained in the transition from (1) to (2), before the fusion of operands can be reached. The general rules for such a transformation are given in formulas (9a–12f, § 25); we give here an example which can be treated by the means developed in this section. We use the formula

$$(x)\{(y)[g(y) \lor h(x, y)] \supset f(x).(y)k(x, y)\} \qquad (28)$$

Since for implication we have only a formula effecting a division of operands, expressed in the transition from (6) to (7), a fusion cannot be directly achieved in (28). We must first change the notation

by putting the letter 'z' instead of the 'y' in the last term of (28).
We thus have

$$(x)\{(y)[g(y) \vee h(x, y)] \supset f(x).(z)k(x, z)\} \tag{29}$$

We now apply the transition from (9) to (11) with respect to 'y', and (12) with respect to 'z'. It is irrelevant in which order this is done because formula (23) applies here. Thus we obtain the one-scope formula

$$(x)(\exists y)(z)[g(y) \vee h(x, y) \supset f(x).k(x, z)] \tag{30}$$

In a transformation of this kind the operations 'and', 'or', and 'implies' remain unchanged; the equivalence operation, however, has first to be dissolved into two implications or to be otherwise eliminated before the operands can be merged. The reason is that according to (11e and 12e, § 25) we have, for this operation, no equivalence between divided and merged operands. Negation lines drawn over operators can always be shortened by the rules expressed in (1 and 2, § 19) and then transformed into shortest negation lines, as shown in (9, § 19). Since the rules of the calculus of propositions apply to functionals, it is even possible to transform the operand of a one-scope formula into the disjunctive normal form, or the conjunctive normal form.

The introduction of propositional functions and operators represents an extension of the formation rules of language. The additions we thus make to the formation rules stated in § 6 can be summarized as follows; we continue the numeration begun with those rules.

4. There are two types of elements: functions and arguments. An element belonging to one type cannot belong to the other. A functional is an expression containing a function outside and one or more arguments inside the parentheses.

5. A functional is a propositional expression.

6. When an all-operator or an existential operator containing a variable 'x' is placed before a propositional expression within which the variable 'x' is not yet bound by another operator, the resulting expression is a propositional expression. The expression before which the operator is placed is called the *operand* of the operator; and the whole expression resulting is called the *scope* of the operator.

In combination with rules 1–3, § 6, these rules lead to a recursive definition of propositional expressions, which reveals the step structure of all such expressions. Rule 6 includes the condition that a variable must not be bound by two operators. On the other hand, it admits expressions of the form '$(x)a$' and '$(\exists x)a$' as meaningful since it does not require that the variable 'x' be contained in the propositional expression before which the operator is placed. The rule stating that in such expressions the operator can be canceled is not a formation rule, but a truth rule, and will be formulated later. [Ex.]

§ 21. Synthetic assertions containing free argument variables

When the variables in a formula are either specialized or bound, we shall say that the formula is *closed in the argument variables*. When, in addition, the functions are specialized, we shall say that the formula is *closed*.[1] Usually, a synthetic proposition will be closed. The closure is necessary because expressions containing free variables, in general, have no determinate truth-value. There is, however, one exception to this rule: when an expression containing a free variable is true for all values of the variable, it can be asserted.

This somewhat peculiar use of free argument variables can be made clear as follows. When we have shown that a certain object x has the property f, we can assert the sentence '$f(x)$'. Now, when it is known that the function f holds for all objects, it is not necessary to say which object we mean by 'x'; the sentence '$f(x)$' will then be true in any case and therefore can be asserted even when a specialization of 'x' has not been given. This sort of assertion is not an invention of mathematicians but is frequently used in conversational language. It is usually expressed through the word 'any'. We say, for instance: anybody knows this secret. This means: pick out a person at random, he, or she, will know the secret. It is obvious that we can venture to say so only when all persons know the secret; therefore here the statement using the word 'any' means the same as a statement using the word 'all'.

Not always, however, do these two sorts of statements have the

[1] We shall later extend the meaning of the term 'closed' to include expressions in which the functional variables are bound; such a procedure leads into the higher calculus of functions. Cf. § 42.

same meaning. When we ask, for instance: 'does anybody know the secret?' we mean: 'does there exist a person who knows the secret?' Here the sentence containing the word 'any' has the meaning of an existential statement. The same holds when the sentence is denied. 'It is false that there is anybody who knows the secret' means it is false that there exists such a person; in other words, it means there is no such person. For this reason we cannot say unconditionally that 'any' means the same as 'all'.

What then is the difference between 'all' and 'any'? It can be shown that this difference consists, not in the *meaning* of the generalization, but only in the *scope* of the generalization. When a free variable like the word 'any' is used to express generality, its scope is always the whole formula. The generality expressed by the all-operator, however, can be restricted to parts of the formula by a suitable indication of the scope. Thus, when we assert '$f(x)$', we mean the same as when saying '$(x)f(x)$'; but asserting '$\overline{f(x)}$' has the meaning '$(x)\overline{f(x)}$' and not '$\overline{(x)f(x)}$'. The last formula is inexpressible by the use of a free variable. We now see the reason that in the example 'it is false that anybody knows the secret' the word 'any' means 'there is': the word 'any' means 'all' but applied to the whole formula; since the statement '$(x)\overline{f(x)}$' means the same as '$\overline{(\exists x)f(x)}$' this meaning is translatable into a denied existence. This shows that actually the word 'any' is not ambiguous. It represents an 'all' used with a different rule as to scope; therefore it is not always possible, in a given statement, to replace the word 'any' by the word 'all', or vice versa.

Let us illustrate our result by another example. Our sentence 'anybody knows the secret' has the form of the assertion

$$f(x) \tag{1}$$

Now consider the sentence: 'if anybody knows the secret, Marjorie talked'. It has the form

$$f(x) \supset a \tag{2}$$

According to the rule of scope holding for the word 'any', (2) means the same as

$$(x)[f(x) \supset a] \tag{3}$$

But, as shown in (10, § 20), (3) is equivalent to

$$(\exists x)f(x) \supset a \tag{4}$$

Expression (4) corresponds to the meaning of our example, since our sentence can be transcribed into the form: 'if there is a person who knows the secret, Marjorie talked'. The statement '$(x)f(x) \supset a$' is not expressible in free variables.

Now consider the statement: 'if anybody knows the secret he will have told it to others'. It has the form

$$f(x) \supset g(x) \tag{5}$$

which is the same as

$$(x)[f(x) \supset g(x)] \tag{6}$$

Expression (6) is not translatable into an existential statement. Consequently, the 'any' in this example has the meaning: 'it holds for all persons that if this person knows the secret he will have told it to others'.

Let us turn now to the task of formulating the relations holding between 'any' and 'all'. First, we have the formula

$$(y)f(y) \supset f(x) \tag{7}$$

which expresses the transition from a bound to a free variable. The reverse transition, however, is not expressible in this way. When we write

$$f(x) \supset (y)f(y) \tag{8}$$

the expression means, according to the rule for the scope of a free variable:

$$(x)[f(x) \supset (y)f(y)] \tag{9}$$

Since the implicate does not depend on 'x' it has the form of a constant 'a'; therefore (9) has the form (3) and is translatable according to (4) into the formula

$$(\exists x)f(x) \supset (y)f(y) \tag{10}$$

which is obviously false. Therefore (8) is equally false. We see that the rule of transition from a free to a bound variable is not expressible in a formula of the object language; if such a formula were stated, the rule would apply to this formula and thus make it false. This paradox is inherent in the nature of free variables. In order to eliminate it we would have to introduce a notation limiting the scope of a free variable to a part of the formula; but such a notation rule would change the free into a bound variable, and therefore the formula constructed would not refer to free variables.

The transition from free to bound variables can therefore be stated only in a rule of the metalanguage, called the rule for free variables: *when the formula '$f(x)$' is true for a nonspecialized 'x', the formula '$(x)f(x)$' is also true, and therefore can be asserted.* For other sorts of expressions this relation is translatable into an implication; but such translation is not possible for the above expressions because they contain a free variable. The power of a free variable permeates all boundaries that can be drawn within a language; it can be limited only by the barrier of quotation marks because they represent the transition to another language.

A relation of this sort is called *equipollence;* the formula '$f(x)$', in which 'x' is a free variable, is *equipollent* to the formula '$(x)f(x)$'. Two formulas are called *equipollent* when each is derivable from the other. Thus from '$f(x)$' we can derive '$(x)f(x)$', and vice versa. This relation can be regarded as a generalization of the relation of *tautological equivalence.* Formulas that are tautologically equivalent are also equipollent, but the converse relation does not hold. Thus the equivalence

$$f(x) \equiv (x)f(x) \tag{11}$$

is not tautologous, but synthetic; it holds only for a function 'f' that is true for all arguments 'x'. The reason is that the implication from left to right is not a tautology. In cases of this sort the relation of equipollence takes the place of the relation of tautological equivalence and thus becomes the logistic expression of the relation of *equisignificance;* two equipollent assertions have the same meaning.

It should be noticed that we can speak of equipollence only with reference to assertions. The *expressions* '$f(x)$' and '$(x)f(x)$' are not equipollent, and do not have the same meaning; equipollence holds only for the corresponding *assertions.* Whereas the expression '$(x)f(x)$' need not be distinguished from the corresponding assertion, such a distinction must be made for the functional '$f(x)$'. An expression in free variables acquires generality only through its assertion. The reason is that only through the assertion the scope of the free variables is delimited. As long as the expression is not asserted it might as well be used as a part of a more comprehensive expression; such use would result in a different scope for the free variable, and thus the first expression would have a different meaning. An expression like '$f(x)$' is therefore an *indefinite* expression;

its meaning is not definite, and depends on the way the expression is used.[1] On the other hand, the expression '$(x)f(x)$', for a determinate 'f', is a *definite* expression.

As to the truth properties of the functional, we must therefore distinguish three cases. When the expression '$f(x)$', or '$\overline{f(x)}$', is assertable, the free variable expresses generality, and the corresponding statements '$(x)f(x)$', or '$(x)\overline{f(x)}$', are true. When neither of the two expressions is assertable we have a third case in which the expression '$f(x)$' is unassertable; it then represents a functional that is ambiguous as to truth-value. Its truth-value depends on the specialization chosen for 'x', which variable, then, represents an uninterpreted constant. It is different with the two all-statements; when they are not assertable they are false. The fact that the functional cannot be incorporated into a two-valued classification represents another feature distinguishing indefinite from definite expressions.

Not all expressions containing free variables are indefinite. If the expression is a tautology, it will be definite even when it contains free variables because it then cannot be made false. Thus the expression '$f(x) \vee \overline{f(x)}$' is definite. It does not change its meaning when it stands as the implicans in a formula like (8); and the formula

$$f(x) \vee \overline{f(x)} \supset (x)[f(x) \vee \overline{f(x)}] \tag{12}$$

is therefore a tautology. This consideration explains the notation introduced by us in § 12. With reference to functions we shall include in definite expressions all tautologies, and all synthetic expressions in which the functions are constants and the arguments are either constants or bound, i.e., which are closed. It is clear that an indefinite expression consisting of functions that are constants can be made definite simply by binding the free argument variables by all-operators, i.e., by closing it.

A second relation between free and bound variables, to be added to (7), is expressible in the formula of the object language

$$f(x) \supset (\exists y)f(y) \tag{13}$$

This formula replaces the false formula (8). It states that, if anything has the property f, there is a thing having this property.[2]

---

[1] Cf. the explanation of indefinite expressions given at the beginning of § 12.
[2] Cf. our remarks concerning (5, § 18).

Free variables are used in the logistic calculus because they simplify all manipulations considerably. The omission of the operator makes it possible to apply the rules of the calculus of propositions to such expressions as (5) directly; the functional '$f(x)$' thus has the character of a propositional variable. Since it is always permissible to deal with free variables as though they were specialized, i.e., to treat them as quasi-specialized variables (cf. § 12), we can regard the free variables of an expression as remaining constant while the bound variables run through all their values. This process is then repeatable for every value of the free variable. Such usage corresponds to the use of constants in mathematical equations. Incidentally, mathematical constants are sometimes free variables, sometimes quasi-specialized variables, depending on whether they are subject to restricting conditions.

A further reason for the use of free variables as arguments is that we employ free variables also in other places of our calculus. The use of free propositional variables 'a', 'b' for the expression of generality, both within tautologies and within synthetic statements, has been mentioned in § 12. Similarly, function symbols like 'f', 'g' represent free variables when used for the expression of generality. Such usage occurs in tautologies; but it may also occur in synthetic formulas. Formulas of the latter sort represent indefinite expressions. We shall deal later with such expressions, which, in general, can be eliminated only by the transition to the higher calculus of functions (cf. § 42). In the present chapter we use free functional variables only in definite expressions, and thus only in tautologies.

On the other hand, the application of free variables cannot make operators dispensable. We need operators for the indication of a scope limited to a part of a formula. In particular, the distinction between a long and a short negation line over an all-statement cannot be made by the use of free variables. Furthermore, an existential operator standing as major operator cannot be expressed through a free variable. A complete calculus will therefore contain free variables by the side of bound variables, and thus employ two different sorts of scope rules.

Conversational language, in addition to the use of the word 'any', possesses various other means of expressing free variables. Thus the pronoun 'he who', sometimes abbreviated to 'who', assumes this function, as in sentences like 'he who trespasses against the law will be punished', or 'who steals my purse steals trash'. Such sentences represent general implications stated in terms of free variables; thus we can construe the first sentence in the form 'x trespasses against the law implies x will be punished'. Furthermore, the indefinite article is sometimes used in the sense of a free variable. Thus we say 'an engineer knows how to use a slide rule'; here both indefinite articles express free variables. Whether the indefinite article is meant to express generality or merely existence (cf. p. 91) is not indicated by the grammatical structure of the sentence but must be understood from the context. Thus it is clear that in the sentence 'an engineer used my slide rule' only existence is meant. Such ambiguities are avoided by the use of the word 'any'.

Expressions for free argument variables are found in most conversational languages; but not many languages possess terms of so consistent a usage as the English word 'any'. The German equivalent of 'any' is given by the pronouns 'man' and 'einer'; but these pronouns have the meaning of a free variable only in some applications, whereas they represent operators in other expressions. Thus 'anybody knows the secret' can be translated by 'man kennt das Geheimnis'; here the pronoun 'man' has the meaning of a free variable, as is apparent from the fact that the negation 'man kennt das Geheimnis nicht' means 'there is nobody who knows the secret', i.e., has the form '$(x)\overline{f(x)}$' or '$\overline{(\exists x)f(x)}$'. The sentence 'if anybody knows the secret, Marjorie talked' is to be translated by 'wenn einer das Geheimnis kennt, hat Marjorie es weiter erzählt'; here we have the existential meaning within the clause. The pronoun 'einer', therefore, acts here as a free variable. But the pronoun 'man' would lead to a different sense. The clause 'wenn man das Geheimnis kennt' means 'if everybody knows the secret'; the pronoun 'man', therefore, has here the limited scope of the all-operator. The pronoun 'einer', on the other hand, has sometimes the existential meaning with reference to the whole statement, and then does not represent a free variable, as in 'irgend einer kennt das Geheimnis', which means 'there is a man who knows the secret'. In certain idioms used in slang, or in archaic phrases, however, the use of 'einer' for simple generality is preserved: for instance, in 'das weiss einer doch', or 'das ist einem doch bekannt', both of which mean 'anybody knows that'. The dative 'einem' and the accusative 'einen' are regarded as more legitimate forms because they are used to replace the pronoun 'man' which has no such cases. In the form (5), moreover, the pronoun 'einer' always expresses the generality of a free variable: for instance, in the sentence 'wenn einer das Geheimnis weiss hat er es weiter erzählt'.

The French 'quelqu'un' is used like the German 'einer' and must be replaced by 'chacun', or 'on', for the expression of simple generality. Only in the form (5) does it have the meaning of 'all'. In this form the use of the free variable cannot be dispensed with, as in the German, because terms like 'chacun' and 'on' have a scope restricted to the clause in which they occur. Therefore (5) must be given by 'si quelqu'un sait le secret il en a parlé'. [Ex.]

§ 22. Some concepts referring to functions

In order to characterize the truth conditions of a propositional function, or the existential conditions of a situational function,[1] we introduce the following diagrams.

We begin with a one-place function. We imagine that all possible argument-objects in the world are represented by marks on the axis of Fig. 3.[2] When an argument satisfies the function $f(\hat{x})$, we put a

FIG. 3.   Matrix of a one-place function.

plus sign in its place; otherwise a minus sign. The resulting arrangement of plus signs and minus signs will be called the *matrix* co-ordinated to the function. This concept, apart from denoting a diagram of signs and not a class of objects, differs from the *extension* of a function in so far as it includes a reference to the arguments which make the function false; it represents the ordered pattern

---

[1] For the concepts developed in this section the distinction between propositional and situational functions is irrelevant. Thus we speak of the extension of the situational function $f(\hat{x})$ or of the propositional function '$f(\hat{x})$', meaning in both cases the class of objects; on the other hand, the matrix coordinated either to $f(\hat{x})$ or to '$f(\hat{x})$' is the diagram consisting of signs. Similarly, we use terms like 'symmetrical' and 'transitive' with reference both to objective and linguistic functions.

[2] For the sake of simplicity we assume for the diagram that the class of possible arguments is denumerable. The matrix is defined, however, also if this is not the case.

determined by both sorts of arguments. The extension is given by the class of objects corresponding to the class of plus signs of the diagram.

For a two-place function $f(\hat{x}, \hat{y})$ we construct a similar diagram, which, however, possesses two axes. On each axis we represent all possible arguments in the same order, and then make a plus sign in the place coordinated to a couple of arguments x and y when this couple satisfies the function in this order. The resulting pattern is the *matrix* of the two-place function $f(\hat{x}, \hat{y})$.

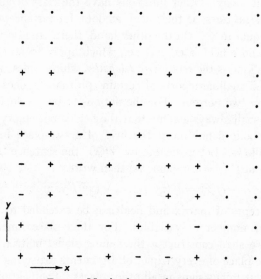

FIG. 4. Matrix of a two-place function.

As before, the extension of the function is determined by the class of plus signs in the whole matrix, since these represent the class of satisfying couples. In addition, we have here another important class which is called the *field* of the function. An argument x belongs to the field of the function $f(\hat{x}, \hat{y})$ when there is an argument y so that '$f(x, y)$' or '$f(y, x)$' holds. In our diagram the field is determined by the class of all horizontal or vertical lines that contain at least one plus sign; thereby lines belonging to the same argument, i.e., having the same distance from the zero point, are counted only once. This class differs from the extension because it does not have couples as its members, but the individual objects. Only for one-place functions are the field and the extension identical.

The field is therefore the class of all argument-objects for which the functional can be made true by a suitable choice of the other variable. Thus the field of the function '$\hat{x}$ is married to $\hat{y}$' is the class of all married men or women; the field of the function '$\hat{x}$ is the father of $\hat{y}$' is the class of all human beings, or, if the word 'father' is also applied to animals, the class of all animals with sexual reproduction. The field of the function '$\hat{x}$ is king of $\hat{y}$' is rather small today. Other functions have the class of all objects, or the *universal class*, as their field, as does, for instance, the function '$\hat{x}$ is equal to $\hat{y}$'. On the other hand, there are also two-place functions whose field is empty, i.e., which apply to no objects at all. Of this kind is the relation of *telepathy*, which, unless we accept the none-too-credible reports of certain spiritualists, does not hold between any two persons. For the notions to be developed in this section we shall always assume that the field is not empty.[1]

The symbolic definition of the field of a two-place function is given as follows. Let us denote by '$cf(x)$' the sentence 'x belongs to the field of f' (c = campus); we then write

$$cf(x) =_{Df} (\exists y)f(x, y) \vee (\exists y)f(y, x) \qquad (1)$$

The concepts of matrix and field can be extended to functions of a higher number of variables. For three-place functions, for instance, we shall construct a three-dimensional matrix, put plus signs in the place of every triplet of arguments satisfying the function, and put minus signs in all other places. The field of a three-place function is the class of things for which there are two other things such that the function holds. In symbols:

$$cf(x) =_{Df} (\exists y)(\exists z)f(x, y, z) \vee (\exists y)(\exists z)f(y, x, z) \vee (\exists y)(\exists z)f(y, z, x) \quad (2)$$

The field of a relation can be restricted by the use of an additional condition stating that the arguments must belong to a certain class. Thus the relation *child* can be restricted to the class of persons constituting the ancestry of a determinate person. Instead of using the function '$f(\hat{x}, \hat{y})$' we then use the function '$r(\hat{x}).f(\hat{x}, \hat{y})$', which is also a two-place function. In our example, $r(\hat{x})$ would represent

---

[1] When the relation is heterogeneous, i.e., when its two places refer to objects of a different type (cf. § 39), the field cannot be defined.

the ancestry of a person x_1; this function, therefore, may here also be written in the form '$r(\hat{x}, x_1)$', where 'x_1' is a constant.

Since two-place functions are more frequently used than others, an elaborate notation referring to them has been developed. In the two-place function $f(x, y)$ the argument x is called the *referent*, the argument y the *relatum*. Thus, with respect to the statement 'Peter is the father of John', Peter is the referent, and John is the relatum. There will be more than one relatum with respect to Peter if Peter has other children. The field includes all referents and all relata of the function. The class of referents is also called the *domain* of the function, and the class of relata is called the *converse domain*. We shall denote the two domains of a function 'f', respectively, by 'df' and 'ef'. These abbreviations are defined as follows:

$$df(x) =_{Df} (\exists y)f(x, y) \tag{3}$$
$$ef(x) =_{Df} (\exists y)f(y, x) \tag{4}$$

There are functions for which an object that is a possible referent cannot be a relatum. Thus, if x is the wife of y, there is no z such that z is the wife of x. In this case the field divides into the two classes of referents and relata; we shall speak here of a *divided field*. The definition of these functions is given by the relation

$$(x)[df(x) \supset \overline{ef(x)}] \tag{5}$$

There are other functions for which everything that is a possible referent is also a possible relatum. Thus if x is the child of y, y must be also the child of somebody; namely, y then must be a human being. When we use the restricted relation *child* introduced above, this condition holds with only two possible exceptions, consisting in a first and a last thing of this sort. Thus x_1 need not have a child, i.e., need not be also a relatum; and furthermore, appropriating a biblical legend for the illustration of a logical relation, we find in Adam a relatum that is not a referent, i.e., is not the child of somebody. In order to make our illustration hold, let us regard Eve as a child of Adam, which may be justified in view of her peculiar descent from him. Similarly, when we regard the relation *smaller* between the integers from 0 to 1000, the number zero is the only referent that cannot be a relatum, and the number 1000 is the only relatum that cannot be a referent. We shall call such functions *uniform.* Apart from two possible exceptions, one among the referents and one among the relata, domain and converse domain are identical for them, and are identical with the field. The definition of uniformity is given by the condition:

$$(\exists u)(\exists v)\{df(u).ef(v).(x)[[df(x) \equiv ef(x)] \lor (x = u) \lor (x = v)]\} \quad (6)$$

The expression in the first brackets states the condition that the possible exceptions u and v are, respectively, in the domain and the converse domain. The equivalence sign may be read as two implications stating that, when x is in the domain, it is also in the converse domain, and vice versa. By putting the existential operators '$(\exists u)$' and '$(\exists v)$' before the all-operator '(x)' we express the condition that there exists at most one exceptional member in the domain, and one in the converse domain, which is not also in the other domain. On the other hand, if there are no such exceptional members, the condition (6) will also be satisfied since then we can choose for 'u' and 'v', respectively, any member of the domain and of the converse domain. An example of this sort is given by the relation *smaller* holding between all positive and negative integers; here there is no exception to the statement that all referents are also relata.

There are also functions which are not uniform, but for which the field is not divided since some arguments at least occur in both domains. We subsume both these functions and the uniform functions under the category of functions with a *nondivided field*. They are defined by the relation

$$(\exists x)df(x).ef(x) \quad (7)$$

It is easily seen that (7) is the negation of (5). An example is the unrestricted relation *ancestor*. Persons who die without children are not ancestors, although they have ancestors; all others both are ancestors and have ancestors.

Another sort of two-place functions, or two-place relations, is characterized by the property that the relation holds for any two different elements of its field in one direction or in the other. Such functions may be called *interconnective*. The name may be illustrated by a telephone network; each subscriber can telephone each other subscriber (though not himself), and the network may be called interconnective. We transfer this name to the relation $\hat{x}$ *can telephone* $\hat{y}$ and call it an interconnective relation.[1] Furthermore, *smaller* is an interconnective relation because, for any two different numbers, one is smaller than the other. The relation *child*, on the other hand, is not of this sort. The definition of interconnective relations is given by the condition:

$$(x)(y)[cf(x).cf(y).(x \neq y) \supset f(x, y) \vee f(y, x)] \tag{8}$$

The sign ' $\neq$ ', meaning nonidentity, was introduced in § 18. An interconnective relation must also be uniform,[2] since if there were two different arguments x and y (or more) which can only be referents, or only be relata, neither ' $f(x, y)$ ' nor ' $f(y, x)$ ' would hold for them. But the converse does not hold, as is shown by the relation *child*, which is uniform when it is used in the restricted sense defined above but is not interconnective.

An important property of two-place functions concerns *univocality*. If a function $f(\hat{x}, \hat{y})$ is univocal with respect to y, i.e., if for a given x there is only one y satisfying $f(\hat{x}, \hat{y})$, whereas for a given y there may be several corresponding objects x, the function is called a *many-one function*. The distribution of such a function has on any vertical line only one plus sign (if it has a plus sign there at all), whereas the horizontal lines may have more than one plus sign. An example is given by the function ' $\hat{x}$ is a native of the country $\hat{y}$ '. Conversely we speak of a *one-many function* if there is only one x corresponding to a given y, whereas there may be several y's having the same x; the matrix then shows never more than one plus sign in the horizontal lines. Of this sort is the function ' $\hat{x}$ is the mother of $\hat{y}$ '. Finally we speak of *one-one functions* when the function is univocal in both directions; its matrix then has never more than one plus sign in any of the vertical or horizontal lines. Thus the function ' $\hat{x}$ is the wife of $\hat{y}$ ' is a one-one function in countries where marriage is restricted to monogamous relations. A function that is not univocal in either direction is called a *many-many function;* of this kind is the function ' $\hat{x}$ is a teacher of $\hat{y}$ '.

A further concept concerning two-place functions is the *converse*

---

[1] The term 'interconnective', applying to relations, must be carefully distinguished from our term 'connective' used for propositional operations. In mathematical literature the term 'connected' is frequently used for our term 'interconnective'.

[2] It is in view of this consequence that we admit two exceptions in the definition of uniform relations. A formal proof of this consequence will be given in § 30.

of $f(\hat{x}, \hat{y})$, which we denote by '$\breve{f}(\hat{x}, \hat{y})$' and define as resulting when the order of the arguments is reversed. Thus the relation *wife* is the converse of the relation *husband*. The definition is:

$$\breve{f}(x, y) =_{Df} f(y, x) \tag{9}$$

This notion is not restricted to one-one functions. Thus 'child' is the converse of 'parent'.

An important property of two-place functions is *symmetry*. A function is called *symmetrical* if its truth properties do not change when the order of the variables is reversed, or, in other words, when it coincides with its converse. These functions are defined by the condition:

$$(x)(y)[f(x, y) \supset f(y, x)] \tag{10}$$

This condition can also be stated in the form

$$(x)(y)[f(x, y) \supset \breve{f}(x, y)] \tag{11}$$

Thus 'equality' is a symmetrical function; if x is equal to y, then always y is equal to x. Other symmetrical functions are given by such predicates as 'similar', 'simultaneous', and 'is married to'. Symmetrical functions are characterized by a matrix in which the plus signs occupy positions symmetrical with respect to the lattice diagonal line; i.e., the pattern on one side of this line is the mirror image of that on the other side. All symmetrical functions are uniform.

The contrary of a symmetrical function is an *asymmetrical* function, i.e., a function which, when it holds between two arguments, excludes the validity of its converse. This type is defined by the condition:

$$(x)(y)[f(x, y) \supset \overline{f(y, x)}] \tag{12}$$

This can also be written:

$$(x)(y)[f(x, y) \supset \overline{\breve{f}(x, y)}] \tag{13}$$

Such functions are given by predicates like 'father', 'taller', and 'earlier'.[1] Between the two kinds we have the *mesosymmetrical*

---

[1] It can be easily shown that the implication in (10) or (11) can be reversed. We prove this by substituting 'x' for 'y' and 'y' for 'x', and then reversing the operators. The implication in (12) or (13), however, cannot be reversed. It is therefore permissible to replace the implication in (10) or (11) by an equivalence, whereas such a replacement is not permissible in (12) or (13). Thus, if y is not the father of x, it does not follow that x is the father of y. On the other hand, if y is not equal to x, it follows that x is not equal to y.

functions, i.e., functions for which an exchange of variables sometimes changes the truth-value and sometimes does not. Of this type
is the function 'brother'; if x is a brother of y, y may be a brother of
x, but it is also possible that y is a sister of x. Other mesosymmetrical
functions are 'x loves y', 'x is a friend of y', 'x sees y'. We see here
why we introduced the requirement that the field be not empty
(cf. p. 114): otherwise both (10) and (12) would hold for the same
function. We include both mesosymmetrical and asymmetrical
functions in the group of *nonsymmetrical* functions.

Another classification is given by the logical property of transitivity, which, however, is defined only for functions with a nondivided field. A function of this kind is *transitive* if it satisfies the
condition

$$(x)(y)(z)[f(x, y).f(y, z) \supset f(x, z)] \qquad (14)$$

An example is the relation *taller*, since with 'x is taller than y' and
'y is taller than z' we always have 'x is taller than z'. Other examples are given by functions like 'earlier', 'ancestor', and 'equal'.
We define the contrary, or *intransitive*, function by the postulate

$$(x)(y)(z)[f(x, y).f(y, z) \supset \overline{f(x, z)}] \qquad (15)$$

Thus 'father' is an intransitive function; if x is the father of y, and
y is the father of z, x is not the father of z, but the grandfather.
Between the two extreme cases we have the *mesotransitive* functions,
for which neither (14) nor (15) holds, such that, if '$f(x, y)$' and
'$f(y, z)$' are true, '$f(x, z)$' is sometimes true and sometimes false.
Examples are 'x is similar to y', 'x sees y', 'x is a friend of y'. The
nontransitive functions include both the intransitive and the mesotransitive functions.

A further division is given by the characteristic of reflexivity.
A function is *reflexive* if we can put the same argument in both
places of the variable, i.e., if the condition

$$(x)f(x, x) \qquad (16)$$

holds. Thus equality is a reflexive relation since x is equal to x.
Another example is similarity. In its matrix a reflexive function
will be recognizable from the fact that all places on the lattice
diagonal line are occupied by plus signs. A function is called *irreflexive* if '$f(x, x)$' is always false, i.e., if

$$(x)\overline{f(x, x)} \qquad (17)$$

holds. Examples are given by the functions 'taller', 'father'. Between the two cases we have as a third case the *mesoreflexive* functions for which '$f(x, x)$' is sometimes true and sometimes false, as in 'x admires y'. The *nonreflexive* functions comprise the irreflexive and the mesoreflexive functions. A function with a divided field is always irreflexive; so is an asymmetrical function. A reflexive function is uniform.

The use of the notion of field permits us to introduce a somewhat restricted reflexivity as follows: if a condition corresponding to (16) holds at least for all arguments that belong to the field of f, we say that the function f is *reflexive within its field*. For two-place functions we can write this condition:

$$(x)[cf(x) \supset f(x, x)] \tag{18}$$

It can be shown that, if a function is symmetrical and transitive, it is also reflexive within its field. This is proved by putting in (14) 'x' for 'z'; a formal presentation of this proof will be given in § 30.

An interesting combination of properties is given by a relation that is interconnective, asymmetrical, and transitive. Such a relation defines a *series*, i.e., a class ordered in a linear arrangement. Thus the relation *smaller*, which has these properties, establishes a linear order among the numbers. The relation *ancestor in the male line*, restricted to the ancestors of a certain man, is also of this kind. The general relation *ancestor*, however, does not constitute a series because, although it is asymmetrical and transitive, it is not interconnective. On the other hand, an interconnective relation that does not establish a linear order is given by the relation of *possible telephone connection*, already mentioned; the reason is that this relation is symmetrical rather than asymmetrical. It may even be conceived as a transitive relation when we extend the meaning of the word 'telephone' in such a way that we may say a man can telephone himself, in order to make the relation reflexive. [Otherwise (14) would not hold for the case that we put $z = x$.] An example of an interconnective relation that is not transitive may be constructed as follows. In a chess tournament each player plays with every other player. Let us now, after the tournament has been closed, regard the relation $\hat{x}$ *was not defeated by* $\hat{y}$, which holds when x defeated y or when the game was a draw. This relation is an interconnective relation since it must hold, in one direction or the other, or in both, between any two players. But it need not be transitive since it may happen that x defeats y and y defeats z, whereas x is defeated by z. If there were no draw games in the tournament the relation would even be asymmetrical. It is clear that we do not arrive here at a linear order of the players. The relation, however, will be uniform, since it is interconnective. For a tournament of the latter sort, which includes no draw games, the uniformity is seen from the fact that every player will be sometimes a winner and sometimes a loser, with the possible exception of two players: one may have always won, and one may have always lost. The result is easily extended to the inclusion of draw games.

The concepts developed here for two-place functions can be given extended meanings for functions with several variables. Thus a three-place function has three domains. It will be called uniform when every member of its field can belong to each domain, with three possible exceptions. It will be interconnective when for any three members of its field the relation holds, at least for one order of the members. The relation *between* applied to points on a line has both these properties. Similarly, also, the concepts of symmetry, transitivity, and reflexivity can be extended. Thus we say that the function '$\hat{y}$ is between $\hat{x}$ and $\hat{z}$' is symmetrical in the variables 'x' and 'z'. The function '$\hat{x}$ is on a straight line with $\hat{y}$ and $\hat{z}$' is transitive in 'y' and 'z'. The function '$\hat{x}$ introduces $\hat{y}$ to $\hat{z}$' is not irreflexive in the variables 'x' and 'y', but irreflexive in the variables 'x' and 'z'.

The extension of the concepts concerning univocality to functions of more than two variables is somewhat complicated since univocality will depend on what arguments we regard as given. We shall call a three-place function a (*one-one*)-*one* function when, given the first two arguments, the third is uniquely determined; and it will be called a *one-*(*one-one*) function when, given the first argument, the two others are determined. Thus the function '$\hat{x}$ is halfway between $\hat{y}$ and $\hat{z}$' is a (one-one)-one function and at the same time a one-(many-many) function. The function '$\hat{x}$ is the child of $\hat{y}$ and $\hat{z}$' is both a (one-one)-one and a one-(one-one) function. This notation, however, characterizes a function only for the given order of the variables and does not include a characterization for another order, whereas the notation used for two-place functions includes a characterization of the converse.

After considering the divisions of functions by place properties, we turn now to another division which concerns the inner structure of functions. A function like '$f(\hat{x})$', '$f(\hat{x}, \hat{y})$', is called an *elementary function* when the symbol 'f' is not reducible to other symbols. There are other expressions in which the symbol 'f' possesses an inner structure, i.e., is defined in terms of other functions or arguments. Such functions are called *complex functions*.

The simplest form of complex function consists in functions defined as a combination of other functions by propositional operations, such as given in

$$g(\hat{x}) =_{Df} f(\hat{x}) \lor h(\hat{x}) \tag{19}$$

Here '$f(\hat{x})$' and '$h(\hat{x})$' are elementary functions; '$g(\hat{x})$' is a complex function. A symbol '$g(\hat{x})$' used for a complex function, and likewise the symbol '$g(x)$' used for the functional, will be called a *contracted symbol*.

Another kind of complex functions comprises functions that have absorbed some of their specialized arguments. We can, for instance, consider the term 'son of William' as a one-place predicate constructed from the two-place predicate 'son of' by specializing one of its arguments. In symbols,

$$g(x) =_{Df} f(x, y_1) \tag{20}$$

We call 'y_1' an *implicit argument* of '$g(x)$'. The symbol '$g(x)$' is a *contracted symbol;* the symbol '$f(x, y_1)$' is not.

Another form of complex functions, or predicates, results when the implicit argument is not a specialized but a bound variable. Of this kind is, for instance, the function '$\hat{x}$ is married', as this means 'there is a y such that $\hat{x}$ is married to y'. Thus the complex predicate 'g' is here introduced by the definition

$$g(x) =_{Df} (\exists y) f(x, y) \tag{21}$$

The contracted symbol '$g(x)$' has here the bound variable 'y' as an implicit argument. Of the same kind is the function '$\hat{x}$ drives', which means 'there is a y, which is a vehicle, which $\hat{x}$ drives'. The function '$\hat{x}$ drives' has therefore absorbed, not only a bound variable, but also a predicate, namely, 'vehicle'.

A similar form of complex functions is given by the *relational product*. This term is introduced by the definition

$$h(x, z) =_{Df} (\exists y)f(x, y).g(y, z) \qquad (22)$$

Thus 'grandfather' is the relational product of the predicates 'father' and 'parent'. Here also two predicates are merged, and a bound variable has been absorbed. In the special case that 'g' is the same predicate as 'f', and 'f' is transitive, the relational product 'h' is the same as the predicate 'f'.

There are complex functions of a similar kind, in which, however, three functions are merged into one. Of this kind are noun compounds such as 'street corner', 'door bell', 'table cloth'. Only two of the functions are named in the composite expression, while the third, meaning 'belonging to', 'connected with', or something similar, is understood. Thus 'door bell' can be defined as follows, when we write 'd' for 'door', 'b' for 'bell', 'cn' for 'connected with', and 'db' for 'door bell':

$$db(x) =_{Df} b(x).(\exists y)d(y).cn(x, y) \qquad (23)$$

Another form of complex function contains a variable bound by an all-operator. Thus '$\dot{x}$ is successful' can be defined as 'for all y, if $\dot{x}$ aspires y, $\dot{x}$ reaches y'; in symbols:

$$s(x) =_{Df} (y)[a(x, y) \supset r(x, y)] \qquad (24)$$

The word 'successful' is therefore a contracted symbol standing for a combination of two predicates and a variable bound by an all-operator.

The introduction of contracted symbols, although, of course, arbitrary to some extent, is subject to certain rules. These are as follows:

1. Free argument variables used in the definiendum must be contained in the definiens.
2. One and the same free variable must not occur more than once in the definiendum; e.g., the definiendum must not have the form '$h(x, x)$'.
3. Argument constants must not be used in the definiendum; they can be introduced later, however, by a specialization of the defined expression.
4. The definiendum must not contain bound variables.

5. The definiendum must not contain a function sign defined previously in the language — in particular, a function sign used in the definiens.

6. The definiendum must not contain propositional operations.

When these rules are violated, contradictions may arise. For instance, the definition

$$(\exists x)h(x).f(x) =_{Df} (x)[f(x) \supset g(x)] \qquad (25)$$

violates rules 4–6. Now from the left-hand side we can derive the statement '$(\exists x)f(x)$', which is not derivable from the right-hand side. This result leads to contradictions.

On the other hand, rules 1–5 appear rather rigorous. They admit as contracted terms only expressions of the form '$f(x, y, \ldots)$', such as those used in our examples. It may well be possible to replace our rules by weaker rules which nonetheless exclude contradictions. But little work has been done so far in this field, and the author feels unable to present here a more satisfactory set of rules. We shall therefore regard the given rules only as defining the term 'contracted symbol', but not as rules to be required for all kinds of definitions.[1] [Ex.]

---

[1] For an analysis of rules for definitions cf. also R. Carnap, *Logical Syntax of Language*, Harcourt, Brace, New York, 1937, p. 24. Our rules 1 and 2 are taken from two rules used by Carnap.

IV. The Simple Calculus of Functions (Continued)

§ 23. Truth characters of one-place functions

One-place propositional functions admit of a rather simple classification of their matrices. They divide into functions that are always true, functions that are always false, and mixed functions. We have, therefore, three *truth characters* of one-place propositional functions, and we denote them by the letters A, E, M. Let us use the term *functional derivatives* for statements derived from propositional functions; they will include all-statements, existential statements, and the functional '$f(x)$'. The truth-values of the functional derivatives are determined by the truth characters of the functions; we express these relations in truth tables II.

II. TRUTH TABLES FOR FUNCTIONAL DERIVATIVES

$f(\hat{x})$	$(x)f(x)$	$(\exists x)f(x)$	$f(x)$
A	T	T	T
M	F	T	T, F
E	F	F	F

These tables resemble the truth tables for propositions. But they are different in that functions are divided into three categories, whereas propositions are classified in two categories. The calculus of one-place functions, therefore, represents a three-valued system. A further difference is found in the fact that the truth-value of the functional derivatives is not always determined: for a function of the mixed kind the truth-value of the functional is indeterminate; it can be true or false.[1] We express this fact by putting both the letters 'T' and 'F' in the respective place in the table. On the other hand, these tables show why free variables can be used in

---

[1] Cf. our discussion of the functional on p. 109.

assertions: for a function of the *A*-kind, the functional '$f(x)$' is true.

We must now introduce a second sort of table. Combining two functions '$f(\hat{x})$' and '$g(\hat{x})$' by propositional operations, we can ask what the truth character of the resulting function will be when those of the constituents are given. We thus arrive at tables III a and III b.

III a. TRUTH TABLES FOR THE NEGATION OF A FUNCTION

$f(\hat{x})$	$\overline{f(\hat{x})}$
A	*E*
M	*M*
E	*A*

Table III a is easily understood. It is clear that, if '$f(\hat{x})$' is always true, '$\overline{f(\hat{x})}$' is always false. For the binary operations we construct tables which refer to nine possible combinations of the truth characters. We see that in general the truth character of the combination is determined when the characters of the constituents are given.

III b. TRUTH TABLES FOR BINARY OPERATIONS BETWEEN FUNCTIONS

$f(\hat{x})$	$g(\hat{x})$	$f(\hat{x}) \vee g(\hat{x})$	$f(\hat{x}).g(\hat{x})$	$f(\hat{x}) \supset g(\hat{x})$	$f(\hat{x}) \equiv g(\hat{x})$
A	*A*	*A*	*A*	*A*	*A*
A	*M*	*A*	*M*	*M*	*M*
A	*E*	*A*	*E*	*E*	*E*
M	*A*	*A*	*M*	*A*	*M*
M	*M*	*A, M*	*M, E*	*A, M*	*A, M, E*
M	*E*	*M*	*E*	*M*	*M*
E	*A*	*A*	*E*	*A*	*E*
E	*M*	*M*	*E*	*A*	*M*
E	*E*	*E*	*E*	*A*	*A*

Thus, when '$f(\hat{x})$' is of the *A*-kind and '$g(\hat{x})$' of the *M*-kind, their disjunction will be of the *A*-kind, and their conjunction will be an *M*-function. Only when both functions are of the mixed kind does an indeterminacy result. We therefore give several possible values in the middle line of the tables. Which of these values holds will depend on the functions used. Thus if the function '$g(\hat{x})$' has a plus sign in its matrix in all those cases where '$f(\hat{x})$' has a minus sign, the or-combination will produce an A-function; if not, the or-combination will only be an M-function. But it cannot be an E-function. The and-combination will be an M-function if at least for one argument both functions have a plus sign; in all other cases it will be an E-function. It can never be an A-function. Similar considerations hold for the other combinations.

Tables III a and III b can be extended by the following device to include operations combining propositions with propositional functions. In correspondence with our rule (cf. § 20) to regard statements of the form '$(x)a$' and '$(\exists x)a$' as meaningful, we shall consider propositions as degenerate cases of propositional functions which can assume only the truth characters A and E, according as the propositions are true or false. This rule allows us to coordinate a truth character, for instance, to expressions of the form '$a \vee f(x)$', which represent nondegenerate functions of 'x'.

Let us now turn to a further application of our results. The truth characters A, M, E of functions can be interpreted as representing the *modalities*, i.e., the notions of *necessity*, *possibility*, and *impossibility*. This interpretation can be constructed as follows.[1]

When we say that it is possible that a physical object is red, we mean to say, in this interpretation, that there are red objects, i.e., that the function 'red' is of the mixed kind. To say that it is impossible that an animal is a sea serpent means, similarly, that there are no sea serpents, i.e., that the function 'sea serpent' is of the E-kind. Furthermore, when we say: it is necessary that a magnet attracts a piece of iron, we mean that all magnets do so, i.e., that the function '$\hat{x}$ is a magnet implies $\hat{x}$ attracts a piece of iron' is of the A-kind.

---

[1] This interpretation was first given by Bertrand Russell, *Introduction to Mathematical Philosophy*, Macmillan, New York, 1919, chapter XV. Tables II, III a, and III b were first constructed by W. Dubislav, *Journal für die reine und angewandte Mathematik*, Bd. 161, 1929, p. 107, and used for case analysis as explained in § 25 of this book. The same tables were derived by the author within the frame of probability logic and used for the interpretation of modalities in *Berichte der Berliner Akademie*, math. Kl., 1932, XXIX.

Tables III a and III b are easily understandable in this interpretation. Thus, if it is impossible that something is a sea serpent, it is necessary that a given thing is not a sea serpent (table III a). Similarly, it will be impossible that a thing is at the same time red and a sea serpent (line 6 of table III b). The indeterminacy of the middle line of table III b is also easily explained. Thus when we toss a coin it is possible to get heads, and possible to get tails; but it is necessary to get heads or tails. Here the A-value of the middle line is the correct one. On the other hand, casting two dice it is possible to get a 'six' on the one, and possible to get a 'six' on the other; however, to get a 'six' on the one or on the other is only possible, not necessary. Therefore here the M-value will be correct. Similar interpretations are easily given for the and-combination.

It should be noticed that the concept *possible* used in conversational language is not always the same as our truth character M; a second meaning is given when we say that what is necessary is also possible. We therefore must distinguish between a wider and a narrower meaning. The wider meaning is given by the disjunction 'A or M', and thus by the statement '$(\exists x)f(x)$'. Conversational language employs both meanings. The use of the narrower meaning is visible in the fact that we do not speak of possibility when we know that something is necessary. Thus it would be unusual to say that it is possible that the sides of a square have equal length. For technical reasons it is advisable to limit possibility to the narrower meaning, i.e., to the category M, and we shall use the word in this sense; if ambiguities arise we shall employ the phrase 'merely possible'.[1]

There is, however, another discrepancy. We may say: although, in fact, there are no sea-serpents in this world, it is possible that there were such animals. We might assert this possibility even if we knew that there actually never have been, nor will be, any sea-serpents. Similarly, we may deny necessity for the statement that all persons in a given room at a given time wear red neckties, even though this statement should be true in a particular case and thus would represent an A-function. The modalities given by the truth characters A, M, E are determined simply by the extensions of the

---

[1] For *possible* in the narrower sense the word *contingent* is employed. But since this word has some further connotations it appears preferable not to use it.

functions and may therefore be called *extensional modalities*. Another type of modality will be defined later (§ 65) under the name of *nomological modalities;* these modalities correspond to the use of the terms 'necessary', 'possible', and 'impossible' in examples like the ones given.

Because of the ambiguity of the middle row in tables III b the tables are not adjunctive, i.e., they cannot throughout be read from left to right. The extensional modalities, therefore, are not adjunctive. This fact must not be regarded as a fault in the logical interpretation of the modalities; it rather is inherent in their nature since, as we saw, the ambiguity of the middle row in tables III b corresponds to the common usage of the terms.

§ 24. Definition of tautologies containing functions

A formula written in functions may contain free or specialized functions, and free or bound or specialized arguments. In addition, it may contain propositions, or propositional variables, whose structure is not indicated. We do not consider, for the present, formulas that contain higher functions having lower functions as their arguments; i.e., we remain in the simple calculus of functions.

Let us assume that the formula contains no contracted symbols for functions, i.e., no abbreviations standing for complex functions (cf. § 22). [The elimination of contracted symbols can easily be achieved by the substitution of their definitions.] The formula then is written entirely in *elementary functions*, i.e., functions that do not include operators or propositional operations. Furthermore, it may contain *elementary propositions* or *elementary propositional variables*, i.e., propositions, or propositional variables, that do not include propositional operations.

When a true statement contains free variables, in the form of functions, or of arguments, or of propositional variables, it is necessary that the statement hold for all values of these free variables. For free functional variables this condition can be expressed by saying that the statement holds for all matrices which can be coordinated to these functions. This formulation includes the condition for arguments of such functions because, for a given argument, the expression to which it belongs can be made true or false by a suitable choice of the matrix. Since by inserting other arguments the expression can be made only true or false, the statement, therefore, will hold also for all other arguments. Only for free argument variables in specialized functions must we add the condition that the statement holds for all values of this variable.

A tautology differs from a merely true statement by the requirement that the condition of generality be extended to include all specialized variables. This requirement means: if we put free variables in the place of corresponding constants occurring as functions, arguments, or propositions, the statement must be true in this form also. Thus when we call the statement 'John's tie is either red or not red' a tautology, we do so because the statement will also be true when we formulate it correspondingly for any other object and for any other property. On the other hand, the statement 'all men are mortal' is not a tautology, although it holds for all arguments, because it does not hold for all functions put in the place of 'men' and 'mortal'. We can therefore define a tautology in propositional functions as follows.

Definition of a tautology in propositional functions. A statement is a tautology if, and only if, it is true for all matrices of its elementary functions, whether they are free or specialized variables, and for all truth-values of its elementary propositional variables, including both free and specialized ones of those variables.

As an example of a tautology let us use the formula

$$(x)[f(x) \vee \overline{f(x)}] \tag{1}$$

Given any matrix of the function '$f(\hat{x})$', it is clear that the function '$\overline{f(\hat{x})}$' has the complementary matrix, i.e., has plus signs in all places where the function '$f(\hat{x})$' has minus signs, and vice versa. The sign of the 'or', therefore, makes (1) true for all possible matrices. By similar considerations we show that the formula

$$(x)(y)[f(x, y) \vee \overline{f(x, y)}] \tag{2}$$

is a tautology.

We have used in these considerations material thinking applied to any infinity of objects, since the number of possible arguments may be regarded as infinite, or at least as practically infinite. It is true that this material thinking is in the metalanguage; for instance, it is of the form: for every x for which '$f(x)$' is true, '$\overline{f(x)}$' is false. Now we know that we cannot entirely eliminate material thinking from the metalanguage. But we can attempt to reduce it to simpler operations involving only a finite, or even a small, number of objects. It is this sort of operation which we mean when we say that the ascertainment of tautologies should be mechanized. We present such methods in the following section. Only for the establishment of such methods do we need material thinking involving an infinity of objects, in the same sense as we needed such thinking for the construction of tables III a and III b. But, once these methods have been formulated as rules, their application can be carried through by mechanical operations.

§ 25. The use of case analysis for the construction of tautologies in propositional functions

The method through which a mechanical ascertainment of tautologies can be achieved makes use of the classification of functions by truth characters. Since truth characters are defined only for one-place functions, this method is restricted to such functions. We shall analyze a given formula in one-place functions by going through all possible cases resulting for different truth characters of its constituents, thus using a method of *case analysis* corresponding to the case analysis used in the calculus of propositions.[1]

Compared with the situation in the calculus of propositions there is, however, one difference. We saw that tables III b, § 23, include an indeterminacy in their middle line; this indeterminacy leads to difficulties which must be discussed now.

There is a group of formulas in which no such difficulties appear because the indeterminacy of the middle line drops out. Of this kind is the formula

$$(x)[f(x).g(x)] \equiv [(x)f(x).(x)g(x)] \tag{1}$$

Assuming for the functions '$f(x)$' and '$g(x)$' one of the three truth characters, and going through all possible cases, we find that this formula is always true and therefore is a tautology. The indeterminate case resulting when the truth character M is assumed for both functions, which on the left-hand side of (1) leads to the two possible values M and E, does not disturb us here because for both these values the left-hand side is false.

---

[1] This method, which is little known, was introduced by W. Dubislav; cf. footnote on p. 127.

However, case analysis meets with difficulties in a second group of formulas. Let us take for instance the formula

$$(x)[\overline{f(x).g(x)}] \equiv (x)[\overline{f(x)} \vee \overline{g(x)}] \tag{2}$$

In those cases where at least one of the two functions is assumed not to be of the mixed kind the truth-value '*T*' for the whole formula is clearly determined; we encounter difficulties, however, for the case that both '$f(\hat{x})$' and '$g(\hat{x})$' are mixed functions, since then the operand on each side can have the truth characters '*M*' or '*A*'. We therefore cannot say immediately whether the equivalence holds. In order to overcome this indeterminacy we must show, by material thinking referring to an infinity of objects, that there is a coupling between the two sides of (2) such that, if we have '*A*' on the left-hand side, we have also '*A*' on the right-hand side, whereas we shall have '*M*' on the right-hand side if we have '*M*' on the left-hand side. Only with this addition can we show that (2) is a tautology, i.e., that the formula holds for all matrices of its elementary functions. We find other examples of proofs needing such additional material thinking by the consideration of formulas like

$$(x)[f(x) \vee \overline{f(x)}] \tag{3}$$
$$(x)f(x) \supset (x)[g(x) \vee f(x).\overline{g(x)}] \tag{4}$$

which are also tautologies. The first formula leads to an indeterminacy if '$f(\hat{x})$' is assumed to be an *M*-function; the second, when '$f(\hat{x})$' is an *A*-function and '$g(\hat{x})$' is an *M*-function.

The aforementioned difficulties are not insurmountable, since the material analysis referring to an infinity of objects can be strictly given, at least for not too complicated formulas. With the use of such material thinking, however, we have abandoned our program of a complete mechanization using only the tables for truth characters. We must therefore restrict the method of case analysis to formulas for which this method is applicable without additional material thinking. We can say: if a formula can be shown to be a tautology by a mechanical use of case analysis, it certainly is a tautology; if the proof leaves the result undetermined, the formula still may be a tautology. Case analysis furnishes here merely a sufficient condition of tautological character, not a necessary condition. We therefore establish the following metatheorem, which we numerate in continuation of the numeration given in § 14:

Metatheorem 11. (*Law of case analysis for functions.*) If a formula can be shown, by the use of case analysis in terms of its elementary functions, to be true for all possible combinations of truth characters, it is a tautology.

Tautologies containing free argument variables are included in our definition of tautologies given in § 24. Now we know that expressions of the form '$f(x)$' can only be true or false, whatever be the matrix of the function $f(\hat{x})$. The relations between the truth character of the function and the truth-value of the functional is given in the last column of table II, § 23; we therefore shall use this column for the case analysis of formulas containing free argument variables. Thus in the case analysis of formula (7, § 21) we see from table II that, if $f(\hat{x})$ is of the A-kind, both implicans and implicate are true; therefore the formula is true. If $f(\hat{x})$ is of the M-kind or E-kind, the implicans is false, and therefore the formula is also true. The indeterminacy expressed in the middle line of table II for the last column, therefore, drops out for this formula. Corresponding considerations show that formula (13, § 21) is a tautology.

The number of formulas which by case analysis can be shown to be tautologies is not small. Thus all formulas in one-place functions given in our list on p. 134, i.e., formulas 9a–14c, can be proved by case analysis. We shall therefore discuss these formulas here; the remaining formulas of the list will be discussed later (§ 28). The numeration continues the numbers used in the list of tautologies of the calculus of propositions, given on p. 38.

Our collection begins with some groups of formulas concerning the scope of operators. Formulas 9a–10f govern the fusion or division of operands containing variables in every term. Some of these formulas, such as 9a or 10b, are equivalences; they state division of operands when read from left to right, and a fusion of operands when read from right to left. The others, which are implications, state that either a fusion or a division of operands is at least implied by, though not equivalent to, the left-hand side; thus in 9b a fusion is implied, in 9d a division is implied. We see that the different propositional operations lead to different rules according as the formula contains all-operators or existential operators.

Some of these formulas were discussed in § 20. Thus 9d was introduced in the transition from (6 to 7, § 20). As to 9e, it should be noticed that the consequence '$(\exists x)f(x) \supset (\exists x)g(x)$' follows only from the implication '$(x)[f(x) \supset g(x)]$'; it does not follow from the form '$(x)f(x) \supset (x)g(x)$'. These relations may be illustrated by an example. From the implication: 'if all men refuse to bear arms, all men will live at peace', which is true, we cannot derive the implication 'if some men refuse to bear arms, some men will live at peace'. The reason is that those men who do bear arms might disturb the peace of those who do not. In order to derive the implication concerning some men, we must start from the statement: 'for all x, if x is a man refusing to bear arms, x will live at peace'. But this statement cannot be maintained as true.

As to the equivalence operation we have a scope formula only for all-operators, stated in 9f. There is no corresponding formula for existential operators; i.e., between the two expressions '$(\exists x)[f(x) \equiv g(x)]$' and '$(\exists x)f(x) \equiv (\exists x)g(x)$' no tautological implication can be established in either direction.[1]

TAUTOLOGIES IN THE CALCULUS OF FUNCTIONS

Formulas concerning fusion or division of operands:

9a. $(x)[f(x).g(x)] \equiv (x)f(x).(x)g(x)$
9b. $(x)f(x) \vee (x)g(x) \supset (x)[f(x) \vee g(x)]$
9c. $(x)[f(x) \vee g(x)] \supset (x)f(x) \vee (\exists x)g(x)$
9d. $(x)[f(x) \urcorner g(x)] \supset [(x)f(x) \supset (x)g(x)]$
9e. $(x)[f(x) \supset g(x)] \supset [(\exists x)f(x) \supset (\exists x)g(x)]$
9f. $(x)[f(x) \equiv g(x)] \supset [(x)f(x) \equiv (x)g(x)]$
9g. $(x)[f(x) \equiv g(x)] \supset [(\exists x)f(x) \equiv (\exists x)g(x)]$
9h. $(x)f(x).(x)[f(x) \supset g(x)] \supset (x)g(x)$

10a. $(\exists x)f(x).g(x) \supset (\exists x)f(x).(\exists x)g(x)$
10b. $(\exists x)[f(x) \vee g(x)] \equiv (\exists x)f(x) \vee (\exists x)g(x)$
10c. $(\exists x)[f(x) \supset g(x)] \equiv (x)f(x) \supset (\exists x)g(x)$
10d. $[(\exists x)f(x) \supset (\exists x)g(x)] \supset (\exists x)[f(x) \supset g(x)]$
10e. $[(\exists x)f(x) \supset (x)g(x)] \supset (x)[f(x) \supset g(x)]$
10f. $(\exists x)f(x).(x)g(x) \supset (\exists x)f(x).g(x)$

[1] To show this, let '$f(\hat{x})$' be always false, and '$g(\hat{x})$' be sometimes true, then the first expression is true and the second is false; and put '$\overline{f(x)}$' for '$g(x)$' and let '$f(\hat{x})$' be a mixed function, then the second expression is true and the first is false.

11a. $(x)[a.f(x)] \equiv a.(x)f(x)$
11b. $(x)[a \vee f(x)] \equiv a \vee (x)f(x)$
11c. $(x)[a \supset f(x)] \equiv a \supset (x)f(x)$
11d. $(x)[f(x) \supset a] \equiv (\exists x)f(x) \supset a$
11e. $(x)[f(x) \equiv a] \supset [(x)f(x) \equiv a]$
11f. $[(x)a] \equiv a$

12a. $(\exists x)[a.f(x)] \equiv a.(\exists x)f(x)$
12b. $(\exists x)[a \vee f(x)] \equiv a \vee (\exists x)f(x)$
12c. $(\exists x)[a \supset f(x)] \equiv a \supset (\exists x)f(x)$
12d. $(\exists x)[f(x) \supset a] \equiv (x)f(x) \supset a$
12e. $[(\exists x)f(x) \equiv a] \supset (\exists x)[f(x) \equiv a]$
12f. $[(\exists x)a] \equiv a$

Formulas concerning negation of operators:

13a. $\overline{(x)f(x)} \equiv (\exists x)\overline{f(x)}$
13b. $\overline{(\exists x)f(x)} \equiv (x)\overline{f(x)}$
13c. $\overline{(x)f(x)} \supset (x)\overline{f(x)}$
13d. $\overline{(\exists x)f(x)} \supset (\exists x)\overline{f(x)}$

Formulas of subalternation:

14a. $(y)f(y) \supset f(x)$
14b. $f(x) \supset (\exists y)f(y)$
14c. $(x)f(x) \supset (\exists x)f(x)$

Formulas concerning two operators:

15a. $(x)(y)f(x, y) \equiv (y)(x)f(x, y)$
15b. $(\exists x)(\exists y)f(x, y) \equiv (\exists y)(\exists x)f(x, y)$

16a. $(\exists x)(y)f(x, y) \supset (y)(\exists x)f(x, y)$
16b. $(\exists x)(y)f(x).g(y) \equiv (y)(\exists x)f(x).g(y)$
16c. $(\exists x)(y)[f(x) \vee g(y)] \equiv (y)(\exists x)[f(x) \vee g(y)]$
16d. $(\exists x)(y)[f(x) \supset g(y)] \equiv (y)(\exists x)[f(x) \supset g(y)]$
16e. $(x)(y)[f(x, y) \vee g(x, y)] \supset (\exists x)(y)f(x, y) \vee (x)(\exists y)g(x, y)$

17a. $(x)(y)f(x, y) \supset (x)f(x, x)$
17b. $(\exists x)(y)f(x, y) \supset (\exists x)f(x, x)$

Formulas 11a–12f concern the extension of a scope over a constant. We add these formulas for the following reason. As will be seen later it is permissible in any of the formulas 9a–10f to replace a function by a constant; we thus can derive formulas containing operators and constants. For the case of constants, however, even stronger formulas can be established; thus 11b is stronger than a formula derived from 9b by putting 'a' for '$g(x)$', because 11b states an equivalence between the two sides whereas 9b states only an implication. [Cf. the considerations added to (4 and 5, § 20).] In spite of the equivalence the two sides of 11b have of course a different logical structure. We see this when we substitute in 11b the form of writing used for the all-sign in (5, § 19); 11b then assumes the form

$$[a \vee f(x_1)].[a \vee f(x_2)] \ldots \equiv a \vee [f(x_1).f(x_2) \ldots] \qquad (5)$$

Applying the same way of writing to 9b, we obtain

$$[f(x_1).f(x_2) \ldots] \vee [g(x_1).g(x_2) \ldots] \supset [f(x_1) \vee g(x_1)].[f(x_2) \vee g(x_2)] \ldots$$
$$(6)$$

It is clear that we have here no equivalence when we apply to the left-hand side the second distributive rule (4d, § 8), whereas (4b, § 8), applied to the right-hand side of (5), furnishes the equivalence. An illustration of the right-hand side of 11b is given by the sentence: 'all things are predetermined or Leibniz is mistaken'; the left-hand side then may be stated: 'consider anything whatsoever, it will be predetermined or Leibniz is mistaken'.

We see that the implications 9b and 10a are transformed into the equivalences 11b and 12a. The equivalences 11c and 11d correspond, respectively, to the implications 9d and 9e. On the other hand, the two equivalences 12c and 12d can be derived from the equivalence 10c by putting 'a', respectively, for '$f(x)$' or '$g(x)$'. Similarly, 11e follows from 9f when we put 'a' for '$g(x)$', whereas 12e has no analogue. Formulas 11f and 12f state the rule for an operator whose scope is a constant. In the case analysis of formulas 11a–12f we apply the rule according to which a proposition can be regarded as a function capable only of the two characters A and E (cf. p. 127).

Next to the group of formulas concerning a fusion or division of operands we find the group 13a–13d dealing with the negation of operators.

For the group 14a–c we have chosen the name of formulas of subalternation, although this name is usually applied only to 14c [cf. (8, § 19)]. But 14a–b can be regarded as intermediate steps of subalternation since from them 14c is derivable by the transitivity of implication. The formulas 14a–b, which we have discussed in (7 and 13, § 21), are written in the free variable 'x'; they can also be written with 'x' as a bound variable when we apply the rule for free variables introduced in § 21. They then have the form, respectively,

$$(x)[(y)f(y) \supset f(x)] \tag{7}$$
$$(x)[f(x) \supset (\exists y)f(y)] \tag{8}$$

These formulas, as well as 14a–c, can be proved tautologous by case analysis. For 14c we use the first two columns of table II, for 14a–b we also use the last column. Formulas (7) and (8) possess two operators; but they include only one-place functions and can therefore also be dealt with by case analysis. Thus when we assume for the function '$f(\hat{x})$' one of the three values A, M, E, the proposition '$(x)f(x)$' in 14a is determined as true or false by the table II; regarding this proposition as a function capable only of the values A and E and applying tables III b to the expression in the brackets we show easily that the formula is always true.

When we wish to construct tautologies that cannot be verified by case analysis we must use the method of derivation, namely, we must derive such tautologies from formulas that are obtained by case analysis. We therefore turn now to an analysis of the procedure of derivation.

§ 26. The rules of substitution and inference in the calculus of functions

The procedure of derivation in the calculus of functions, though employing generalized methods, is fundamentally similar to the procedure developed for the calculus of propositions. It is devised, likewise, for the purpose of deriving true formulas from true formulas. This statement of the aim can be regarded as a definition of the term 'derivation'; every method which, applied to true formulas, must lead to true formulas, is called a derivation.

The method of derivation can be applied both to synthetic and tautological formulas. In both cases, a stock of formulas must be given as the starting point of the derivations. As for synthetic formulas contained in this stock, their truth must be shown by empirical methods. In order to establish a stock of tautologies we use the method of case analysis. This method, which applies both to the calculus of propositions and to that of functions, is not a derivation because it does not relate tautologies to other true formulas; rather it may be called a *criterion*, since it is applied to a given formula directly and determines its result in terms of the structural properties of the formula alone. From the stock of tautologies constructed by case analysis we proceed by derivations to further tautologies. In fact, all known tautologies can be derived in this way. It is clear, of course, that synthetic premises are not necessary for the derivation of tautologies.

The rules of derivation established for the calculus of propositions, namely, the rules of substitution (cf. § 12) and inference (cf. § 14), can be transferred to the calculus of functions. The rule of substitution, however, requires some additions originating from the use of functions; we shall therefore introduce a generalized rule of substitution which is to be applied in the place of the rule of substitution for propositional variables. The rule of inference remains unchanged. In contradistinction to the calculus of propositions, we need for functions a third rule; this rule states the transition from free to bound variables as discussed in § 21.

As before, the justification of these rules, which are directives, will be stated in metatheorems. For the derivation of these metatheorems we need material thinking in the metalanguage, including a reference to an infinity of objects. For the application of the rules to derivations within the object language, however, we shall use only the simplest form of material thinking in the metalanguage, referring to such obvious facts as that a certain symbol has certain properties. This application, therefore, is completely mechanized.

We now turn to the formulation of the rule of substitution. Whereas, in the calculus of propositions, substitutions are made in the place of elementary propositional expressions, the substitutions in the calculus of functions apply, separately, to functional variables and to argument variables. We therefore use here the circumflex sign. When the circumflex is put on top of only one of two argument variables of a functional, the other argument variable is regarded as part of the function, i.e., we regard this expression as a complex function having absorbed an argument. Thus the propositional function '$f(\hat{x}, y)$' is a one-place complex function of the variable 'x'.

Rule of substitution for propositional functions. In a true formula it is permissible to make the following substitutions for elementary variables (cf. § 24) provided that each substitution is made in all places where the original term occurs:

(α)   For an elementary function '$f(\hat{x})$', occurring as a free variable, we may substitute another function '$g(\hat{x})$'; or a complex function such as '$(y)[g(\hat{x}, y) \lor h(\hat{x})]$'; or a function containing other free arguments, such as '$g(\hat{x}, y)$'; or a functional that does not depend on 'x', such as '$g(y)$'; or a propositional variable 'a'. The substitution is permissible even when the new function occurs in other parts of the original formula, and it applies equally to a free or a bound variable 'x'. The argument variable 'x' remains unchanged in this substitution, except for the case where an expression not containing 'x' is substituted. New free or bound variables introduced by the substitution must not be identical with 'x' and must be given names different from the name of any bound variable of the original formula in whose scope they will be after the substitution. Since expressions like '$f(\hat{x})$' and '$f(\hat{u})$' mean the same function, the substitution must be made in all places where these expressions occur; the names of the variables 'x' and 'u' occurring in the respective functionals '$f(x)$' and '$f(u)$' then remain unchanged. Corresponding substitutions can be made for functions of more variables.

(β)   For a free argument variable 'x' another free argument variable 'y' may be substituted even when 'y' occurs as a free variable in other parts of the formula. When the substitution is made within the scope of a bound variable 'y', the new variable must be different from 'y'.

(γ)   For a free argument or function variable 'x' or 'f', special values 'x_1' or 'f_1', respectively, may be substituted.

(δ)   For an elementary propositional variable 'a' any propositional expression may be substituted. When the substitution is made within the scope of a variable 'x', the variables used in the substituted expressions must be different from 'x'.

(ϵ)   For a bound argument variable 'x' another argument variable may be substituted. It is permissible to make this substitution only within the scope of 'x', and to leave an 'x' within another scope unchanged. When the substitution is made within the scope of a bound variable 'y' the new variable must be different from 'y'.

We shall make this rule clear by some examples. In formula (9d, § 25):

$$(x)[f(x) \supset g(x)] \supset [(x)f(x) \supset (x)g(x)] \tag{1}$$

we can substitute '$h(\hat{x}) \vee k(\hat{x}, y)$' for '$f(\hat{x})$', using part α of the rule, and thus obtain

$$(x)[h(x) \vee k(x, y) \supset g(x)] \supset \{(x)[h(x) \vee k(x, y)] \supset (x)g(x)\} \tag{2}$$

Using part ϵ of the rule, we can substitute, in (1), 'y' for the 'x' in the second scope, 'z' for the 'x' in the third scope; we then obtain:

$$(x)[f(x) \supset g(x)] \supset [(y)f(y) \supset (z)g(z)] \tag{3}$$

Now when we make the same substitution as applied to (1) it must be done in all places where the sign 'f' occurs, since '$f(\hat{x})$', '$f(\hat{y})$', and '$f(\hat{z})$' are the same function. For the expression '$k(x, y)$' constructed from '$k(\hat{x}, y)$' we then write '$k(x, u)$' in order to avoid a conflict with the bound variable 'y' on the right-hand side of (3). We thus arrive at

$$(x)[h(x) \vee k(x, u) \supset g(x)] \supset \{(y)[h(y) \vee k(y, u)] \supset (z)g(z)\} \tag{4}$$

This expression differs from (2) only in the names used for the variables.

We may also substitute in (1) the functional '$h(y, z)$' for '$f(\hat{x})$'; we then obtain

$$(x)[h(y, z) \supset g(x)] \supset [(x)h(y, z) \supset (x)g(x)] \tag{5}$$

Since the term '$h(y, z)$' does not contain the variable 'x', the binding existing for '$f(x)$' in (1) has been eliminated by the substitution. Such elimination is permissible. The first operator '(x)' in the second brackets in (5) can be canceled because the term '$h(y, z)$' constituting its operand does not contain 'x'. We thus obtain

$$(x)[h(y, z) \supset g(x)] \supset [h(y, z) \supset (x)g(x)] \tag{6}$$

On the other hand, it would be impermissible to substitute in (1) for '$f(\hat{x})$' the expression '$(x)f(x)$' because this would introduce a double binding. Instead, we may substitute the expression '$(y)f(y)$'. Furthermore, we cannot substitute a function '$h(\hat{x})$' for the function '$f(\hat{x}) \supset g(\hat{x})$' because the latter is not an elementary function.

As a further example let us use the formula (11d, § 25):

$$(x)[f(x) \supset a] \equiv (\exists x)f(x) \supset a \tag{7}$$

Here we may substitute for '$f(\hat{x})$' the expression '$g(\hat{x}, y) \supset h(z)$', and obtain

$$(x)\{[g(x, y) \supset h(z)] \supset a\} \equiv (\exists x)[g(x, y) \supset h(z)] \supset a \tag{8}$$

Furthermore, using part δ of the rule, we can substitute in (7) for 'a' the propositional expression '$(y)k(y, z)$'. We thus obtain

$$(x)[f(x) \supset (y)k(y, z)] \equiv (\exists x)f(x) \supset (y)k(y, z) \tag{9}$$

But it would be impermissible to substitute for 'a' in (7) the expression '$g(x)$' because on both sides this would introduce a binding of the 'x' in '$g(x)$' which did not exist before the substitution, since 'a' does not contain 'x'.

As a further illustration of part ϵ of the rule, we may substitute 'u' for 'y' on the left-hand side of (9). But it would be impermissible to substitute 'x' for 'y' there because this 'x' would be in the scope of two operators. Part β is applied when we substitute 't' for 'z' in (9) on both sides, or when we substitute 'y' for 'z' in (8) on both sides. However, we cannot put 'y' for 'z' in (9).

From (9) we can go back to (7) by putting 'a' for '$k(\hat{y}, \hat{z})$'; then the operator '(y)' can be dropped.

The justification of the rule of substitution for functions is given by the following metatheorem:

Metatheorem 12. (Law of substitution for functions.) When a substitution of function or argument variables is made in a true formula, the resulting formula is true; and when such a substitution is made in a tautology, the resulting formula is a tautology.

This theorem is proved by the following considerations. Part α of the rule applies to elementary functions which are free variables and thus do not represent complex functions. These functions are therefore capable of all matrices, and the original formula must thus be true for all matrices. The substitutions that are admitted can produce, for any given value of the free variables introduced, only one of the possible matrices; therefore the formula must remain true. Substitutions according to parts β and γ cannot make the formula false because the original formula was true for all values of the free variables. Part δ is clear because the original formula is true whether '*a*' is true or false, and the substituted expression can have only one of these values. Part ε represents only a change in the names of bound variables.

It follows that a true formula will be transformed into a true formula. If the original formula is tautologous it must be true for all matrices of the functions occurring in it, including specialized functions, and for all truth-values of propositions, including specialized propositions. This character cannot be changed by the substitutions. Therefore a tautology will be transformed into a tautology. Incidentally, since we consider in this chapter only synthetic expressions in which the functions are constant, part α of the rule is here used only with respect to tautologies.

We said previously that we do not need to state any addition to the rule of inference and that we can use it in its propositional form for functions. The necessary extension of this rule follows automatically when we apply part δ of the rule of substitution and use the term 'propositional expression' in the wider sense defined in § 20. Thus the propositional schema

$$\frac{\begin{array}{c} a \supset b \\ a \end{array}}{b} \tag{10}$$

can be used for functionals:

$$\frac{\begin{array}{c} f(x) \supset g(x) \\ f(x) \end{array}}{g(x)} \tag{11}$$

Similarly, we can substitute functionals containing several variables. These variables need not be the same in all expressions; for instance, we have

$$\frac{\begin{array}{c} f(x, y) \supset g(y, z) \\ f(x, y) \end{array}}{g(y, z)} \tag{12}$$

We can also use expressions containing bound variables:

$$\frac{\begin{array}{c} (x)f(x) \supset (\exists y)g(y) \\ (x)f(x) \end{array}}{(\exists y)g(y)} \tag{13}$$

It can easily be seen that metatheorems 6–8, § 14, hold also for inferences of this sort, which include functional expressions.

§ 27. The rule for free variables

The calculus of functions, in addition to the rules of inference and substitution, requires a third fundamental rule — the rule for the transition from free to bound variables introduced in § 21. We formulate it as follows.

Rule for free variables. When a formula containing free variables is true, it is permissible to put all-operators referring to some or all of these variables before the whole formula as their scope, in any chosen order.

The formulation of this rule includes an addition stating that the order of the all-operators is arbitrary. The meaning of this addition is made clear in § 28. The justification of the rule follows from the considerations given in § 21. If the original formula in free variables is true, the resulting all-statement must also be true; and if the original formula is a tautology, so will be the resulting all-statement. We therefore have the metatheorem:

Metatheorem 13. (Law of free variables.) When a transition from free to bound variables is made in a true formula, in accordance with the corresponding rule, the resulting formula is true; and when it is made in a tautology the resulting formula is a tautology.

Combining this theorem with metatheorem 12, § 26, and metatheorem 7, § 14, we can now state the general theorem:

Metatheorem 14. (Law of the transfer of tautological character.) When a formula is derivable from tautologies alone, it is a tautology.

We now turn to the problem of tautological derivability. First, it is clear that there exists an inference from tautological implication to derivability:

Metatheorem 15. When a formula 'q' is tautologically implied by a formula 'p', i.e., when the relation '$p \supset q$' is a tautology, 'q' is derivable from 'p'.

This theorem follows because, if 'p' is given, we can always add the formula '$p \supset q$' and use the rule of inference. The converse relation must be treated in connection with the considerations given in § 21. We know from (8, § 21) that we cannot generally assert a tautological derivability when the rule for free variables is used in a derivation. On the other hand, this rule enables us to eliminate all free argument variables, both in tautological and in synthetic expressions. Furthermore, when we restrict synthetic expressions used as premises to expressions in which propositional and function symbols occur only as constants, we shall have definite expressions alone as premises when we bind all free argument variables by all-operators, i.e., when we close the expressions. Since in this case no change of meaning will take place when the premise is put into the implicans of another formula, and since, if '*p*' is true, the derived formula '*q*' will be true, we have the theorem:

Metatheorem 16. (*Law of tautological derivability*, also called *theorem of deduction*.[1]) When a formula '*q*' is derivable from a set '*p*' of definite formulas, the formula '*p* ⊃ *q*' is a tautology.

If the set '*p*' contains both synthetic and analytic formulas, the analytic formulas can be omitted; the implication '*p* ⊃ *q*' then remains tautological. This result follows from the considerations presented in § 16. The term 'definite' is understood here in the sense defined in § 21. The omission of expressions which are indefinite because of free argument variables is practically no limitation on the generality of the theorem because such expressions can be made definite by binding the variables. The exclusion of indefinite expressions containing free propositional or function variables, however, does represent an important limitation; variables of this kind can be eliminated, in general, only by a transition to the higher calculus of functions. We shall therefore postpone the treatment of such expressions until we turn to the discussion of that calculus (cf. § 42).

§ 28. Derivation of tautologies

As the basis for the derivations to be given here we shall use the tautologies of the calculus of propositions, and, in addition, formulas (9a–14c, § 25). We explained in § 25 that all these formulas can be proved by case analysis.

As a first example let us use formula (3, § 25), which cannot be proved by case analysis. We start from the tautology in propositions

$$a \vee \bar{a} \tag{1}$$

---

[1] Cf. D. Hilbert and P. Bernays, *Grundlagen der Mathematik*, Vol. I, Berlin 1934, Springer, pp. 155–156.

Substituting '$f(x)$' for 'a' we obtain

$$f(x) \vee \overline{f(x)} \tag{2}$$

When we apply the rule for free variables (2) assumes the form

$$(x)[f(x) \vee \overline{f(x)}] \tag{3}$$

This example illustrates the necessity for the use of free variables. Since the propositional variables 'a', 'b', . . . are free variables we must first introduce other propositional variables containing free argument variables, namely, functionals. Only after this has been done can we introduce an all-operator.

A second example is given by formula (4, § 25), which also cannot be proved by case analysis. The formula

$$a \supset b \vee a.\bar{b} \tag{4}$$

is a tautology in propositions, since by the application of the second distributive rule (4b, § 8) it can be transformed into '$a \supset a \vee b$'. Substituting '$f(x)$' for 'a', and '$g(x)$' for 'b', we have

$$f(x) \supset g(x) \vee f(x).\overline{g(x)} \tag{5}$$

With the use of the rule for free variables this assumes the form

$$(x)[f(\hat{x}) \supset g(x) \vee f(x).\overline{g(x)}] \tag{6}$$

Now we use formula (9d, § 25). Substituting here the function '$g(\hat{x}) \vee f(\hat{x}).\overline{g(\hat{x})}$' for '$g(\hat{x})$' we arrive at the formula

$$(x)[f(x) \supset g(x) \vee f(x).\overline{g(x)}] \supset \{(x)f(x) \supset (x)[g(x) \vee f(x).\overline{g(x)}]\} \tag{7}$$

Abbreviating the expression in the major implicans by 'a', and the expression in the braces by 'b', and using (6), we can apply the schema (10, § 26) of an inference holding for propositions. We thus derive

$$(x)f(x) \supset (x)[g(x) \vee f(x).\overline{g(x)}] \tag{8}$$

which is formula (4, § 25). This formula must be a tautology because it has been derived from tautologous premises only (metatheorem 14, § 27).

A formula which is thus proved to be a tautology, i.e., to be true *for all matrices*, will be true also *for all truth characters* even when case analysis does not directly prove this result. With reference to (3 and 4, § 25) we said that, for such formulas, material thinking could be added to a case analysis with the result that there is a coupling between the functions so that the indeterminacy of the middle line of table III b is eliminated. The existence of such a coupling is indirectly proved when the formula is shown to be a tautology by the method of derivation.

The extension of the method of derivation to functions of several variables involves no difficulties. For instance, starting from the tautology (13a, § 25), we can substitute for the function '$f(\hat{x})$' the function '$f(\hat{x}, y)$'; thus we obtain

$$\overline{(x)f(x, y)} \equiv (\exists x)\overline{f(x, y)} \tag{9}$$

Using the rule for free variables we arrive at

$$(y)[\overline{(x)f(x, y)} \equiv (\exists x)\overline{f(x, y)}] \tag{10}$$

Now the tautology (9f, § 25) can be written in the form

$$(y)[f(y) \equiv g(y)] \supset [(y)f(y) \equiv (y)g(y)] \tag{11}$$

When we substitute in (11) the function '$\overline{(x)f(x, \hat{y})}$' for '$f(\hat{y})$', and '$(\exists x)\overline{f(x, \hat{y})}$' for '$g(\hat{y})$', the left-hand side becomes identical with (10); the right-hand side then assumes the form

$$(y)\overline{(x)f(x, y)} \equiv (y)(\exists x)\overline{f(x, y)} \tag{12}$$

Since this is derived by an inference from tautological premises it must be a tautology, according to metatheorem 14, § 27.[1]

Another example is as follows. From the tautology (8b, § 8) we derive

$$f(x).g(x) \supset f(x) \tag{13}$$

Using the rule for free variables we have

$$(x)[f(x).g(x) \supset f(x)] \tag{14}$$

Applying (9d, § 25) we derive the tautology

$$(x)f(x).g(x) \supset (x)f(x) \tag{15}$$

Applying (9e, § 25) to (14) we derive also the tautology

$$(\exists x)f(x).g(x) \supset (\exists x)f(x) \tag{16}$$

Notice that (16) cannot be derived from (15); for the derivation we must go back to (14).

---

[1] For a simple derivation of (12) and similar formulas cf. the beginning of § 29.

Let us now turn to problems involving the order of operators. We start from formula (14a, § 25), which can be written in the form

$$(x)f(x) \supset f(u) \tag{17}$$

We now substitute the function '$(y)f(\hat{x}, y)$' for '$f(\hat{x})$'; then the same must be done for the function '$f(\hat{u})$' with the difference that here 'u' takes the place of 'x'. We thus arrive at

$$(x)(y)f(x, y) \supset (y)f(u, y) \tag{18}$$

Now (14a, § 25) can also be written in the form

$$(y)f(y) \supset f(v) \tag{19}$$

Substituting '$f(u, \hat{y})$' for '$f(\hat{y})$' we have

$$(y)f(u, y) \supset f(u, v) \tag{20}$$

(18) and (20) furnish the formula

$$(x)(y)f(x, y) \supset f(u, v) \tag{21}$$

We now apply the rule for free variables to the variable 'u' and obtain

$$(u)[(x)(y)f(x, y) \supset f(u, v)] \tag{22}$$

Now formula (11c, § 25) can be written in the form

$$(u)[a \supset f(u)] \equiv a \supset (u)f(u) \tag{23}$$

Substituting here the expression '$(x)(y)f(x, y)$' for 'a', and '$f(\hat{u}, v)$' for '$f(\hat{u})$', we derive from (22)

$$(x)(y)f(x, y) \supset (u)f(u, v) \tag{24}$$

We now apply the rule for free variables to the variable 'v' and obtain

$$(v)[(x)(y)f(x, y) \supset (u)f(u, v)] \tag{25}$$

Using (23) once more we arrive at

$$(x)(y)f(x, y) \supset (v)(u)f(u, v) \tag{26}$$

Applying part ϵ of the rule of substitution, we substitute 'y' for 'v', and 'x' for 'u'; we thus obtain

$$(x)(y)f(x, y) \supset (y)(x)f(x, y) \tag{27}$$

Now this formula holds for all functions 'f' and therefore also for the converse '$\breve{f}$'. Substituting '$\breve{f}$' for 'f' we have

$$(x)(y)\breve{f}(x, y) \supset (y)(x)\breve{f}(x, y) \tag{28}$$

Putting for '*f*' its definition according to (9, § 22), we obtain

$$(x)(y)f(y, x) \supset (y)(x)f(y, x) \tag{29}$$

Substituting '*y*' for '*x*', and '*x*' for '*y*', we arrive at

$$(y)(x)f(x, y) \supset (x)(y)f(x, y) \tag{30}$$

Replacing the two implications of (27) and (30) by an equivalence, we obtain

$$(x)(y)f(x, y) \equiv (y)(x)f(x, y) \tag{31}$$

This formula states the *commutativity of all-operators*. In order to derive it we made use of the addition given to the rule for free variables according to which the rule may be applied to free variables in any order. It is important to realize that, without such a liberty in the application of the rule, the commutativity of all-operators could not be derived. Thus if we were bound to apply the rule to all free variables simultaneously and in such a way that the order of all-operators introduced corresponded to the order of the free variables as they appear in the formula, we would have to go from (21) to the formula

$$(u)(v)[(x)(y)f(x, y) \supset f(u, v)] \tag{32}$$

From this we can derive only the identity

$$(x)(y)f(x, y) \equiv (x)(y)f(x, y) \tag{33}$$

but not the commutativity of the all-operators. This commutativity, therefore, is expressed in the addition given to the rule for free variables.[1]

This result shows the far-reaching content of this addition. Applied to the matrix of a two-place function, the commutativity

---

[1] A very simple proof of the implication (27) can be constructed as follows. We start with the synthetic expression

$$(x)(y)f(x,y)$$

Omitting the all-operators, according to the rule for free variables, we obtain

$$f(x,y)$$

Now we apply the rule for free variables again, this time adding the all-operators in the reversed order, and arrive at

$$(y)(x)f(x,y)$$

Since this formula is derived from the first formula, it is tautologically implied by it, according to metatheorem 16, § 27; we thus obtain implication (27). The reversed implication can be derived in the same way. This derivation, however, is not constructed in terms of the rules of substitution, inference, and free variables alone but applies a further rule, derived from metatheorem 16, § 27.

of the all-operators means that, if all horizontal lines are fully occupied by plus signs, so must be all vertical lines. This inference appears strictly evident; but we should not forget that we apply this inference to an infinite number of things and that graphic representation by means of a diagram cannot strictly prove it. With this extension to infinity, the principle of commutativity of all-operators reminds us very much of the axiom of the parallels in Euclidean geometry; the historical fate of this axiom should serve as a warning not to find any synthetic character in the principle. It should rather be considered as a definition determining the meaning of 'all', stating that feature which we express by the words 'without exception'.

As soon as we drop this feature in the meaning of 'all', the commutativity of all-operators no longer holds. Thus if we interpret the all-operator as meaning 'all, with a finite number of exceptions',

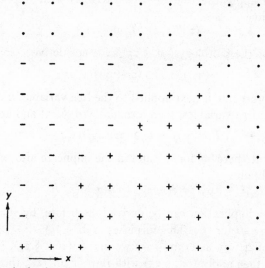

Fig. 5. Matrix showing the noncommutativity of operators meaning *practically all*.

which phrase can be abbreviated to 'practically all', the commutativity breaks down. This consequence is shown in Fig. 5. We construct here a matrix which has plus signs to the right of the diagonal line, but minus signs on the other side. Each horizontal line has only a finite number of minus signs; therefore the statement

'$(y)(x)f(x, y)$' will be true when the operator '(x)' is interpreted as meaning 'practically all'. But each vertical line has only a finite number of plus signs; therefore the statement '$(x)(y)f(x, y)$' will be false.

This shows that by the use of the addition to the rule for free variables we rule out interpretations of the 'all' in terms of approximative concepts. Incidentally, this restriction includes a prohibition to identify the term 'all' with the term 'probability *one*'. The latter probability states only that the limit of the frequency is *one*. The horizontal lines in Fig. 5 therefore represent sequences in which the probability of a plus sign is *one*, whereas the vertical lines furnish the probability *zero* for plus signs.[1]

After the commutativity of all-operators, stated in (15a, § 25), has been demonstrated, that of existential operators is easily proved. We start with the tautology (14b, § 25), substitute '$f(\hat{x}, z)$' for '$f(\hat{x})$', and apply the rule for free variables with respect to 'z'. Thus we have

$$(z)[f(x, z) \supset (\exists y)f(y, z)] \tag{34}$$

Now we use the tautology (9e, § 25) and thus derive

$$(\exists z)f(x, z) \supset (\exists z)(\exists y)f(y, z) \tag{35}$$

A similar procedure is next applied to the free variable 'x'. We use the rule for free variables, then formula (11d, § 25) and arrive at

$$(\exists x)(\exists z)f(x, z) \supset (\exists z)(\exists y)f(y, z) \tag{36}$$

We substitute here 'y' for 'z', and in the implicate also 'x' for 'y'. We thus obtain

$$(\exists x)(\exists y)f(x, y) \supset (\exists y)(\exists x)f(x, y) \tag{37}$$

The reverse implication can be derived, as before, by substituting the converse '$\breve{f}$' for 'f'. This furnishes (15b, § 25).

In a similar way as (15b, § 25) we derive (16a, § 25). We start from (34), then apply (9d, § 25) with respect to '(z)', then use the rule for free variables with respect to 'x' and apply (11d, § 25) to this variable. In the resulting formula we make the same substitutions as applied to (36). In this case the implication in the opposite direction cannot be proved, and we therefore have no equivalence.

[1] Cf. the author's *Wahrscheinlichkeitslehre*, A. W. Sijthoff, Leiden, 1935, p. 276.

Formulas (16b–d, § 25) state the rule that two different operators are commutative in the special case that the two-place function splits into a conjunction, disjunction, or implication of two one-place functions. Formula (16b, § 25) can be derived as follows. Since (16a, § 25) states that the implication runs in one direction, we must add only a proof for the implication from right to left. This proof can be easily given by considering the expression '$(\exists x)f(x)$' on the right-hand side of (16b, § 25) as a constant 'a' and applying formula (11a, § 25). The then resulting expression '$(\exists x)f(x).(y)g(y)$' can be transformed with (12a, § 25) and (11a, § 25) into the left-hand side of (16b, § 25). The proof of formulas (16c–d, § 25) is similar. A proof of (16d, § 25) has been given in (23, § 20). It can be shown, furthermore, that, for a combination of '$f(x)$' and '$g(y)$' by an equivalence operation, commutativity of two different operators does not hold. Formula (16e, § 25) is proved when we apply the formula (9c, § 25) to the left-hand side twice, but change the order of the terms of the disjunction for the second application.

Formula (17a, § 25) is proved as follows. In (14a, § 25) we substitute '$f(u, \hat{y})$' for '$f(\hat{y})$', obtaining

$$(y)f(u, y) \supset f(u, x) \qquad (38)$$

We now put 'x' for 'u':

$$(y)f(x, y) \supset f(x, x) \qquad (39)$$

Using the rule for free variables we put an '(x)' before the whole formula, and then apply (9d, § 25); this furnishes

$$(x)(y)f(x, y) \supset (x)f(x, x) \qquad (40)$$

This is formula (17a, § 25). Similarly, putting the operator '(x)' before (39) and then applying (9e, § 25) we derive (17b, § 25).

We do not add further formulas to our list on p. 134 because a great number of tautologies can immediately be derived from the list of the tautologies in the calculus of propositions, given on p. 38. A simple technique for such a transcription of tautologies in propositional variables into tautologies in functional variables is given by the following rule, which is, of course, a secondary rule, i.e., which represents only a short cut and can be eliminated (cf. § 13).

Rule of transcription. If a tautology in propositional variables is given, it is permissible to replace every letter '*a*', '*b*', etc., by functionals '*f(x)*', '*g(x, y)*', etc., and to put an all-operator for each argument variable before the whole formula as operand. If the major operation of the original formula is a conjunction or an implication or an equivalence, it is permissible to put the all-operators before each of the major terms as operand. If a major term consists of a conjunction of other terms, the all-operators may be put before each of these other terms as operand.

That the formula thus resulting must be a tautology is clear from the rule for free variables and the structure of formulas (9a, d, f, § 25).

As an example let us use formula (6c, § 8):

$$a \supset b \equiv \bar{b} \supset \bar{a} \tag{41}$$

Applying the rule of transcription we have

$$(x)[f(x) \supset g(x)] \equiv (x)[\overline{g(x)} \supset \overline{f(x)}] \tag{42}$$

This formula states the law of contraposition in the calculus of functions.

A further example is given by (8l, § 8):

$$(a \supset b).(b \supset c) \supset (a \supset c) \tag{43}$$

This is transcribed into the law of transitivity in the calculus of functions:

$$(x)[f(x) \supset g(x)].(x)[g(x) \supset h(x)] \supset (x)[f(x) \supset h(x)] \tag{44}$$

Similarly, the inferential implication, which we stated in (9h, § 25), follows by transcription from the corresponding formula (8e, § 8). Furthermore we thus can derive the transitivity of equivalence for functions when we start from the tautology (8m, § 8)

$$(a \equiv b).(b \equiv c) \supset (a \equiv c) \tag{45}$$

Applying the rule of transcription we arrive at

$$(x)[f(x) \equiv g(x)].(x)[g(x) \equiv h(x)] \supset (x)[f(x) \equiv h(x)] \tag{46}$$

The derivations of tautologies given here are greatly simplified by the fact that we begin with a rather large stock of tautologies in functions, furnished by case analysis and stated in formulas (9a–14c, § 25). It can be shown that it is not necessary to use all these formulas. It is sufficient to use as initial stock merely the two

formulas (14a–b, § 25); when we add to these the tautologies of
the calculus of propositions, all other tautologies in functions can
be derived.[1]

In § 16 we mentioned the fact that the tautologies in propositions are reducible to the four formulas (1–4, § 16). It follows that
all tautologies of the simple calculus of functions are derivable
from only six axioms, given by formulas (1–4, § 16) and formulas
(14a–b, § 25). In addition to the axioms, only the rules of substitution and inference and the rule for free variables are to be used.
This result demonstrates the power of the method of derivation,
leading from a small set of simple axioms to all the numerous and
complicated formulas which the calculus includes.

§ 29. Secondary rules

In § 13 and § 15 we introduced some secondary rules for the calculus of propositions. We now turn to the analogues of these rules
in the calculus of functions.

The rule of replacement (§ 13) can be taken over in the calculus
of functions without any change. Only the proof of the corresponding metatheorems 2–5 requires a supplementation because the term
'propositional expression', in this calculus, has a wider meaning.
This supplementation will be added to formulas (2–6, § 13) and
will consist in statements formulating that corresponding implications hold when the step to a more comprehensive propositional
expression is made by the use of operators. It is not possible, however, to write these statements as formulas of the object language;
the occurrence of free variables makes it necessary to use the metalanguage, for reasons similar to those explained with respect to
(8, § 21). We therefore write these statements in the form:

If '$f(x) \equiv g(x)$' is true, then '$(x)f(x) \equiv (x)g(x)$' is true   (1)

If '$f(x) \equiv g(x)$' is true, then '$(\exists x)f(x) \equiv (\exists x)g(x)$' is true   (2)

Using these two relations in addition to (2–6, § 13), and following

---

[1] For the proof of this statement we refer the reader to D. Hilbert and W. Ackermann,
Grundzüge der theoretischen Logik, J. Springer, Berlin, 1928, pp. 53–60. This axiomatic
system has been constructed by Hilbert and Bernays. In this system, however, the
rule for free variables is used in a more complicated form which represents a transcription of the formulas listed by us as (11c–d, § 25). When the rule is applied in
the usual form, such as given by us in § 27, the latter two formulas must be added as
axioms.

the step structure of a formula as in the proof given for meta-theorem 2, § 13, we can prove this theorem in the wider meaning including propositional functions. In the same way we prove metatheorems 3–5, § 13.

Using the rule of replacement we can, for instance, simplify the proof of (12, § 28). Because of the equivalence (9, § 28) we can replace the expression '$\overline{(x)f(x, y)}$' on the left-hand side by the equivalent expression '$(\exists x)\overline{f(x, y)}$'; we thus arrive at the right-hand side of (12, § 28). Similarly, other formulas concerning the shortening of the negation line over expressions with two or more operators can be easily derived by means of the rule of replacement. On the other hand, the derivation of (12, § 28) given in § 28 shows that the rule of replacement is dispensable and represents only a logical short cut, as in the calculus of propositions. Another example is given by the formula

$$(\exists y)\overline{(x)[f(x, y) \supset g(x, y)]} \equiv (\exists y)(\exists x)f(x, y)\cdot\overline{g(x, y)} \qquad (3)$$

which follows by the use of the equivalence (9, § 19). The use of the rule of replacement in combination with (9, § 28) allows us also to construct a simple proof for the commutativity of existential operators. We then take in (31, § 28) the negation on both sides and shorten the negation line according to (9, § 28). The latter operation represents an application of the rule of replacement.

Another sort of secondary rule is derived from the tautology (14a, § 25) and states the transition from bound to free variables. It is as follows:

Rule for the canceling of all-operators. When a true formula contains an all-operator whose scope is the whole formula, this all-operator may be canceled, and the resulting formula may be asserted.

Such a rule, of course, does not apply to existential operators; nor does it apply to an all-operator preceded by an existential operator, since the scope of such an all-operator is not the whole formula. Only if a set consisting merely of all-operators has the whole formula as its scope, can it be canceled by successive application of the rule.

The secondary rules developed for the rule of inference in § 15 can be taken over in the calculus of functions. In addition we can introduce some further secondary rules of inference which simplify the procedure of inference applied to functional expressions.

Using the formula (9h, § 25) we can immediately establish the schema

$$\frac{\begin{array}{c}(x)[f(x) \supset g(x)] \\ (x)f(x)\end{array}}{(x)g(x)} \tag{4}$$

The proof of this schema follows from the fundamental rule of inference when we write it in the form

$$\frac{\{(x)[f(x) \supset g(x)].(x)f(x)\} \supset (x)g(x)}{(x)[f(x) \supset g(x)].(x)f(x)}}{(x)g(x)} \tag{5}$$

The second line of this schema representing the modus ponens corresponds to the first two lines of (4); the first line is the tautology (9h, § 25). Since a tautology may always be added to given formulas, (4) states the same as (5), only in a simplified form.

Similarly, some further schemas containing all-operators can be derived. It is clear that every schema in which the conclusion is tautologically implied by the premises will be valid; the schema then can be derived, by analogy with (5), from the fundamental rule of inference. Thus using the tautology (44, § 28) we derive the schema:

$$\frac{\begin{array}{c}(x)[f(x) \supset g(x)] \\ (x)[g(x) \supset h(x)]\end{array}}{(x)[f(x) \supset h(x)]} \tag{6}$$

which expresses the transitivity of implication.[1] A corresponding schema is derivable for the equivalence with the use of (46, § 28):

$$\frac{\begin{array}{c}(x)[f(x) \equiv g(x)] \\ (x)[g(x) \equiv h(x)]\end{array}}{(x)[f(x) \equiv h(x)]} \tag{7}$$

Both these schemas are frequently used in derivations.

Let us now turn to schemas containing existential operators. From (10f, § 25) we derive the schema

$$\frac{\begin{array}{c}(x)g(x) \\ (\exists x)f(x)\end{array}}{(\exists x)f(x).g(x)} \tag{8}$$

---

[1] As to the application of this schema in the theory of the syllogism, cf. § 36, where some examples are given.

Substituting here for '$g(\hat{x})$' the function '$f(\hat{x}) \supset g(\hat{x})$', and using the tautology '$a.(a \supset b) \equiv a.b$', we derive the further schema

$$\frac{\begin{array}{c}(x)[f(x) \supset g(x)] \\ (\exists x)f(x)\end{array}}{(\exists x)f(x).g(x)} \qquad (9)$$

An example is given by the inference

$$\frac{\begin{array}{c}\text{All Sioux Indians are Americans.} \\ \text{There are Sioux Indians.}\end{array}}{\text{There are Sioux Indian Americans.}} \qquad (10)$$

A further schema of this group, though of rather trivial nature, can be derived from the tautology (16, § 28):

$$\frac{(\exists x)f(x).g(x)}{(\exists x)f(x)} \qquad (11)$$

This schema has only one premise.

We can use this schema to derive from (8) the following schema:

$$\frac{\begin{array}{c}(x)[f(x) \supset g(x)] \\ (\exists x)f(x).h(x)\end{array}}{(\exists x)h(x).g(x)} \qquad (12)$$

This follows when we use (8), replace the conclusion '$(\exists x)h(x).f(x).$ $[f(x) \supset g(x)]$' by the equivalent expression '$(\exists x)h(x).f(x).g(x)$', and apply (11).[1]

Finally we must mention two forms of inference which lead from general statements to specialized statements. Like (11) the first has only one premise:

$$\frac{(x)f(x)}{f(x_1)} \qquad (13)$$

When we use for '$f(\hat{x})$' the complex function 'if $\hat{x}$ is a man $\hat{x}$ is mortal', the inference is illustrated by

$$\frac{\text{All men are mortal.}}{\text{If Socrates is a man he is mortal.}} \qquad (14)$$

The other schema of this sort has the form

$$\frac{\begin{array}{c}(x)[f(x) \supset g(x)] \\ f(x_1)\end{array}}{g(x_1)} \qquad (15)$$

---

[1] Like (6) this schema is used in the theory of the syllogism, where we illustrate it by examples. Cf. § 36.

It can be illustrated by the inference

$$\frac{\text{All men are mortal.}}{\text{Socrates is a man.}} \qquad (16)$$
Socrates is mortal.

Schema (13) follows directly from (14a, § 25). Schema (15) follows when we apply (13) to the first premise of (15), obtaining '$f(x_1) \supset g(x_1)$', and then use the fundamental rule of inference.

As to inferences containing functions of several variables, such forms can be derived by substitutions made in the preceding schemas. We then can put all-operators before each premise separately, referring to the free variables introduced. This procedure is permissible because, when we replace the inference by the implication between the premises and the conclusion, we can apply (9d and 9a, § 25). Thus from (4) we can derive the schema, substituting '$f(\hat{x}, y)$' for '$f(\hat{x})$', and '$g(\hat{x}, y)$' for '$g(\hat{x})$':

$$\frac{(x)(y)[f(x, y) \supset g(x, y)]}{(x)(y)f(x, y)} \qquad (17)$$
$$(x)(y)g(x, y)$$

There are, however, some schemas that cannot be derived in this way, but require a separate derivation.

An important schema of this sort is derivable by the use of the tautology (16e, § 25). We may call it the *inference by crossed operators:*

$$\frac{(x)(y)[f(x, y) \vee g(x, y)]}{(\exists x)(y)f(x, y) \vee (x)(\exists y)g(x, y)} \qquad (18)$$

An example is given by the inference:

$$\frac{\text{Consider any two men; they will either love each}}{\text{other or hate each other.}} \qquad (19)$$
There is a man who loves all men, or for every man there is some man whom he hates.

Another schema is as follows. When we substitute two-place functions in (8) we arrive at a schema each line of which begins with an all-operator; but we cannot thus derive a schema in which some lines begin with existential operators. Therefore we shall now construct a more general schema. We write it by leaving, in the first and third line, the places of the operators unoccupied; in these blanks we can insert existential operators or all-operators ad libitum, but with the qualification that the set used in the conclusion is the same as that used in the first premise. The operators of the second premise, however, are throughout all-operators. We thus obtain the schema

$$\frac{\substack{(\ )\ldots(\ )g(x\ldots z)\\(x)\ldots(z)f(x\ldots z)}}{(\ )\ldots(\ )[f(x\ldots z).g(x\ldots z)]} \tag{20}$$

This schema can be proved by the following consideration. Starting from the tautology

$$f(x) \supset [g(x) \supset f(x).g(x)] \tag{21}$$

which represents a tautology in propositional variables, we derive by the use of the rule for free variables (§ 27) and the relation (9d, § 25) the formula

$$(x)f(x) \supset (x)[g(x) \supset f(x).g(x)] \tag{22}$$

Applying (9e, § 25) to the all-operator before the brackets, we derive, using the transitivity of implication,

$$(x)f(x) \supset [(\exists x)g(x) \supset (\exists x)f(x).g(x)] \tag{23}$$

When we use (6d, § 8) the formula becomes

$$(r)f(x).(\exists x)g(x) \supset (\exists x)f(x).g(x) \tag{24}$$

This is the relation (10f, § 25). Starting from (22) we can generalize this relation for functions of several variables, including the use of both existential and all-operators, as follows. We substitute in (22) for '$f(\hat{x})$' the function '$f(\hat{x}, y)$', for '$g(\hat{x})$' the function '$g(\hat{x}, y)$', use the rule for free variables with respect to the variable 'y', and then apply (9d, § 25) to the first implication sign. We thus obtain:

$$(x)(y)f(x, y) \supset (x)(y)[g(x, y) \supset f(x, y).g(x, y)] \tag{25}$$

The right-hand expression can be changed by a procedure similar to that applied in the transition from (22) to (23). Using for each of the operators one of the relations (9d or 9e, § 25), we derive the three relations

$$(x)(y)f(x, y) \supset [(\exists x)(y)g(x, y) \supset (\exists x)(y)f(x, y).g(x, y)] \tag{26}$$
$$(x)(y)f(x, y) \supset [(x)(\exists y)g(x, y) \supset (x)(\exists y)f(x, y).g(x, y)] \tag{27}$$
$$(x)(y)f(x, y) \supset [(\exists x)(\exists y)g(x, y) \supset (\exists x)(\exists y)f(x, y).g(x, y)] \tag{28}$$

These relations can be transformed by the use of (6d, § 8), by analogy with (24). It is clear that corresponding formulas result when
we use more than two operators. These relations state that the
conclusion of (20) is tautologically implied by the conjunction of
the two premises. According to our remark preceding (6), schema
(20) is thus proved.

In a similar way we can generalize (11). We start here from the
tautology

$$(x) \ldots (z)[f(x \ldots z).g(x \ldots z) \supset f(x \ldots z)] \qquad (29)$$

which is derivable from (8b, § 8). Applying (9d or 9e, § 25) separately to each operator, according to choice, we derive the tautology

$$(\ ) \ldots (\ )[f(x \ldots z), g(x \ldots z)] \supset (\ ) \ldots (\ )f(x \ldots z) \qquad (30)$$

We now construct the schema

$$\frac{(\ ) \ldots (\ )[f(x \ldots z).g(x \ldots z)]}{(\ ) \ldots (\ )f(x \ldots z)} \qquad (31)$$

which states generally that, if the major operation is the conjunction, one of the factors can be omitted.

Using (20) and (31) we derive another schema which represents a
generalization of (6) and therefore expresses the *general transitivity
of implication:*

$$\frac{(\ ) \ldots (\ )[f(x \ldots z) \supset g(x \ldots z)]}{(x) \ldots (z)[g(x \ldots z) \supset h(x \ldots z)]} \qquad (32)$$
$$\overline{(\ ) \ldots (\ )[f(x \ldots z) \supset h(x \ldots z)]}$$

The proof is as follows. First applying (20) to the premises we
derive the conclusion

$$(\ ) \ldots (\ )[f(x \ldots z) \supset g(x \ldots z)].[g(x \ldots z) \supset h(x \ldots z)] \qquad (33)$$

We now add the tautology

$$(x) \ldots (z)\{[f(x \ldots z) \supset g(x \ldots z)].[g(x \ldots z) \supset h(x \ldots z)] \supset [f(x \ldots z) \supset h(x \ldots z)]\} \qquad (34)$$

and apply (20) a second time; the conclusion then can be transformed by the use of the tautology '$a.(a \supset b) \equiv a.b$'. Using (31) we
derive the schema (32). It is the significance of this schema that
it extends the transitivity of implication to cases in which the
implication is preceded by any set of operators.

By similar considerations we can prove a generalization of (17):

$$
\begin{array}{l}
(x) \ldots (z)[f(x \ldots z) \supset g(x \ldots z)] \\
\underline{(\ ) \ldots (\ )f(x \ldots z)} \\
(\ ) \ldots (\ )g(x \ldots z)
\end{array}
\qquad (35)
$$

This schema states that, if the functional '$f(x \ldots z)$' is preceded by any set of operators, we can put the functional '$g(x \ldots z)$' in its place if '$g(x \ldots z)$' is implied by '$f(x \ldots z)$' for all values of the variables.

§ 30. Derivations from synthetic premises

As an example of a derivation let us give here the proof that an interconnective relation must also be uniform, i.e., that (6, § 22) is derivable from (8, § 22). The latter condition can be written, when we use the definitions (3 and 4, § 22), as

$$(x)(y)\{[df(x) \lor ef(x)].[df(y) \lor ef(y)].(x \neq y) \supset f(x, y) \lor f(y, x)\} \quad (1)$$

Using the first distributive rule in the form (4c, § 8), we transform the implicans into a disjunction of four products. Applying (8h, § 8) and the schema (17, § 29) twice, we derive the two conditions:

$$(x)(y)[df(x).df(y).(x \neq y) \supset f(x, y) \lor f(y, x)] \qquad (2)$$
$$(x)(y)[ef(x).ef(y).(x \neq y) \supset f(x, y) \lor f(y, x)] \qquad (3)$$

We begin with derivations from (2). We use the tautology

$$a.b \supset c \lor d \equiv a.(b \supset c) \lor (a \supset d) \qquad (4)$$

which is easily proved when we remove the implication signs by the use of (6a, § 8) and then transform each side, applying repeatedly, on the right-hand side, the relation (5e, § 8). When we transform the operand of (2) according to (4), (2) assumes the form:

$$(x)(y)\{df(x).[df(y).(x \neq y) \supset f(x, y)] \lor [df(x) \supset f(y, x)]\} \qquad (5)$$

Applying the schema (18, § 29) to the disjunctive operand, we obtain:

$$(\exists x)(y)\{df(x).[df(y).(x \neq y) \supset f(x, y)]\} \lor (x)(\exists y)[df(x) \supset f(y, x)] \qquad (6)$$

When we replace in (14b, § 25) 'y' by 'z', substitute '$f(\hat{x}, y)$' for '$f(\hat{x})$', and add the two all-operators according to the rule for free variables, we arrive at the tautology:

$$(x)(y)[f(x, y) \supset (\exists z)f(z, y)] \qquad (7)$$

The implicate can here be written in the form '$ef(y)$', according to (4, § 22). Combining the first term of the disjunction (6) with (7), we can apply the schema (32, § 29); we thus show that from this first term the relation

$$(\exists x)(y)\{df(x).[df(y).(x \neq y) \supset ef(y)]\} \tag{8}$$

is derivable. According to metatheorem 16, § 27, relation (8) will be tautologically implied by the first term of (6). We now combine the second term of (6) with (7), interchanging 'x' and 'y'; applying again schema (32, § 29) we obtain

$$(x)[df(x) \supset ef(x)] \tag{9}$$

As before we infer that this expression is tautologically implied by the second term of (6). Since the disjunction (6) was derived from (2) we now infer, by the use of the dilemma (17, § 15), that the disjunction of (8) and (9) is derivable.

Furthermore, it can easily be seen that (8) is derivable from (9). We show this by putting 'y' for 'x' in (9), adding the true statement '$(\exists x)df(x)$', and then adding to the implicans in the brackets the term '$x \neq y$', which is permissible because of (8g, § 8). We infer as before that therefore (8) is tautologically implied by (9). Using now the schema (25, § 15) we infer that (9) is redundant, and that (8), taken alone, is derivable from (2) and thus from (1).

The same derivation can be applied to (3). Since (3) differs from (2) only in that 'df' is replaced by 'ef', we arrive at a formula which results from (8) by the interchange of the expressions 'df' and 'ef'. This is the formula:

$$(\exists x)(y)\{ef(x).[ef(y).(x \neq y) \supset df(y)]\} \tag{10}$$

Since both (8) and (10) are derived from (1), their conjunction is also derivable. In order to write this conjunction as a one-scope, formula we put 'v' for 'x' in (10); then (12a, § 25) can be used. Transforming the expressions in the brackets by the use of (6d, 8j, and 7a, § 8), and shifting the operator '(y)' to the right by the use of (11a, § 25), we thus arrive at the condition:

$$(\exists x)(\exists v)\{df(x).ef(v).(y)[(x \neq y).(v \neq y) \supset [df(y) \equiv ef(y)]]\} \tag{11}$$

This is the same as (6, § 22) when we put 'u' for 'x' and 'x' for 'y' and remove the implication in the operand by the use of (6a, § 8).

We now turn to other problems. In the construction of derivations we are frequently concerned with proofs limited to objects of a certain class, such as all men, or all numbers. This sort of restriction is expressible in our calculus by the use of a general implication whose implicans refers to the class of objects considered. The actual construction of the derivations, however, appears unnecessarily complicated since all the theorems include a common implicans and all transformations concern only the expressions following this implicans. We shall therefore introduce a simplified notation by the use of *restricted operators*.

We indicate these operators by the addition of a prime mark, and define them as follows:

$$(x)'f(x) =_{Df} (x)[k(x) \supset f(x)] \tag{12}$$
$$(\exists x)'f(x) =_{Df} (\exists x)k(x).f(x) \tag{13}$$

The class K coordinated to the function $k(x)$ represents the class of objects to which the operators are restricted; we call K the range of the operators. We read the restricted operators in the form: 'for all x in K'; and 'there is an x in K'.

It can easily be seen that all tautologies for functions hold equally with the use of restricted operators if the class K is not empty. Thus (13a, § 25) is translated into the formula

$$\overline{(x)'f(x)} \equiv (\exists x)'\overline{f(x)} \tag{14}$$

Inserting the definitions (12) and (13), we obtain

$$\overline{(x)[k(x) \supset f(x)]} \equiv (\exists x)k(x).\overline{f(x)} \tag{15}$$

which is a tautology. Similarly, (14c, § 25) is translated into

$$(x)'f(x) \supset (\exists x)'f(x) \tag{16}$$

which means:

$$(x)[k(x) \supset f(x)] \supset (\exists x)[k(x).f(x)] \tag{17}$$

This is correct if K is not empty; i.e., the formula

$$(\exists x)k(x) \supset \{(x)[k(x) \supset f(x)] \supset (\exists x)[k(x).f(x)]\} \tag{18}$$

is a tautology. When restricted operators are used, we shall always assume that the range is not empty.[1]

---

[1] Expression (18) gives the correct formulation of the inference from 'all' to 'some' used in traditional logic; cf. § 19. The use of restricted operators may be regarded as a correct version of what the scholastic notation intended to represent.

Furthermore, it can be shown that derivations made by the use of the rules of substitution and inference hold also for restricted operators. We can therefore carry through all sorts of derivations with restricted operators when we follow the rule that all the operators used in the derivations must be given the mark of restriction. Finally it is also permissible to use free variables of a restricted range. We shall add a prime mark to such variables, and write 'x'', in order to indicate the restriction. The operations with such variables are the same as those with unrestricted free variables. In the transition from free to bound variables we then must, of course, introduce a restricted all-operator.

As an example, let us give here the proof previously mentioned, showing that a function $f(x, y)$, which is symmetrical and transitive, is also reflexive within its field. We use the field cf of the function as the range of the operators and free variables. Because of the definition (1, § 22) of the field and the condition of symmetry we have

$$(x)'(\exists y)'f(x, y) \tag{19}$$

This can be written in the form

$$(\exists y)'f(x', y) \tag{20}$$

The condition of symmetry according to (10, § 22) furnishes

$$(y)'[f(x', y) \supset f(y, x')] \tag{21}$$

Applying the schema (9, § 29) to (21) and (20) we derive

$$(\exists y)'f(x', y).f(y, x') \tag{22}$$

Now the definition of transitivity (14, § 22) can be written in the form

$$(y)'[f(x', y).f(y, z') \supset f(x', z')] \tag{23}$$

Using the rule of substitution we specialize this formula by putting x' for z', and obtain

$$(y)'[f(x', y).f(y, x') \supset f(x', x')] \tag{24}$$

Applying the schema (35, § 29) to (22) and (24) we arrive at the conclusion

$$(\exists y)'f(x', x') \tag{25}$$

According to the rule that '$(\exists y)a$' is the same as 'a' we derive:

$$f(x', x') \tag{26}$$

Now we apply the rule for the transition from free to bound variables and arrive at the result

$$(x)'f(x, x) \qquad (27)$$

This states, according to (18, § 22), that the function $f(x, y)$ is reflexive within its field.

§ 31. The formal and the material conception of language

We have finished the exposition of the method of derivation and now must enter into a critical discussion of this method. We said that it is the aim of a derivation to construct a true formula; are we sure that this aim has been reached? In order to answer this question, let us turn to a closer consideration of the method.

It is a general characteristic of the method of derivation that it is completely *formalized*. Let us make clear the meaning of this term by the following illustration. When we consider a formula, such as

$$a \lor \bar{a} \qquad (1)$$

we usually conceive of it as having a meaning: thus (1) is regarded as meaning:

There is a certain situation or there is not. $\qquad (2)$

We then say that we use the *material conception* of the formula. In this conception the symbol '*a*' is conceived of as an abbreviation standing for a general statement about a situation; the symbols '∨' and '⁻' are considered as representing the familiar logical operations 'or' and 'not'. The symbols of formula (1) are then *interpreted;* and the *material conception* of a formula is therefore characterized by the use of an *interpretation* of the symbols that occur.

We need the interpretation, and thus *material thinking*, when we want to *understand* a formula. However, when we are concerned only with *manipulations* of the formula within the logistic calculus, no interpretation and no material thinking are necessary. It is sufficient, for this purpose, to know that the symbol '*a*' represents a variable of a certain kind, called a propositional variable, and that the symbols '∨' and '⁻' represent operations whose rules are given in the truth tables. For instance, when we show by case analysis that (1) is a tautology, it is not necessary to realize the meaning of the symbols of the propositional operations; it is sufficient to apply the rules of the truth tables mechanically, making use only of the fact that in the first line of the column headed by the symbol 'v' there is a '*T*', that there is a '*T*' also in the second line, etc., and of similar external properties of the truth table of negation. We then speak of the *formal conception* of the formula; this conception is characterized by the use of the *uninterpreted* formula, regarded only as an aggregate of symbols equipped with certain structural properties.

It is of fundamental importance that derivations, too, can be completely carried through within the formal conception. This is clear from the examples of derivations presented in the preceding sections. We compared given formulas with tautologies printed in a list of such formulas, checking whether there existed a letter-to-letter correspondence; we substituted, for given letters, combinations of other letters; we wrote down a new formula following the mechanical rules expressed in one of the inferential schemas. In derivations, therefore, it is not necessary to realize the meaning of the symbols; it is sufficient to examine the structure of the formulas.

Material thinking is thus replaced by formal thinking; mechanical manipulations with symbols, distinguished only by their geometrical shapes, take the place of thought operations based on realizing the meaning of the symbols. Instead of saying, for instance: 'if neither a certain situation nor a certain other one exists, then the first situation does not exist, and the second also does not exist', we say with formula (5b, § 8): 'if two letters separated by the hook-symbol stand below a long horizontal line, this expression is equivalent to an expression consisting of the same two letters below short horizontal lines, separated by a dot'. It is an advantage of this mechanization that it greatly facilitates the process of derivation and at the same time makes thinking more dependable.

The opinion has been expressed that such a mechanization means an impoverishment of thought, that it represents the intrusion of machine methods into a domain that should be reserved for mental processes proper. It seems to us that such a criticism must be regarded as an expression of intellectual shortsightedness. If those mental processes that admit of a machine technique are mechanized, the human mind will be freer to pursue other activities that withstand mechanization, and will be capable of further advances and deeper insights. The use of a technique, be it the technique of logistic, or the technique of arithmetic, or any of the techniques of the machine age, carries with itself the possibility of progressive application; whether this possibility will be actualized for the good of civilization will depend on what those who control the technique do with it.

Our discussion has shown that the use of the formal conception is associated with a transition from object language to metalanguage. Instead of speaking of a situation we speak of a propositional variable; instead of saying 'v' we say that there is a hook-shaped symbol, i.e., we say "'v'". It is this transition to the metalanguage that represents the mechanization, since it is the very focusing on symbols, apart from their meanings, that makes the analysis mechanical and precise. On the other hand, we see that this transition could not be made if the statements of the metalanguage were not thought materially; we must realize the meaning at least of what is stated in the metalanguage when we want to manipulate the symbols of the object language and to determine their structural properties. Thus we must understand the meaning of the statement 'there is a hook-shaped symbol between two letters', or of the statement 'there is a '*T*' in the second line of the column of the tables', in order to derive consequences stating, for instance, that a certain formula is a tautology, or that it is implied by another formula. In the derivation of metatheorems we have made wide use of material thinking in the metalanguage, sometimes of a rather complicated form. The formal conception of the object language, therefore, involves a material conception of the metalanguage; *formal manipulations with formulas of the object language are made possible through material thinking in the metalanguage.* This is one of the fundamental laws of symbolic logic.

This law makes it clear why the rules of language must be stated in the metalanguage. In order to apply these rules we must understand what they mean; therefore these rules must be thought materially even when the formulas with which we operate are considered formally, i.e., as groups of symbols. This is the reason that we have stated these rules in words, as is usual, and not in symbols. We could, of course, introduce a symbolic technique expressing these rules; but these symbols would belong to the metalanguage, and the symbolic technique would be of limited value because it could not be considered formally as long as we are concerned with a formal consideration of the object language. The schema (1, § 14) by which we have expressed the fundamental rule of inference can be considered, for instance, as a symbolism belonging to the metalanguage. This symbolism must be distinguished from the symbolism of the object language. It therefore cannot be replaced by the inferential implication

$$(a \supset b).a \supset b \qquad\qquad (3)$$

since this formula, in the formal conception, does not tell us anything. The schema (1, § 14) tells us something because it is not included in the formalization performed upon the object language.

Let us see to what extent the distinction between the formal and the material conception of language helps us to answer the question raised in the beginning of this section.

§ 32. The proof of consistency

It is the question of the truth of derivable formulas that we had raised. Now we must realize that the notion of *truth* refers, not to the formal conception, but to the material conception, of the object language, since it applies only when the object language is interpreted. To say, for instance, that '$a \lor b$' is true has a meaning only when the symbols 'a' and 'b' are interpreted as propositions and the hook is interpreted as the operation 'or'. The reason is that truth is a *semantical* concept and applies to signs only when they are coordinated to objects, i.e., when they are given a meaning. Within the formal system, the notion of truth is represented merely by the letter 'T' used in the truth tables, a symbol to which this system gives no meaning. We therefore do not know, within the formal conception, to which synthetic formulas the letter 'T' is to be coordinated. The formal system provides us merely with rules for the *transfer* of the T-symbol; i.e., it tells us which formulas will be of the T-character *if* certain other formulas are of this character. Furthermore, it tells us which formulas will always have a T-character. Both these results are achieved by the rules of derivation and the rules for the establishment of tautologies. As long as the T-symbol is given no meaning, these are both *syntactical* rules; and it is clear, therefore, that the formal conception makes only a *syntactical* use of the T-symbol. The semantical properties of the T-symbol are connected with its interpretation and cannot be dealt with in the formal conception.

It must be realized that within the formal conception all rules of language have the nature of *conventions*. They are rules of the game, so to speak; and there is no question of their justification. The problem of justification comes in only when symbols are interpreted, i.e., within the material conception of language. Now there exists always a number of possible interpretations for a system of signs; but this does not mean that *every* interpretation is permissible. When a particular interpretation is used we must first ask whether the formal properties of the system allow us to introduce this interpretation. Thus when we interpret the T-symbol as meaning *truth* we must ask whether the properties that this symbol has within the formal system make such an interpretation possible. We must transcribe the question raised in the beginning of § 31 into this form when we wish to answer it for a formalized language.

The chief property of truth is its exclusiveness; truth excludes falsehood. If a sentence is true, it cannot be false. When the T-symbol is to be interpreted as truth we must be able to show that it possesses the same exclusiveness, i.e., that the syntactical properties of the T-symbol establish such exclusiveness. In other words: when a 'T' has been coordinated to some sentences and it has been shown then that to certain other sentences also a 'T' must be coordinated, it must be impossible to coordinate an 'F' to the latter sentences without a change of the first coordination. This means: we must prove that *if a sentence is derivable from given statements, it is impossible to derive the negation of this sentence from the same statements.* This is what is called the *theorem of consistency.* We see that this theorem represents a merely syntactical problem, which concerns the rules for the transfer of the letter 'T' within the formal system.

Let us illustrate these considerations by the presentation of an analogy which has been used repeatedly for this purpose. The formal consideration of a language may be compared to the way we regard a game of chess. The pieces are moved on the board according to certain rules which lead from one position to another position; we have a choice as to which move to make, but the choice is limited by the rules of the game. A position, i.e., a certain arrangement of the pieces on the board, corresponds to a formula, in our comparison; and the rules of the game correspond to the rules of derivation. The initial position of the pieces, by which each game begins, corresponds to the set of axioms used for a derivation. Within the process of a derivation, the symbols used do not possess any more meaning for us than the chess pieces; they are merely concrete objects which we manipulate in accordance with the rules of the procedure.

Now imagine that somebody takes the chess pieces out of the box and arranges them in an arbitrary fashion on the board. Will the arrangement represent a position which can be obtained within a regular game? That is, is it possible to reach this position when we start from the usual initial position and follow the rules of the game, disregarding any question as to the playing abilities of the opponents? Some general statements can be made here. For instance, if in the arbitrary arrangement the two bishops of the same player occupy fields of the same color while all the pawns of this player are on the board, the question must be negatively answered. It is impossible to reach such a position, because in the initial position the two bishops occupy fields of different colors, and because they will preserve this property in all their moves, as is implied by the rules of the game. Furthermore, the second bishop cannot have been introduced through the fact that a pawn reached the eighth line, because all pawns are still on the board. We thus judge from a purely formal criterion that the position is unattainable.

In the same sense we wish to prove that, given certain axioms, it is impossible to reach, by derivations, a 'position' in which both a formula 'q' and its negation '$\bar{q}$' appear among the derived formulas. We require that this proof be given by the use of formal criteria alone, such as were represented in the analogy by the use of the color of the field as the criterion. The criterion to be looked for must consist in a property of the initial formulas which is transferred to every derived formula, in the same way as in a chess game the property of being on a white field is transferred in every move of a bishop that has this property.

The significance of the problem of consistency was first seen by D. Hilbert; and he and his collaborators have been able to solve this problem for a number of cases. Thus a proof of consistency was given for the calculus of propositions and for the simple calculus of functions. The proof we give here differs from Hilbert's proof in its form, though not in essential features; it is constructed in continuation of a method introduced by W. Dubislav.[1] As to the proof of consistency for the higher calculus of functions and the whole of mathematics, important steps have been made; but a complete solution has so far not been possible.[2]

We give the proof first for the calculus of propositions and then for the simple calculus of functions. Each of these proofs is subdivided according as 'q' has been derived from tautologies alone, or from synthetic formulas.

We have shown in metatheorem 9, § 14, that a formula 'q' derived from tautologies alone is a tautology. Now a tautology in the calculus of propositions has the property that, when it is subjected to case analysis on the basis of the truth tables, it always furnishes the result 'T'. This is a syntactical property, which can be defined and verified within the formal conception. Consequently, if 'q' is derivable from tautologies alone, it must have this syntactical property. It follows that '$\bar{q}$' cannot be derivable, since, as the negation of 'q', it will always furnish the result 'F' when subjected to case analysis.

Let us illustrate this inference by reference to the chess game. The property of being a tautology is the characteristic mark which is transferred from the initial formulas to every derived formula. It corresponds to the mark which, in the chess example, is given by the color of the field that the bishop occupies, which mark is transferred in every move. As in the case of the moves of the bishop in the chess game, we infer that no derivable formula can have a mark different from the one that is transferred in all derivations.

Now let us regard the case that 'q' is derived from a set 'p' of synthetic formulas containing only constants. According to metatheorem 10, § 14, the formula '$p \supset q$' then is a tautology. Now if

---

[1] Cf. footnote on p. 127. Hilbert does not use the truth-table method, but coordinates to every formula one of the numbers 1 and 0, and shows that tautologies always have the number 0. Cf. D. Hilbert and W. Ackermann, *Grundzüge der theoretischen Logik*, Springer, Berlin, 1928, pp. 29, 65. A more comprehensive presentation is given in D. Hilbert and P. Bernays, *Grundlagen der Mathematik*, Vol. I, Springer, Berlin, 1934.

[2] Cf. D. Hilbert and P. Bernays, *op. cit.*, Vol. II, Berlin, 1939.

the formula '$\bar{q}$' were also derivable from 'p', the formula '$p \supset \bar{q}$' would also be a tautology. From (6e, § 8), we then infer that the formula

$$p \supset q.\bar{q} \tag{1}$$

holds. Since it is derived from tautologies, this formula must be a tautology. Now (1) can be written in the form '$\bar{p} \vee q.\bar{q}$', which by use of (5d, § 8) can be transformed into '$\bar{p}$'. Therefore 'p' is a contradiction. It follows that when 'p' is synthetic this result cannot hold, i.e., that in this case it is impossible to derive both 'q' and '$\bar{q}$'.

Consequently, in order to be sure that the use of a set 'p' of definite axioms does not lead to contradictions, we need only show that 'p' is not contradictory. In the calculus of propositions such demonstration can always be given, in principle, by case analysis. The proof of consistency for the calculus of propositions is thus concluded.

The proofs for the two cases considered, viz., for tautologies and for synthetic formulas as starting points of the derivations, can be transferred to the calculus of functions, since we have metatheorems 14 and 16, § 27. These theorems state that in the calculus of functions also a formula derived from tautologies is a tautology, and a formula derived from definite synthetic axioms is tautologically implied by these axioms. For such synthetic formulas the proof of consistency is therefore given. In order to extend it to formulas containing free argument variables we introduce equipollent definite formulas by binding the variables by an all-operator. Since the equipollent expressions so introduced cannot lead to contradictions, the same must hold for the original expressions. Otherwise, if 'p' is an indefinite expression and 'p'' an equipollent definite expression, and if we could derive a contradiction from 'p', we could derive the same contradiction from 'p'' because 'p' is derivable from 'p''.[1] The theorem of consistency is thus also proved for the calculus of functions.

There is, however, an important difference between the calculus of propositions and the calculus of functions which bears upon the

---

[1] This argument applies also to the case in which we use as premises indefinite expressions containing free propositional and functional variables, with which we shall deal in § 42.

proofs given. In the calculus of propositions, the concept of tautology is definable by a purely mechanical criterion, given through case analysis. We saw that in the calculus of functions case analysis furnishes merely a sufficient, not a necessary, criterion of tautological character (cf. § 25). Therefore we cannot give here a general criterion for tautologies which consists in the use of mechanical rules applicable to any given formula. The property of being a tautology is here an abstract property; in general we know of its existence only because the formula under consideration is derived from other tautologies to which the mechanical criterion of case analysis can be applied.

This qualification does not make our proof invalid. But we prefer a proof which uses a mechanical criterion and thus can be carried through without reference to the meaning of the term 'tautology' as stated in the definition given in § 24. Now it can be shown that such a mechanical criterion can be constructed, and that it can be used for the proof of consistency. The criterion resembles the one used for the calculus of propositions and is therefore of the same concrete kind as the color criterion applied in the game of chess.

The criterion to be used, in contradistinction to the criterion of case analysis as stated in the rule of case analysis for functions (cf. § 25), is only a *necessary* criterion of tautological character, not a *sufficient* one. It is as follows. Given any formula written in functions, we first cancel all the operators contained in it, and replace all functionals by propositional variables '*a*', '*b*', etc.; the formula thus resulting may be called the *root-formula* belonging to the original formula. Thus from the formula

$$(\exists x)(y)f(x, y) \supset (y)(\exists x)f(x, y) \qquad (2)$$

we proceed to the root-formula

$$a \supset a \qquad (3)$$

Now it can easily be shown that, if a formula is a tautology, the corresponding root-formula must be a tautology in the calculus of propositions. This is clear because, in the transition to the root-formula, we have substituted special matrices in the place of the functional variables; since a tautology is to hold for all matrices of the functions occurring, it must hold also for these matrices. The special matrices we use here have either plus signs in all places of the matrix, or they have only minus signs; i.e., they represent the degenerate case of functions which do not vary with the argument. It is therefore a necessary criterion of tautological character that the corresponding root-formula is a tautology in the calculus of propositions.[1] That this is not a sufficient criterion is clear when we reverse the two major terms of our example (2); the resulting formula then is not a tautology (although not a contradiction), whereas the corresponding root-formula is the same as before, namely, (3).

Let us introduce the name 'R-T-formula' for a formula whose root-formula is a tautology. We now can prove that, if a formula is an R-T-formula, so will be every formula derivable from it. This is shown by proving it separately for the use of the rules of substitution, of inference, and of free variables, to which all derivations are reducible. A substitution for a functional variable will appear in the corresponding root-formula as the substitution of a new expression in propositional variables for a propositional variable, i.e., as a substitution conforming to the rule of substitution used in the calculus of propositions. Similarly, the derivation of a new formula by an inference will be expressed, in the corresponding root-formula, by an inference in terms of the fundamental rule of inference. Finally, the use of the rule for free variables (cf. § 27) will not change the root-formula at all because an operator does not appear in that formula.

On the other hand, it is clear that, if a formula in functions is a contradiction, i.e., the negation of a tautology, its root-formula must also be a contradiction, since otherwise the original formula would hold at least for some special matrices. These results enable us to give the proof of consistency for the derivation of tautologies in the same form as before. If a formula 'q' is derived from tau-

---

[1] Incidentally, this result confers also some practical value to the use of root-formulas in the technique of derivations. When we are presented a certain formula and wish to know whether it is a tautology, we can construct the corresponding root-formula; if this is *not* a tautology the original formula certainly cannot be tautologous. This criterion, of course, cannot be used for a positive proof of a tautological character of the original formula. But if the latter is a tautology, it will be possible to derive it from its root-formula by the substitution of functionals, the use of the rule for free variables (§ 27), and the application of the formulas governing the fusion and division of operands, given in formulas (9a–12f, § 25), if necessary by the addition of formulas (13a–14c, § 25).

tologies which are verified by case analysis, its root-formula must be a tautology. The negation '$\bar{q}$' then must have a root-formula which is a contradiction; therefore '$\bar{q}$' is not derivable. We use here merely the fact that all tautologies verified by case analysis are *R-T*-formulas, and that this character is transferred in all derivations.

The class of *R-T*-formulas is a wider class than the class of tautologies; it contains the tautologies as a subclass. This wider class admits of an interesting interpretation. It comprises all those formulas that hold for a universe of objects that contains only one individual. For such a world the distinction between all-operators and existential operators can be dropped, since the formula

$$(x)f(x) \equiv (\exists x)f(x) \qquad (4)$$

will be true in this world. This formula is an *R-T*-formula. Furthermore, it is here unnecessary to use operators at all; instead, we can always use free variables. This follows because the formula

$$f(x) \equiv (x)f(x) \qquad (5)$$

which is an *R-T*-formula, will be true. The reason is that the distinction between (8 and 9, § 20) drops out, formula (9, § 20) being equivalent to (10, § 20). In this very simple world the calculus of functions, therefore, assumes the same form as the calculus of propositions; the two calculi are here *isomorphous*.

On the other hand, we can interpret the given proof of consistency for the tautological part of the functional calculus as follows: the consistency of this calculus is proved by reduction to the consistency of the calculus of propositions. For what we proved is this: if there were a contradiction in the tautological part of the calculus of functions, it would be translatable into a contradiction in the calculus of the *R-T*-formulas, and thus also into the calculus of propositions.

The extension of this method to the case of synthetic premises of a derivation involves some further complications. We said that in a completely mechanized conception the concept of tautology defined in § 24 must not be used. Instead, we have here only what may be called *T*-formulas. They include the initial *T*-formulas, which are verified by case analysis as having a '*T*' in all cases, and, in addition, all formulas which are derivable from these initial *T*-formulas.

Now it can be shown that, if a formula 'q' is derivable from a set 'p' of definite formulas, the formula '$p \supset q$' is a T-formula. This theorem replaces our metatheorem 16, § 27. The proof is given by the following device. Let 'r' be the first formula derived on the path of the derivation leading from 'p' to 'q'; then 'r' will be obtained either by the rule of substitution or by the rule of inference. The rule for free variables cannot have been used because 'p' contains no free variables. In the first case, since 'p' contains only constants or bound variables, the substitution must have been made by putting 'p' in the place of an elementary propositional variable contained in a T-formula; therefore '$p \supset r$' must be a T-formula. In the second case, 'r' must be derived in the form

$$
\begin{array}{c}
p \supset r \\
\underline{p} \\
r
\end{array}
\qquad (6)
$$

Here '$p \supset r$' must also be a T-formula; otherwise the first premise could not be established. Let 's' be the next formula reached in the derivation. Now we cannot generally prove that '$r \supset s$' will be a T-formula because 'r' may contain free variables. For instance, 's' may be obtained by the rule for free variables, i.e., by binding a free variable by an all-operator; then '$r \supset s$' is not a T-formula. But it is clear that then, at least, '$p \supset s$' will be a T-formula since 'p' does not contain the free variables. On the other hand, if '$r \supset s$' should be a T-formula, so also will be the formula '$p \supset s$' because of the transitivity of implication. Continuing this method step by step until we reach the formula 'q' we prove that '$p \supset q$' is a T-formula.[1]

Using this result we can infer as in the discussion of (1) that, if 'p' is synthetic, it is impossible that both 'q' and '$\bar{q}$' are derivable. This proof is based on the fact that, as shown above, the system of T-formulas is consistent. The extension above given (p. 171), according to which no contradictory formulas can be derived when the initial formulas contain free variables, applies equally to the present result. The proof of consistency is thus given by the use of mechanical criteria alone.

---

[1] We take this interesting demonstration from D. Hilbert and P. Bernays, *op. cit.*, Vol. I, p. 151.

The practical use of this proof of consistency for derivations from synthetic formulas should not be overestimated. It shows merely that the derivation of contradictions is impossible if the set 'p' of axioms is not contradictory. In every practical application it remains to prove that this is the case. We cannot give a general rule allowing us to construct a proof in all cases. We possess only some special rules which, if applicable, will prove the noncontradictory character of 'p'. Let us collect these rules in a short summary.

First, if 'p' contains only one-place functions, it may be possible to prove by case analysis that at least for one choice of the truth characters the formula obtains a 'T'. Of this sort is the formula

$$(x)[f(x) \supset g(x)] \tag{7}$$

which leads to a 'T' if '$g(\hat{x})$' is of the A-kind, or '$f(\hat{x})$' is of the E-kind. A second means is given by the construction of the root-formula; if the root-formula is not contradictory, the original formula cannot be contradictory either. Thus the formula

$$(x)(\exists y)[f(x, y) \supset g(x, y)] \tag{8}$$

which we cannot test by case analysis, cannot be contradictory because its root-formula '$a \supset b$' is not contradictory. Third, we can sometimes prove a noncontradictory character by showing that for one specific matrix the formula is true. Thus the formula

$$(\exists x)f(x).g(x).(\exists x)\overline{f(x)}.\overline{g(x)} \tag{9}$$

is not contradictory because it will be true when we choose for '$f(\hat{x})$' and '$g(\hat{x})$' matrices which contain at least one plus sign for the same argument 'x', and also at least one minus sign for the same argument 'x'. This formula cannot be dealt with by the previous methods. Applying case analysis we obtain an F' in all cases where one of the two functions is of the A-kind or E-kind, and for the case of two M-functions the result will be indeterminate. Furthermore, the method of the root-formula does not help us here because the root-formula '$a.b.\bar{a}.\bar{b}$' is here a contradiction. The method of constructing a specific matrix which satisfies the formula is also called the method of the *construction of a model*. In our example, we have constructed the model in the sphere of signs, by designing a suitable matrix. Sometimes it is also possible to construct the model in the sphere of objects. Thus we can construct for (9) a model by choosing for $f(\hat{x})$ the property of being black, and for $g(\hat{x})$ the property of being a swan. It is clear that, if a formula is shown to be empirically true in a certain interpretation, it cannot be contradictory.

Finally, we can combine these methods with our results as to consistency. If a formula 'p' is derivable from a formula 'p''' of a sort for which the noncontradictory character can be proved by one of the preceding methods, 'p' cannot be contradictory. This allows us sometimes to prove the consistency of a set of axioms to which the direct methods mentioned are not applicable. On the other hand, if it is possible to derive from 'p' both a formula 'q' and its negation '$\bar{q}$', the formula 'p' must be contradictory.

These methods can be used frequently. But it is clear that they do not represent a general rule applicable to all cases, and that there will be many sets of synthetic axioms for which we know no means of giving a proof of consistency.

§ 33. Two objections against the proof of consistency

Two serious objections have been raised against the proof of consistency. The first states that this proof is superfluous; the second, that it is circular. Let us discuss these objections in this order.

The adherents of the first objection admit that within a formal conception of the object language a proof of consistency is necessary; but they argue that, when the object language is used, it is always interpreted, and that an interpreted language does not need a proof of consistency. They say that to use a language we must know whether its axioms are true, and whether the process of derivation is so constructed that it leads from true formulas to true formulas. But, if a formula is true, it cannot be false; therefore the process of derivation, if it possesses this property, cannot lead both to a formula and its negation. The proof of consistency thus appears superfluous. On the other hand, if we do not know whether the process of derivation leads from true formulas to true formulas, the proof of consistency cannot help us because it proves only a necessary, not a sufficient, condition for this property. The language then cannot be applied.

This objection can be answered as follows. When we say, in the material conception of language, that the process of derivation leads from true formulas to true formulas, it will be possible only in some simple cases to see immediately that this statement is true. But since truth rules and formation rules, in combination, extend the domain of assertable formulas in an unlimited way, it is impossible for us to realize the extent of a statement of this sort in its full generality. The statement about the truth-preserving property of derivational processes, therefore, does not represent, to its full extent, a cognitive statement. Instead, it must be regarded as including a *convention*, even in the material conception of language — a convention that determines that all formulas resulting from derivational processes shall be considered meaningful. What can be proved is only that, *if* a derivable statement is meaningful, it must be true. But *that* such statements are always meaningful is not a matter accessible to demonstrative argument.

Consider, for instance, the statement

$$(x)f(x) \tag{1}$$

If this statement is true, we can specialize the variable 'x' by inserting any particular value. It therefore might appear plausible to choose as argument the function 'f' itself; we thus derive from (1) the statement

$$f(f) \tag{2}$$

It is clear that this statement cannot be false when (1) is true. But may we regard it as meaningful? We have no way of knowing this directly. It is shown later (§ 40) that a statement of the form (2) is not permissible because it would lead to contradictions. This makes clear that the question of the meaningfulness of derived statements is not a cognitive question, i.e., cannot be answered by a statement of truth character. Rather it must be answered by a convention. We may set up the rule that derivable statements be meaningful; but then we must be able to prove that this convention will never lead to contradictions. That is to say: we must have a proof of consistency. In fact, the rules of substitution given in § 26 are so constructed that they do not allow for such substitutions as used in (2). Moreover, an expression like (2) cannot be derived from expressions like '$(g)f(g)$' because such expressions are excluded by the formation rules: in particular, by the definition of the func-

tional given in rule 4, § 20. Both the rules of formation and substitution were so delimited for the very reason that an expression
like (2) would lead to contradictions. We owe this result to the
analysis of the antinomies (§ 40); this interesting chapter of logic
represents a case where a contradictory language was remedied by
a change, not only in the rules of derivation, but also in the formation rules.

Are the narrower rules used by us free from contradictions? We
cannot see that directly. The bearing of the formation and substitution rules on possible extensions of the language is incalculable.
We do not know into what maze of complications the recursive
definition of the term 'propositional expression' may lead, when
the formation rules, given in § 6 and § 20, are combined with the
rule of substitution, formulated in § 26. It is only the proof of consistency which authorizes us to carry the notion of meaning this
far, and to speak of meaning with respect to expressions which
evade any immediate understanding. We see that the proof of
consistency makes legitimate not only the rules of derivation but
likewise the formation rules and the truth rules of a language.

Let us present a simple example showing the bearing of the truth
rules upon consistency. Assume for a moment that the fourth horizontal line in truth table I b, § 7, is canceled. The formula '$a \vee b$'
then would be a tautology since it would have a 'T' in all three
places of its column. We therefore could make substitutions in this
formula. Putting 'a' for 'b' and applying (1b, § 8) we could derive
'a'; and putting '$\bar{a}$' first for 'a', then for 'b', we similarly could
derive '$\bar{a}$'. We thus would arrive at a contradiction. Now a procedure of this sort is obviously impermissible; we see that the fourth
line in table I b cannot be canceled, that it is possible that both 'a'
and 'b' are false. Though this emendation appears satisfactory,
there remains the question how we know that the four horizontal
lines of the tables are sufficient to exclude similar inconsistencies.
Why must all propositions be true or false? Should we not assume
that there are propositions which are neither true nor false? The
only tenable answer to this question appears to be that we do not
regard linguistic utterances of such a sort as propositions, that we
decide to delimit the domain of sentence-meaning to linguistic forms
which are either true or false. In other words, the restriction to two

truth-values represents a convention. But then it remains to see whether this convention does not lead to contradictions. If every proposition were capable of three truth-values, including an intermediate case of *indeterminacy*, table I b would have to be constructed with nine horizontal lines, accounting for all possible combinations of the three truth-values.[1] We therefore must make sure that the omission of further horizontal lines in table I b does not lead to inconsistencies like those caused by the omission of the fourth line.

These considerations show that we must distinguish between the *statement of contradiction* and the *statement of consistency*, and that the latter cannot be inferred from the former. The statement of contradiction has the form '$\overline{a.\bar{a}}$'; or, when the metalinguistic equivalent of this object statement is used, it says that not both '*a*' and '*ā*' can be true. From this statement we immediately derive a postulate, namely, that a contradiction *should* not be derivable. That a language will satisfy this postulate, i.e., that a contradiction *will* not be derivable, constitutes the content of the statement of consistency. Its truth will depend on the totality of the rules of the language, including the conventions contained in such rules, and must be ascertained by means of a proof of consistency.

Let us now turn to the second objection. It may be explained by the use of ideas which originally were advanced in favor of a proof of consistency.

The proof of consistency has been set forth by those who insist upon its necessity with the claim that it establishes the reliability of the language for which it is given. It is argued that, if a language is inconsistent, every statement can be proved in it. Indeed, if both a certain sentence '*q*' and its negation '*q̄*' are derivable, the conjunction '*q.q̄*' is also derivable; and since for every sentence '*s*' the implication '$q.\bar{q} \supset s$' holds, even as a tautology, every sentence '*s*' can be derived in such a language by applying the rule of inference to the two premises '*q.q̄*' and

---

[1] A logic of this kind, which is called *three-valued*, was constructed by J. Lucasiewicz and A. Tarski, and, independently, by E. L. Post. Such logical systems are consistent and offer certain advantages when they are used for the language of the physics of the microcosm. Cf. the author's *Philosophic Foundations of Quantum Mechanics*, University of California Press, Berkeley, 1944, §§ 30–33; this exposition includes an introduction to the rules of three-valued logic.

'$q.\bar{q} \supset s$'. It follows that, when we do not know whether a language is consistent, it is useless to construct proofs of any theorems in it. Reliance on derivations made in a language presupposes the consistency of the language.

Against this consideration an objection has been raised which turns the very basis of the argument into an argument to the contrary. It is true that — so argues the objection — the object language, as long as it is conceived formally, does not guarantee consistency and therefore requires a particular proof achieving this aim. But since the proof is given by a material use of the metalanguage, its validity presupposes that the metalanguage is free from contradictions; otherwise the objections made above against the use of proofs in a language whose consistency is not known will apply within the metalanguage to the proof given for the consistency of the object language. The proof of consistency, therefore, presupposes what it is aimed at. It is true that it does not presuppose the content of its thesis, namely, the consistency of the object language; but it does presuppose the analogous thesis regarding the metalanguage. What we have, therefore, is not a circularity, strictly speaking, but an infinite regress. It is clear, however, that one is as bad as the other.

A way out of this difficulty might be attempted as follows. Suppose that it should be possible to give the proof of consistency, not in the metalanguage, but in the object language itself. Then we would have proved at the same time the consistency of the language and the dependability of the proof, since the proof was given within the very language to which it refers. Against this sort of solution, however, an objection has been constructed by K. Gödel.[1] He showed that it is impossible to give the proof of consistency within the language to which it refers.

Gödel's theorem about the proof of consistency has been regarded by some logicians as a serious argument against the value of any proof of consistency. Such judgment can scarcely be con-

---

[1] A good exposition of Gödel's theorem is given in D. Hilbert and P. Bernays, *Grundlagen der Mathematik*, Vol. II, Springer, Berlin, 1939, § 5. Gödel's original publication was made in 1930. His theorem goes even beyond the statement that the proof of consistency must be given in the metalanguage; he proves that the metalanguage, in a certain sense, must be richer in logical means than the object language. The answer to this criticism will be included in our analysis in § 34.

sidered a correct evaluation of Gödel's theorem. Let us assume for a moment that Gödel had proved the contrary, that he had proved that a demonstration of consistency could be given within the language itself. Would this improve the situation? Obviously not. The proof then would be valid only if we knew the language were consistent. But if it were not, the proof of consistency could also be given within this language since in an inconsistent language every theorem can be derived. The fact that the proof was given, therefore, would not constitute a reason to believe that the language was consistent. We should have here a circularity rather than an infinite regress. We see that it is not the transition to the metalanguage which makes a proof of consistency objectionable. The objection has deeper reasons and requires a solution by other means.

Serious as appears the criticism expressed in the second objection, it seems that it, too, can be overcome. It is based on a misunderstanding of the nature of logic, namely, on the claim that logic should give us absolute certainty. We must enter, therefore, into an analysis of the general problem of the reliability of logic.

§ 34. Logical evidence

The question of the reliability of logic is connected with the problem of *logical necessity*. The analysis of this problem follows the same pattern as the analysis of the problem of consistency: here, too, the division into formal and material conception of language furnishes us first with an apparent solution and then with a criticism that reduces the so-called solution to a partial contribution, leaving the ultimate solution to considerations of a different nature.

We frequently use the notion of logical necessity in the discussions of the logical calculus. We say, for instance, that a tautology is *necessarily* true, or that the conclusion of an inference is *necessarily* true if the premises are true. Why can we speak here of necessity? The solution offered by the formal conception of language is this: necessity is but the expression of the formal structure of language. By the method of case analysis we have shown that a tautology is true for all truth-values of the elementary propositions; therefore it is not possible to construct any propositions for which the tau-

tology is false. Similarly, we see from the truth tables that, whenever '$a \supset b$' and 'a' are true, 'b' is true; therefore it is not possible
to construct propositions for which the rule of inference does not
hold. By means of such considerations, logical necessity is shown
to be a structural property of the respective formulas, or manipulations, of the object language, when it is regarded in the formal
conception. We need not refer to the content of the propositions
in order to prove logical necessity; the algorism of the uninterpreted
symbols carries necessity with it.

This elucidation of logical necessity, however convincing, must
be subject to the same criticism as presented before. We saw that
the formal conception of the object language is associated with
material thinking in the metalanguage. In considerations of the
given kind, therefore, we rely on the validity of material thinking
in the metalanguage. We take it for granted, for instance, that
more than four combinations of the letters 'T' and 'F' cannot be
constructed in the first two columns of truth table I b; we rely upon
the result derived from the truth table according to which the only
case where both '$a \supset b$' and 'a' are true is given in the first horizontal
line. In this procedure we apply tautologies of the metalanguage,
for instance the tertium non datur, in the form: a proposition is
either true or not true, the latter case being identified with being
false. In other words, when we reduce the necessity of tautologies
or inferences in the object language to properties of a formal calculus, we presuppose the validity of some tautologies or inferences
for the metalanguage. It may even happen that we use in the
metalanguage tautologies or inferences of the same form as those
whose necessary character in the object language we point out.

It is true that the process of formalization can be iterated, that
we can formalize the metalanguage and thus derive the necessary
character of the statements used in the metalanguage by formal
considerations. In this derivation, however, we shall be dependent
on material thinking in the meta-metalanguage. The attempt to
formalize *all* thinking leads into an infinite regress. We must stop
on one level, at which we shall rely on material thinking.

This result is clear also from the form in which we gave the proof
of consistency. Although the object language was here completely
formalized, we had to use material thinking in the metalanguage.

Furthermore, we saw that consistency of a calculus is not a sufficient condition for the truth of the formulas contained in it. Truth means more than incorporation into a consistent system. Thus the class of R-T-formulas (cf. § 32) satisfies the criterion of consistency with respect to the usual rules of derivation; but we reserve the name of tautology for a narrower class. The reason is that only the formulas of this narrower class are true for all matrices of the functions occurring in them. But that this is so can be shown only by the use of a rather complicated form of material thinking, which refers to an infinity of objects. We explained this in the discussion of the tables for truth characters in § 23.

What then tells us that material thinking is true? We have no answer other than an appeal to *logical evidence*. It is *self-evident*, for instance, that, if a proposition is regarded as either true or false, there are only four combinations of letters 'T' and 'F' in the first two columns of the truth tables. We *see* this; we *know* it in realizing the meaning of words like 'not', 'or', 'four'. Without relying on evidence of this kind we could not think. It is true that we have banished self-evidence from the object language by using the formal conception; but self-evidence reappears in the metalanguage, since in the metalanguage we must use material thinking.

We must therefore admit that symbolic logic cannot rid us from the use of logical evidence. This tool bequeathed to us by traditional logic cannot be dispensed with. We can speak of progress inasmuch as self-evidence is applied, in symbolic logic, on a smaller scale and with restriction to thought processes of a very simple kind. It is certainly an advantage that we use as self-evident only very simple statements, and then derive the necessity of complicated statements by steps each of which requires, once more, only the evidence of simple thought operations. There are degrees of evidence; and we shall be glad if we can replace questionable evidence by chains of thought operations each link of which represents strong evidence. Furthermore, there are many results which do not appear evident at all, but which can be shown to hold when we use such chains of evidence. It is not evident, for instance, that the theorem of Pythagoras follows from Euclid's axioms; but we can derive it by a chain of inferences each of which carries a high degree of self-evidence.

Of this sort is also the contribution which is represented by the proof of consistency. The statement that logic is consistent is by no means evident. The erroneous contention that the consistency of logic is evident seems to be grounded in a confusion of the statements of *consistency* and of *contradiction*, distinguished above (cf. p. 180). It is self-evident that, if 'q' is true, it cannot be false. But there is no self-evidence to the statement that if 'q' has been derived from true premises it will be impossible to construct a chain of derivations leading from the same premises to the formula '$\bar{q}$'. Such a statement cannot be self-evident because it refers to all possible derivations and thus to a comprehensive totality which evades a judgment by immediate insight. When we wish to assert this statement, it must be proved, like the theorem of Pythagoras. Long before Pythagoras constructed his deductive proof, however, the theorem was regarded as valid on the basis of practical experience; thus the Egyptians used Pythagoras' theorem in their methods of surveying. Compared with such empirical knowledge, the deductive proof constituted an essential improvement: it showed that what so far had been assumed on the basis of empirical induction could be reduced to a reliance on a number of self-evident steps of reasoning. Similarly, the belief in the consistency of logic and mathematics had so far been grounded in empirical induction, based on the fact that no serious contradictions had been encountered. When it is possible to construct a deductive proof of consistency, such empirical induction is eliminated and replaced by steps of self-evident reasoning. Hilbert's proof of consistency, therefore, plays a role similar to that of any proof of a complicated mathematical theorem; historically speaking, it perhaps occupies in the development of logic a position comparable to that of Pythagoras' theorem in the development of geometry.

The specific reason why a proof of consistency must be required, therefore, is that without such deductive proof we would be dependent on the use of empirical induction. Logical evidence appears superior to this principle, the reliability of which has so often been questioned. But we must not think that the use of deductive proof will eliminate all sources of doubt, i.e., that self-evidence must be associated with absolute certainty. We know that evidence can be misleading: that what appears evident at one time may later turn out to be false. Since logic cannot be constructed without dependence on self-evidence, we therefore must admit that even logic is not absolutely reliable. Such criticism, however, requires some qualification, which will show it to be less destructive than it may appear in the beginning.

The statements that we make in the metalanguage refer to logical statements; but they are not themselves logical statements. They are empirical statements. This is clear from the fact that they refer to sign combinations which are empirically given, which we see with our eyes. We say, for instance, that the formulas of the list in § 8 are tautologies. This statement refers to a group of empirically given symbols; it states properties of these symbols in the same sense as, for instance, the statement that these formulas are classified in eight groups, and has therefore only the truth of empirical knowledge. *The statement that a given formula is a tautology, is not a tautology but an empirical statement.*

To make this clear, let us consider the two statements:

$$a \lor \bar{a} \qquad\qquad (1)$$

$$\text{statement (1) is a tautology} \qquad\qquad (2)$$

Statement (1) is a tautology, as is said in (2). Statement (2), however, is not a tautology; it is an empirical statement about structural properties of the symbols occurring in (1).

The reason that (2) is empirical originates from the fact that reference to the other statement is made in (2) by means of the name '(1)', which we have introduced by the number symbol printed to the right of the statement. We must look at the number symbol, then at the statement, and convince ourselves by the use of our eyes that the symbols printed in the statement have the properties of a tautology. It will be the same, when, instead of using the number symbol, we point to a formula and say 'this formula is a tautology'. When we wish to avoid the use of empirical observation, we must write instead of (2) the statement:

> a statement consisting of a propositional variable, a hook-shaped sign, and the same propositional variable covered by a horizontal line, is a tautology   (3)

Here we have given a syntactical description of statement (1), instead of the use of a name. In this case it is correct to say

$$\text{statement (3) is a tautology} \qquad (4)$$

Now we must add, however, that statement (4) is not a tautology, but an empirical statement, because reference to statement (3) is given in (4) by the use of a name. Furthermore, statement (3) cannot be used to replace (2). It could be used for this purpose only if we knew that statement (1) has the properties described in (3); but such a correspondence must be asserted in a new statement, which will be empirical.

These considerations show that statements about logical formulas and their necessary validity, including the statement of consistency, are made in an empirical language and have only the validity of empirical statements. Now we know that empirical statements are never absolutely reliable.[1] The uncertainty introduced through the use of logical evidence in the metalanguage, for this reason, is superimposed on the general uncertainty of empirical statements. To require that the metalanguage be proved consistent before we use it for a proof of consistency of the object language would be demanding too much, because we use this metalanguage only for the proof of an empirical statement.

We come to the result that the discussion of the reliability of statements in the metalanguage about logic must be given within the theory of empirical knowledge. This theory has led to the result that empirical statements, since we are never sure that they are true, can be maintained only in the sense of *posits*. We posit these statements, i.e., we deal with them as true statements on the basis of given evidence, as long as we have no new evidence against them.[2] Should such evidence turn up, however, we are ready to correct our posits, and to posit other statements. This method of trial and error, known from the use of inductive inference, must be transferred to the use of logical evidence. We rely upon logical evidence as long as we have no evidence to the contrary. Even logical evidence cannot claim absolute truth, but leads only to posits. All statements of the metalanguage asserting properties of

---

[1] This follows from the fact that all empirical statements can be tested by later observations; cf. the author's *Experience and Prediction*, University of Chicago Press, Chicago, 1938, § 9, § 21.   [2] *Op. cit.*, p. 313.

logical formulas or procedures in the object language must therefore be considered as posits. It is permissible to say that a given formula is a tautology and thus necessarily true; but that this statement about its necessary truth is true can only be posited, with the proviso of later correction. Absolute truth, although a property of logical formulas, is unknowable, since we never know whether we have it.[1]

When we say that we choose our posits in following logical evidence, we may be asked whether we wish to formulate with such a statement a *rule of logical evidence* which is to represent the ultimate rule of thought. Let us answer this question by examining the possible signification of such a rule.

The rule of logical evidence will differ from all other rules of language in that it uses a psychological criterion, since evidence is a psychological phenomenon. When we consider the rule of substitution, or the rule of inference, we see that these rules make a certain permission to act dependent on properties of symbols; but they do not refer to the psychological state produced by the symbols in the mind of the person confronted by them. It is true that all rules are pragmatic in so far as they are commands, since a command addresses a person; but the rule of logical evidence, in addition, has a *pragmatic reference*, whereas the other rules have a *semantical reference*.

Now if we look at the language which we consider, and which may be called the object language, we see that for this language the rule of logical evidence can be dispensed with. We have, instead, the rules of substitution and inference, and in addition a set of tautological formulas from which all others can be derived. These rules of semantical reference replace a rule of pragmatic reference. In operating in the object language we therefore need

---

[1] The fact that a proposition stating that a formula is logically necessary is in itself an empirical statement seems to have been first pointed out by J. F. Fries (*Neue Kritik der Vernunft*, 1807). Fries, who was a Kantian, applied this distinction to Kant's synthetic a priori. The form in which we present the idea differs from Fries's conception because we do not recognize a synthetic a priori and apply the distinction to analytic statements. Of course, Fries does not connect his distinction with the distinction of object language and metalanguage, and of formal and material conception of language, ideas unknown at his time. He believed that the empirical character of statements in the metalanguage (in our sense) originated from a psychological nature of these statements.

not watch for the psychological criterion of evidence; although the rules used *will be* evident, we shall not *refer* to this psychological fact as substantiating our use of the rules. We may say that, for the object language, evidence has crystallized into a set of rules with semantical reference.

It is only when we consider a nonformalized language, such as the metalanguage, that we speak of the rule of logical evidence. We then need the psychological criterion of evidence because criteria in terms of the symbols have not yet been formulated. The rule of evidence, therefore, appears as a provisional rule used only until rules of semantical reference have been established, and discarded as soon as this is done. It follows that when we speak of one specific language we need not speak of a rule of evidence, since the language considered can always be assumed to be formalized. To speak of a rule of logical evidence appears necessary only when we speak, not of the object language alone, but at the same time of the meta-language in which we are speaking. In other words, a rule of evidence appears necessary only when we want to include the language we are speaking in our considerations.

Now it is known from the analysis of antinomies that there must be some restrictions as to a self-referent language, i.e., a language a part of which speaks about that language. Independently of these questions, the following can be stated with respect to the rule of evidence: it is possible to state the rule of logical evidence in a self-referent sense; but it is not possible to question or criticize the rule of logical evidence so stated.

We realize the impossibility when we consider the problem of a *justification* of the rule of evidence. We said that we use the rule of evidence to make posits. Now in making posits we are interested in finding the best posit attainable. We therefore require that a rule determining posits be justified, i.e., be shown to lead to the best posits. Such demonstration is possible, however, only for a rule which is not self-referent, since otherwise the procedure would be circular. Thus it is possible to give a justification for the inductive rule since this justification can be given within analytical thinking, i.e., without the use of the inductive rule.[1] It is also possible to give

---

[1] Cf. the author's *Experience and Prediction*, University of Chicago Press, Chicago, 1938, § 39.

a justification of the rules of substitution and inference, as in § 12, § 14, and § 26, because in the metalanguage in which we give this justification we use, explicitly, not these rules, but the rule of evidence. Only when we wish to proceed to a justification of the rule of evidence are we led into circularity, since such a justification would concern the language in which it is given and therefore presuppose the use of the rule itself.

Inquiring further into this problem we see that not only the justification will be circular; even the requirement that a justification should be given involves a circularity. For this requirement is based on a statement asserting that the validity of evidence is not evident, i.e., can be questioned; but such a statement cannot be made without the use of evidence. Here again we see the difference between the criticism of the principle of induction and a criticism of the rule of logical evidence. A statement questioning the rule of induction can be made within a noninductive language, i.e., a language that does not use the rule of induction. It is impossible, however, to question the rule of evidence in a language that does not use this rule. In consequence of the self-referent nature of the rule, the statement questioning the rule will therefore be included in what is questioned. Whereas a criticism of the rule of induction is legitimate, and requires a specific settlement, it is shown by the considerations presented that the attempt to criticize logical evidence cannot be carried through.

We have reached a point here where thinking about logic must stop. We can question logical evidence only when we rely upon it. That is to say: any general statement questioning logical evidence is contradictory because the statement itself will be included in what is questioned. It follows that the statement cannot be meaningfully asserted.

The desire to ask a question about the validity of logical evidence in general seems to originate from an impermissible extension of a question which can be meaningfully asked only for a special case. With respect to any individual statement obtained by the use of logical evidence we can always ask: is this statement reliable? The answer is given by considerations in the next higher language, and may possibly assert that the statement considered is false. Extending questions of this kind to a question concerning the method of evidence in general, including the language in which we think, we believe we arrive at a reasonable question; but actually we arrive at a question which cannot be meaningfully asked.

Such *unlimited extension* of questions originally restricted to special cases represents a procedure which repeatedly has led to pseudo-problems. Thus the. question 'why' can reasonably be asked with respect to every physical phenomenon; but it cannot be meaningfully asked with respect to all phenomena simultaneously. In doing so, certain philosophers have formulated the question: 'why is there something at all, and not nothing'? Those who claim that with this combination of words a philosophical question is asked do not realize that they ask nothing. It seems that the mechanism of thought, starting from meaningful questions concerning individual cases and turning to more and more general questions, is carried on by a certain inertia to run idling when there is nothing left to be asked. It is such an idling of thought, such an empty extension of questions meaningful only with respect to limited objects, that leads to questioning the method of logical evidence as a general method. What is asked is nothing; and it is no breakdown of human capacities if to such a question no answer can be given.

V. The Calculus of Classes

§ 35. Classes

We have repeatedly used the concept of class in the preceding considerations but have not yet entered into an analysis of this concept. We now turn to this problem, beginning with the introduction of a specific notation.

In § 17 we explained that a class F can be defined as the totality of objects having a certain property f, i.e., of the objects x for which the expression '$f(x)$' is true. Following the rule stated in that section we denote the class by a capital letter corresponding to the letter denoting the function. When an object x *belongs* to a class F, or *is a member* of F, we write

$$x \in F \tag{1}$$

We read this: 'x epsilon F'. The Greek letter 'ϵ', used here in a notation introduced by Peano, stands as an abbreviation for the word '$\dot{\epsilon}\sigma\tau\iota$', meaning the copula 'is'. The definition of classes in terms of functions can be expressed by the rule that (1) means the same as '$f(x)$'; i.e., we have

$$x \in F =_{Df} f(x) \tag{2}$$

From this definition it is clear that classes can be dispensed with; they can be eliminated by the use of functions. It is, however, convenient for many purposes to have a class notation. Moreover, this notation corresponds more closely to those forms of conversational language in which the copula 'is' is combined with a noun expressing a one-place function. When we say 'John is an American' we are stating that John is a member of the class of Americans.[1]

The definition (2) defines both the class and the relation of class membership at the same time. That is to say: only the combination of these two concepts is defined. Whenever classes are used,

---

[1] Russell denotes the class coordinated to the function $f(\hat{x})$ by the symbol '$\hat{x}[f(x)]$'. We thus have, combining our notation with Russell's notation: $F =_{Df} \hat{x}[f(x)]$.

it is possible to transcribe the statements occurring into statements of the form (1). We speak here of a *definition in use* of the class concept.

When a class has been defined by the use of a propositional function, as in (2), we sometimes say that it is *intensionally* defined. This kind of definition is distinguished from an *extensional* definition, i.e., a definition given by listing the names of the members, or by pointing to every member successively. The extensional definition, however, can always be transcribed in such a way that it is made intensional. Thus we can regard the fact that the name of an object is written in a certain list, or that the object is pointed to, as a property of that object, which can be conceived as a function $f(\hat{x})$. The definition (2) can therefore be considered to include all possible forms of defining classes.

On the other hand, this interpretation shows that the definition of classes is not limited by a restriction to 'reasonable' properties $f(\hat{x})$. We can define a class consisting of the moon, my dog, and Napoleon; the defining property then is that these things are named on this sheet of paper. The only limitation to class definitions is that the objects must be of the same type; i.e., they must be so chosen that an expression of the form '$f(x)$' is meaningful for all of them. Thus we must not add the property of redness as a member to the above class. This limitation is discussed in § 40.

The distinction between objective and linguistic functions (cf. § 17) has its analogue in the distinction between objective and linguistic classes. We thus distinguish between a class of objects, or *object class*, and a class of argument-signs. In contradistinction to functions, the object class plays a more important part in all applications than the linguistic class. The reason is that the *thing*, which constitutes the element of the class, is frequently referred to in conversational language, whereas the *situation* remains a somewhat unusual object of thought and in general is evaded by reference to propositions.

The application of propositional operations leads to the definition of operations with classes. First, the negation can be used to define the *complement* $\bar{F}$ of a class F, or *complementary class*, by the relation

$$x \in \bar{F} =_{Df} \overline{x \in F} \qquad (3)$$

Thus the complement of the class of Americans is the class of non-Americans. In the definition of this class we must remember that the rule of meaning restricts this class to those objects for which the sentence '$x \, \epsilon \, F$' is false and thus excludes objects for which this expression is meaningless. Thus the set of words 'temperature is a member of the class of non-Americans' is as meaningless as the corresponding positive set of words.

Furthermore, the combination of two propositions by propositional operations can be used for the definition of operations with classes. In the same way as, for instance, the combination '$f(\hat{x}) \vee g(\hat{x})$' can be regarded as representing one function of 'x', being the disjunction of the functions '$f(\hat{x})$' and '$g(\hat{x})$', we can regard the corresponding combination '$F \vee G$' as one class, which is the disjunct of the classes 'F' and 'G'. The indication of the argument 'x' is here usually omitted; such omission is possible because classes usually correspond to functions of only one argument (cf. § 38). We thus define the *disjunct of two classes,* or their *joint class,* by the relation

$$x \, \epsilon \, F \vee G \;\; =_{Df} \;\; (x \, \epsilon \, F) \vee (x \, \epsilon \, G) \qquad (4)$$

The sign of the 'or' on the right-hand side is the usual sign since it stands between propositions. The corresponding sign on the left-hand side has a somewhat different meaning because it stands between classes. In some notations, therefore, a different sign is used for this purpose. Such a sign, however, can be dispensed with when a special notation, such as our capital letters, is used for classes.

The meaning of the disjunct of two classes can be illustrated by the diagram presented in Fig. 6. Here the class F is given by the area shaded by horizontal stripes; G, by the area shaded by vertical stripes. The joint class $F \vee G$ is then given by the whole shaded area, including the cross-shaded part. The cross-shaded area, of course, is counted only once. Thus, when F and G represent two societies, their

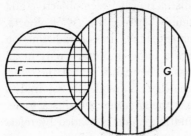

Fig. 6.   Joint class and common class

joint class is given by all those persons who are members of at least one of the two societies; if a person is a member of both societies, he is counted only once within the joint class. When the two societies are planning a joint meeting, the persons entitled to attend it are given by the joint class of the two societies.

Similarly, we define the *conjunct*, or *common class*, of two classes F and G by the relation:

$$x \in F.G \ =_{Df} \ (x \in F).(x \in G) \tag{5}$$

In the diagram of Fig. 6 the common class is given by the cross-shaded area. Like the joint class, the concept of the common class is frequently used in conversational language. Thus when we speak of red flowers we refer to the common class of flowers and of red things.

It is clear from the definition of the term that the common class is in general smaller than either of the two classes for which it is constructed; only in exceptional cases will it be equal to one of them, but it certainly cannot be larger. Similarly, the joint class will in general be larger than either of the two classes, and only in exceptional cases will it be equal to one of them; it cannot be smaller. These two statements fill a place which, in traditional logic, is occupied by the so-called *law of extension and intension*. By 'intension' the meaning of a term is meant; and the law states that, the greater the intension, the smaller the extension, and vice versa. Thus the concept 'red flower' has a greater intension than the concept 'flower'; the attribute 'red' has been added to the latter concept. Correspondingly, the extension of the concept 'red flower' is smaller than that of the concept 'flower'. What is called here an increase in intension, or in content, is obviously the addition of a second class, or property, by means of the operation 'and'. Therefore the law states that a transition to the common class of two concepts is in general accompanied by a decrease in extension. One should not forget, however, that two concepts may also be connected by the operation 'or', and that then the extension will increase because the joint class is introduced. Thus the class of things which are red *or* flowers is larger than either of the two. In order to make the traditional law hold we must therefore regard an or-combination of two concepts as a diminution of intension. For instance, the concept 'parent' has a smaller intension than the concept 'mother' because it is definable as 'mother or father'. When we add this qualification to the traditional law it becomes identical with the extensional relations stated above for the common class and the joint class.

In the same way as the operations 'or' and 'and', all other operations can be used for the definition of compound classes. Thus the *implication class* is defined as

$$x \in F \supset G =_{Df} (x \in F) \supset (x \in G) \qquad (6)$$

Since '$F \supset G$', according to this definition, is the same as '$\bar{F} \vee G$', we can illustrate this class by the help of Fig. 6 as follows. $\bar{F}$ is represented by the area which is not horizontally shaded; therefore $\bar{F} \vee G$ is given by the unshaded area and, in addition, the vertically shaded area. This means that $F \supset G$ is given by all the area of the plane of drawing, with the exception of the crescent-like part which is horizontally shaded but is not crossed by vertical stripes.

The *equivalence class* $F \equiv G$ is defined as

$$x \in F \equiv G =_{Df} (x \in F) \equiv (x \in G) \qquad (7)$$

Since this is the same as the class $F.G \vee \bar{F}.\bar{G}$, it is the joint class of the two common classes $F.G$ and $\bar{F}.\bar{G}$. The first is given by the cross-shaded area of the diagram; the second by the whole of the unshaded area. The sum of these two areas is the class $F \equiv G$.

The class that contains all physical things or events is called the *universal class*.[1] It is denoted by 'V'. Its complement is the *null-class*, i.e., the class that has no member. It is denoted by 'Λ'.

It should be noticed that expressions of the form '$F \vee G$', '$F \equiv G$', denote *classes* but do not constitute *propositions*. When we wish to write propositions we must either add the 'x' and the 'ϵ', as has been done above, or express the proposition under consideration in the form of a statement of the identity of two classes. We shall use here the symbol '$=$' introduced in § 18, thus extending its application to arguments of higher levels. Whereas so far we have used the identity of physical things as a primitive term, we can easily define the identity of classes by the following relation:

$$(F = G) =_{Df} (x)[(x \in F) \equiv (x \in G)] \qquad (8)$$

[1] Universal classes for objects of higher types are discussed later; cf. p. 223.

This definition states: two classes are identical when they have the same members. The identity of classes is thus defined in terms of an equivalence of propositions.

Using the identity relation, we can write, for instance,

$$F \supset G = \bar{F} \vee G \tag{9}$$

This expression is a proposition; according to (8) it means the same as

$$(x)[(x \, \epsilon \, F) \supset (x \, \epsilon \, G) \equiv \overline{(x \, \epsilon \, F)} \vee (x \, \epsilon \, G)] \tag{10}$$

or

$$(x)[f(x) \supset g(x) \equiv \overline{f(x)} \vee g(x)] \tag{11}$$

The all-operator is expressed by the use of the universal class. Thus the formula

$$(x)[f(x) \vee \overline{f(x)}] \tag{12}$$

can be written in the form

$$F \vee \bar{F} = V \tag{13}$$

The formula

$$(x)\overline{f(x) . \overline{f(x)}} \tag{14}$$

can be similarly written

$$\overline{F . \bar{F}} = V \tag{15}$$

or more simply

$$F . \bar{F} = \Lambda \tag{16}$$

The existential operator can be expressed by the negation of a statement saying that a class is identical with the null-class. Thus

$$F \neq \Lambda \tag{17}$$

means

$$(\exists x) f(x) \tag{18}$$

A class F for which (17) holds, i.e., which has members, is called *non-empty*.

An important case obtains when the implication class is identical with the universal class. We then have

$$F \supset G = V \tag{19}$$

According to (6) this means, when we put an all-operator before the whole formula,

$$(x)[f(x) \supset g(x)] \tag{20}$$

or

$$(x)[(x \, \epsilon \, F) \supset (x \, \epsilon \, G)] \tag{21}$$

As an abbreviation for (19) or (21), the notation is used

$$F \subset G \qquad (22)$$

This expression represents a proposition and not a class; the sign '⊂', therefore, has a similar use as the sign ' = '. We say that the

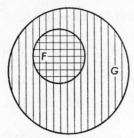

sign '⊂' expresses the relation of *class inclusion*, and we read (22) in the form: '*F* is included in *G*'. A class *F* is *included* in a class *G*, or is a *subclass* of *G*, if all members of *F* are also members of *G*. In the diagram of Fig. 7 the relation of class inclusion finds a graphic presentation.

The propositional equivalent of class inclusion is the general implication, as expressed in the equivalence of (22) and (20).

FIG. 7.   Class inclusion

By employing these two symbolic forms for the same logical relation we gain the advantage of being able to portray more closely the wordings of conversational language. Whereas the 'if-then' form of the implication is used for some kinds of sentences, the relation of class inclusion corresponds to other forms. Thus sentences of the form 'all *F* are *G*', e.g., all persons born in the United States are American citizens, follow the pattern of class inclusion and do not explicitly state an implication.

The relation of class inclusion must be carefully distinguished from the relation of class membership. Class inclusion holds between two classes; membership holds between a class and a thing. Thus the classes of dogs and animals are combined by the relation of class inclusion; an individual dog, however, is a member of both classes.

For technical reasons, we set up the rule that every class is included in itself, or is a subclass of itself. Otherwise the relation (20) could not be identified with class inclusion. It is clear, furthermore, that the null-class is included in every class, since (20) will be true for every function '*g*' if '*f*' is always false. It follows that the null-class is both a subclass of *G* and of *Ḡ*, i.e., that both the statements '$\Lambda \subset G$' and '$\Lambda \subset \bar{G}$' are true. The statement '$\overline{\Lambda \subset G}$', however, is false; this statement is not the same as '$\Lambda \subset \bar{G}$', in contradistinction to the corresponding statements (3) of class membership. For nonempty classes F, too, the statements '$\overline{F \subset G}$' and '$F \subset \bar{G}$' are not equivalent, but for such statements, at least, the first is derivable from the second; namely, it is the subaltern of the second, in the terminology of the square of opposition (13, § 19).

The distinction between adjunctive and connective operations, used within the calculus of propositions, must be transferred to class operations. Since the latter operations are definable in terms of the former, the adjunctive or connective class operations obtain when the respective operations are used for the propositions occurring in the definitions. Thus we shall have an adjunctive or connective implication class given by (6) according as the implication on the right-hand side of (6) is adjunctively or connectively interpreted. Correspondingly, we distinguish an adjunctive from a connective class inclusion according as we use in (20) an adjunctive or a connective implication. Adjunctive class inclusion means that all members of F *are* members of G; connective class inclusion means that all members of F *must be* members of G.

Whereas implication between propositions is usually connectively interpreted in conversational language, it is different with class inclusion; this relation is usually interpreted adjunctively. For instance, we do not hesitate to say, if no colored sailors operate on board a ship, that the crew of the ship is included in the class of white men; but we would not say that being a member of the crew implies being a white man, since we have here only an adjunctive implication. This usage of language is the reason that the term 'extensional' has acquired a meaning given by our term 'adjunctive'. An implication between two functionals is called 'extensional' if it is conceived as meaning that the extension of one function is included in that of the other. It is understood that this is to mean 'adjunctively included'. If we were to use here connective class inclusion, we would of course obtain a connective implication. For this reason we prefer not to use the word 'extensional', but to replace it by the word 'adjunctive'. A correct definition of the above sort would distinguish between 'adjunctively extensional' and 'connectively extensional'; but then the word 'extensional' is redundant.

These considerations can be transferred to the definition (8) of the identity of classes. Since the equivalence in (8) need not be a tautology, the identity of classes thus defined does not presuppose that the classes are determined by functions having the same meaning; all that is required is that the extensions coincide. We may therefore speak here of an *extensional identity*, or an *adjunctive identity*. It is to be distinguished from a *connective identity*, which holds when the equivalence in (8) is of a connective nature. Two examples may illustrate this distinction. The two classes of human beings and featherless bipeds are adjunctively identical. The class of even numbers, however, is connectively identical with the class defined by the predicate 'number differing by 1 from an odd number'.

§ 36. The syllogism

Historically, the class calculus represents the first form of symbolic logic. It was developed in the theory of the syllogism, which goes back to Aristotle. The notation used in this theory was only in part developed by him; additions were made by his followers and later in medieval logic.[1]

In the theory of the syllogism, relations between three classes, S, M, P, are considered. Two relations are given; they constitute the premises of the inference. A third relation is inferred; it represents the conclusion. The relations considered are restricted to the four kinds defined in (11 and 12, § 19), stated in terms of the letters 'A', 'E', 'I', 'O'. Therefore each line of the syllogism contains only two classes. The premises give the relations holding between P and M, and between S and M; the conclusion states the relation holding between S and P. The class M, which thus occurs twice in the premises, is called *middle term*. There exist only four possible arrangements of the letters 'S', 'M', 'P' in the premises:

I	II	III	IV
$M\ P$	$P\ M$	$M\ P$	$P\ M$
$S\ M$	$S\ M$	$M\ S$	$M\ S$
$S\ P$	$S\ P$	$S\ P$	$S\ P$

These four arrangements are called the four *figures* of the syllogism. One of the letters 'A', 'E', 'I', 'O' is inserted in each of the prem-

---

[1] Aristotle (384–322 B.C.) speaks only of three figures. The fourth was constructed two centuries after him, by Galenos (129–99 B.C.). The use of the letters 'A', 'E', 'I', 'O' in the meanings (11, § 19) was introduced by Psellus (born A.D. 1020).

ises; and it is asked whether a conclusion exists and, if it does, which letter must be inserted in it. A valid form thus constructed is called a *mood*. Since each premise can be given four different connecting letters, each figure admits of $4 \times 4 = 16$ forms of the premises; therefore we have 64 forms in all. Of these, only a few lead to a conclusion.

Let us consider the simplest case, resulting when the letter '*A*' is inserted in the premises of the first figure. It has the conclusion '*S A P*' and is represented by

$$\frac{\begin{array}{c} M\ A\ P \\ S\ A\ M \end{array}}{S\ A\ P} \tag{1}$$

This form is traditionally called mood *Barbara*. The other moods have similar names whose vowels correspond to the connecting letters used in the inference. An example for (1) is given by the inference

$$\frac{\begin{array}{l} \text{All men are mortal.} \\ \text{All heroes are men.} \end{array}}{\text{All heroes are mortal.}} \tag{2}$$

Using (12, § 19) for the translation of (1) into modern symbolism, we see that this mood can be written in the form

$$\frac{\begin{array}{c} (x)[m(x) \supset p(x)] \\ (x)[s(x) \supset m(x)] \end{array}}{(x)[s(x) \supset p(x)]} \tag{3}$$

This inference, which we have introduced in (6, § 29), is an expression of the *transitivity of implication*. When we use the relation of class inclusion, (1) can also be written in the form

$$\frac{\begin{array}{c} M \subset P \\ S \subset M \end{array}}{S \subset P} \tag{4}$$

In this form it expresses the *transitivity of class inclusion*.

When we use the distinction between *universals* (letters '*A*', '*E*') and *particulars* (letters '*I*', '*O*'), all moods can be divided into three groups:

   1. Moods that contain only the two connecting letters '*A*' and '*E*'.

2. Moods that contain also letters 'I' or 'O' in the premises, irrespective of the letters contained in the conclusion.

3. Moods that do not contain the letters 'I' or 'O' in the premises but do contain them in the conclusion.

The first group can easily be brought into the form (3) by the use of contraposition as stated in (42, § 28). Let us consider the mood *Cesare* of the second figure

$$
\begin{array}{c}
P\,E\,M \\
\underline{S\,A\,M} \\
S\,E\,P
\end{array}
\tag{5}
$$

An example of this form is given by the inference

$$
\begin{array}{l}
\text{No feathered animals are mammals.} \\
\underline{\text{All dogs are mammals.}} \\
\text{No dog is feathered.}
\end{array}
\tag{6}
$$

Now (5) can be written, when the definitions (12, § 19) are used:

$$
\begin{array}{c}
(x)[p(x) \supset \overline{m(x)}] \\
\underline{(x)[s(x) \supset m(x)]} \\
(x)[s(x) \supset \overline{p(x)}]
\end{array}
\tag{7}
$$

When we apply contraposition to the first premise, (7) is the same form as (3). It is, of course, irrelevant that the function '$p(x)$' occurs here with a negation line on top of it. All inferences of this group differ from mood Barbara only in that they make use of contraposition. There are altogether five valid moods in this group.

In the second group one of the premises contains an existential operator, and the schema to be used is our schema (12, § 29). Thus the mood *Datisi* of the third figure is represented by the schema

$$
\begin{array}{c}
M\,A\,P \\
\underline{M\,I\,S} \\
S\,I\,P
\end{array}
\tag{8}
$$

An example is given by the inference

$$
\begin{array}{l}
\text{All mammals breathe through lungs.} \\
\underline{\text{Some mammals are water animals.}} \\
\text{Some water animals breathe through lungs.}
\end{array}
\tag{9}
$$

With the use of (12, § 19), (8) can be transcribed into the form

$$(x)[m(x) \supset p(x)]$$
$$\underline{(\exists x)m(x).s(x)} \qquad (10)$$
$$(\exists x)s(x).p(x)$$

This is the schema (12, § 29). All inferences of this group are reducible to (10). Sometimes contraposition is to be used thereby for the universal premise; and, if one of the premises is negative, a functional with a negation line above it will occur.

There is no mood in which both premises contain existential operators, because in this case no conclusion follows. This fact is expressed by the rule that two particular premises do not furnish a conclusion. On the other hand, it is clear from (10) that the conclusion of an inference of this group will contain an existential operator, i.e., must be particular. This fact is stated in another of the traditional rules of the syllogism. The second group can be shown to contain ten valid moods.

The third group consists of inferences that have only an 'A' or an 'E' in the premises but contain an 'I' or 'O' in the conclusion. These inferences, however, cannot be regarded as correct when we use the logistic interpretation of all-statements, for which the A-statement or E-statement does not include an existential assertion. We saw in our discussion of the square of opposition (13, § 19) that no inference from 'all' to 'some' can be made unless the condition is added that the first class of the A-statement is not empty. Thus we must regard the mood *Darapti* of the third figure as incorrect:

$$M\ A\ P$$
$$\underline{M\ A\ S} \qquad (11)$$
$$S\ I\ P$$

The invalidity is shown by the example:

All sea serpents are water animals.
All sea serpents are serpents. (12)
Some serpents are water animals.

The conclusion is false because there are no sea serpents. Now the structure of (11) can be written when (12, § 19) is used:

$$(x)[m(x) \supset p(x)]$$
$$\underline{(x)[m(x) \supset s(x)]} \qquad (13)$$
$$(\exists x)s(x).p(x)$$

It is clear that the conclusion is inferable only when we add the third premise '$(\exists x)m(x)$' and write the inference in the form

$$
\begin{array}{c}
(x)[m(x) \supset p(x)] \\
(x)[m(x) \supset s(x)] \\
\underline{(\exists x)m(x)} \\
(\exists x)s(x).p(x)
\end{array}
\tag{14}
$$

This schema is derivable from the general schema (8, § 29) by a procedure similar to that by which we derived the schema (12, § 29). We first combine the two all-statements in the premises of (14) into one statement, according to (9a, § 25), apply the schema (8, § 29), use the tautology '$a.(a \supset b) \equiv a.b$' twice, and finally apply (11, § 29). An example for (14) is given by (12) when we replace the term 'sea serpents' by 'ducks', the term 'serpents' by 'birds', and add the premise 'there exist ducks'.

The schema (14) concerns inferences in which a universal conclusion cannot be drawn from the first two premises. There are some other inferences of this group in which such a conclusion can be drawn. Of this type is the mood Barbari, belonging to the first figure. It is given by the example (2) when we replace the conclusion by the statement 'some heroes are mortal'. A conclusion of this sort is obtainable when we use the conclusion of the schema (3) as the first premise in a schema whose second premise is the existential premise '$(\exists x)s(x)$'. We thus arrive at the schema

$$
\begin{array}{c}
(x)[s(x) \supset p(x)] \\
\underline{(\exists x)s(x)} \\
(\exists x)s(x).p(x)
\end{array}
\tag{15}
$$

We derived this schema in (9, § 29) and illustrated it by the example (10, § 29). The inferences of this sort, which are included in the third group and which are also called *weakened moods*, thus can be conceived as comprehending two steps; the first is given by the schema (3), the second by the schema (15).

In the third group there are 9 valid moods, i.e., moods which can be made valid by the addition of an existential premise. On the whole we therefore have 24 valid moods. Since 5 of the inferences of the third group, namely, the weakened moods, have the same premises as inferences of the first group, we have, among the 64 possible arrangements of premises, only 19 arrangements which lead to a conclusion.

In our classification, the moods of the syllogism are arranged as follows:

	Figure I	Figure II	Figure III	Figure IV
1st group	Barbara Celarent	Camestres Cesare		Camenes
2nd group	Darii Ferio	Baroco Festino	Datisi Ferison Disamis Bocardo	Dimaris Fresison
3rd group	Barbari Celaront	Camestros Cesaro	Darapti Felapton	Bramantip Camenos Fesapo

The weakened moods are indicated by a name which differs from that of the corresponding stronger mood only in the last vowel.

Our analysis leads to the result that the theory of the syllogism essentially contains only two schemas: the schema (3), which expresses the transitivity of implication, or of class inclusion; and the schema (8, § 29), which leads to an existential conclusion. The latter schema is used in the three special forms (10), (14), and (15). All syllogisms may be divided into three groups: the first uses the schema (3); the second, the schema (10); and the third, either the schema (14), or (3) in combination with (15). All we need for the transformation of an inference into one of these schemas is contraposition. It is understood, however, that in the inferences of the third group, which lead from universals to particulars, an existential premise is to be added, a requirement which the traditional theory did not state explicitly. Had this been done, it would have been discovered that this interpretation of universal statements, first, is not necessary for the inferences of the other two groups, and, second, is incompatible with the relations expressed in the square of opposition (cf. § 19 and the footnote on p. 95).

It is a further inadequacy of the traditional syllogistic theory that it does not distinguish between the relations of class inclusion and membership in a class. The statement '$x \, \epsilon \, M$' is regarded as expressible in the form 'xAM'. Now we do have an inference of the form:

$$M \subset P$$
$$\underline{x_1 \in M} \qquad\qquad (16)$$
$$x_1 \in P$$

This holds because of schema (15, § 29). The example (16, § 29), therefore, applies also to (16), which schema shows that this example has a structure different from the example (2). When traditional logic regards both examples as expressible through the schema (1), this incorrectness does not lead to false consequences because of the parallelism between (1) and (16). In other cases, however, we cannot make an inference when we replace class inclusion by class membership; for instance, the two premises

$$M \in P$$
$$\underline{x_1 \in M} \qquad\qquad (17)$$

do not furnish a conclusion. We must postpone the discussion of such expressions as used in the first premise because they lead to classes of a higher type. Only on the lowest level is it permissible, in syllogisms, to replace class inclusion by class membership.

From a technical viewpoint, the traditional theory of the syllogism appears unnecessarily complicated and unelegant. The use of the negative statements containing the letters 'E' and 'O' is superfluous. They can be eliminated when the complements of the respective classes are introduced. Thus the form '$S\ E\ M$' can be replaced by '$S\ A\ \overline{M}$'; similarly, '$S\ O\ M$' is replaceable by '$S\ I\ \overline{M}$'. It is easily seen that then only three forms of inference remain, corresponding to our division into three groups. The first group contains two A-statements as premises and furnishes an A-statement as conclusion; the second group contains an A-statement and an I-statement as premises, and furnishes an I-statement as conclusion; the third group contains two A-statements and an I-statement as premises and furnishes an I-statement as conclusion. The schemas used are the same as given by us in the preceding discussion of the three groups.

On the other hand, when the use of a separate notation for negative statements appears desirable, the traditional notation must be

regarded as incomplete. Statements of the form '$(x)[\overline{m(x)} \supset p(x)]$'
and '$(\exists x)\overline{s(x)}.\overline{m(x)}$' should then be given special letters, for instance in the form '$M \, U \, P$' and '$S \, J \, M$'. In the traditional notation these statements can be expressed only when the classes $\overline{M}$ and
$\overline{S}$ are introduced; they then have the form '$\overline{M} \, A \, P$' and '$\overline{S} \, O \, M$'.
For instance, the valid syllogism

> All nonsmokers save money.
> No vegetarian is a smoker.   (18)
> All vegetarians save money.

cannot be written in the traditional notation when the class M
of smokers is to be used in the inference; instead we must introduce
the class $\overline{M}$ which comprises the nonsmokers. The inference then
has the form of mood Barbara:

$$\frac{\begin{array}{l} \overline{M} \, A \, P \\ S \, A \, \overline{M} \end{array}}{S \, A \, P} \qquad (19)$$

Here the second premise is transformed from an E-statement into
an A-statement. But when this sort of transformation of statements
is admitted, it appears logical to admit these transformations quite
generally and to discard the letters 'E' and 'O' entirely.[1]

Speaking in terms of modern symbolic logic we must classify the
theory of the syllogism as a special chapter in which some secondary
rules of inference, referring to class relations, are developed. It is
clear that these rules are by no means sufficient to cover all sorts
of inference. Thus the inference: 'if Abraham is the father of
Isaac, Isaac is the son of Abraham', cannot be expressed by a
syllogism. This inference has the form

$$\frac{\begin{array}{l} f(x_1, y_1) \supset \breve{f}(y_1, x_1) \\ f(x_1, y_1) \end{array}}{\breve{f}(y_1, x_1)} \qquad (20)$$

This schema corresponds to the fundamental rule of inference (§ 14),
but in addition makes use of the definition of the converse of a

---

[1] Incidentally, example (18) shows that the rule 'If one premise is negative, the
conclusion is negative' does not generally hold; this rule, which has been correctly
derived for the traditional theory, holds only with respect to the scholastic notation
which has no symbolic expression for certain negative statements, and deals with them
as positive statements.

function; it cannot be transformed into a syllogism because it contains a function of two variables. Similarly, most of the inferences used in mathematics cannot be interpreted in terms of syllogisms. Thus the inference: 'if B is between A and C, B is also between C and A' is no syllogism because it refers to a three-place function.

It seems that the limitations of the syllogism were recognized rather early. The modus ponens was introduced by Aristotle's pupils Theophrastos and Eudemos, who thus were the first to begin the construction of a logic of propositions, in contradistinction to the logic of classes; their work was continued by the Stoics. But the logic of propositions did not make any significant progress until the time of Boole (1854). Furthermore, it is only the combination of a logic of functions of several variables with the logic of propositions that leads to the full understanding of conversational language as well as of mathematics. This step was made in our time, mainly through the work of Russell.

This short summary shows that the theory of the syllogism occupies a rather subordinate place in modern logic. We do not wish to belittle the historical significance of this theory; it certainly constitutes the first attempt at formalizing deductive thought operations, and thus represents the historical origin of a scientific logic. But we see today how little was achieved thereby in comparison with the development of deductive operations which were in use, even at the time of Aristotle, in Greek mathematics. The inferences applied in the theorems of Greek geometry cannot be covered by the syllogism, since they include many-place functions such as occur in the above examples.

It appears that early logicians committed themselves to a mistake which ever since has been repeated in many variations: the mistake of planning a logic of knowledge without consideration of the most rigorous and most creative instrument of thought, without an analysis of mathematics. The two thousand years that lie between Aristotle and the origin of modern symbolic logic could not improve Aristotle's logic because they perpetuated his original mistake. Let the inglorious end of the logic of the syllogism be a warning to those who even in our day wish to construct a philosophy which disregards the results of modern science.

§ 37. The principle of abstraction

The notion of class finds an important application in the interpretation of a logical operation which traditionally has been called *abstraction*.

Using traditional terminology we would describe this process as follows. Given a group of objects, we sometimes wish to speak of a common property of all these objects. We then *abstract* this property from its concrete expression in the various objects; i.e., we disregard the differences between the objects and select the common property as the focus of our attention. The term 'abstraction' has here the meaning of 'separation from'. Thus the property of redness can be defined by abstracting it from a group of red objects.

Russell has shown that this psychological analysis of the process can be replaced by a purely logical procedure. First we must define the group of objects from which the abstraction is to be made. We then find that all these objects are connected by a symmetrical relation: in our example, the relation *of color-similarity*. Starting from a particular object, say a certain rose, we thus can define the class of objects having this relation to the selected object, for instance, the class of objects which are color-similar to this rose. Instead of speaking of the common property we then simply speak of the class so defined; in other words, instead of saying that an object is red, we shall say that an object is a member of the class of things color-similar to this rose.

This logical version of the process of abstraction shows clearly that it represents an *ostensive* definition of a property, i.e., a definition given by the act of pointing to a physical object. But it shows also that we need not regard the property as something removable from the concrete object. Instead of imagining a separation between property and object we rather regard the property as given by a union of objects. The notion of property thus can be replaced by the notion of class; the abstraction of the common feature of the group of objects is replaced by the transition to the totality of the group.

It is clear that every interconnective and symmetrical relation will give rise to a class. The interpretation of properties so obtained is very satisfactory because it eliminates unnecessary

entities, in Occam's sense. It shows that abstract properties are
reducible to concrete relations, and it corresponds to the actual
procedure used for the determination of properties. Thus the
definition of a certain shade of color by the use of a color catalogue
is made by reference to a given sample. When we order a suit of
this color we do so by saying: a suit having the same color as this
sample. A similar interpretation can be given to the abstract
property of *spatial direction*. Using Russell's method we can reduce
it to the interconnective and symmetrical relation of *parallelism
between straight lines*. The direction of a line is then interpreted as
the class of lines which are parallel to this line. In fact, every determination of a direction is made by reference to a selected line;
and going in this direction means going parallel with this line. Another example is given by the notion of weight. What is the weight
of a body? It is usually conceived as an abstract property of the
body, recognizable from certain physical effects. Using Russell's
principle of abstraction we can reduce the concept *weight* to the
relation *having the same weight*. The weight of a body is the class
of all objects having the same weight as this body. An adept in
traditional logic would object that in order to define the same
weight we must first define the weight, and then proceed by the
addition of the differentia specifica to the genus. But there is no
reason to insist on this impractical method. It is admissible to
conceive the notion of the same weight as prior to that of weight,
and to define the latter in terms of the former. This conception
corresponds to the actual procedure used in the empirical ascertainment of the weight of a body. The balance is a device which
indicates, not the weight, but equality of weight. Saying that a
body weighs two pounds, therefore, means as much as saying that
the body has the same weight as a certain standard.

Russell has pointed out that it was Aristotle's class logic which
prevented logicians from seeing this simple and natural interpretation of definitions by abstraction.[1] A logic that does not know relations is unable to interpret abstract properties in terms of the

---

[1] It was Leibniz who first saw that the definition of equality of a property is logically
prior to the definition of that property. He applied this principle in his definition of
spatial position, which he gave in terms of the relation 'the same spatial position'.
Cf. the author's paper, 'Die Bewegungslehre bei Newton, Leibniz und Huyghens',
Kantstudien 29, 1924, p. 424.

concrete empirical operations by which these properties are ascertained.

§ 38. Classes of couples, triplets, and so on

The calculus of classes is the analogue of the calculus of one-place functions. If we wish to construct a similar counterpart to the calculus of two-place functions, we must introduce *classes of couples*, since the extension of a two-place function is given by a class of this sort (cf. § 17). Similarly, *classes of triplets* would have to be introduced for three-place functions; and so on for many-place functions.

In the history of logistic such a calculus has proved of little use. The calculus of functions is a satisfactory instrument for all purposes involving two-place, or many-place, functions. We saw that even for one-place functions a class calculus can be dispensed with; but here the notation so constructed offers some advantages. There are, however, special forms of two-place functions for which a class notation may appear useful, and we shall now give a short presentation of the calculus constructed for them.

These special forms are given by many-place functions which consist of several one-place functions, for instance by functions of the kind '$f(\hat{x}) \vee g(\hat{y})$' and '$f(\hat{x}).g(\hat{y})$'. Using capital letters for the classes corresponding to the one-place functions, we can introduce the classes of couples by means of a class operation; but we must then indicate that the expressions refer to different variables. We shall therefore add here the variable as a subscript. Thus we obtain classes of the form $F_x \vee G_y$ and $F_x.G_y$, and define:

$$x, y \in F_x \vee G_y =_{Df} (x \in F) \vee (y \in G) \qquad (1)$$
$$x, y \in F_x.G_y =_{Df} (x \in F).(y \in G) \qquad (2)$$

These definitions constitute the analogues of (4 and 5, § 35). The class $F_x.G_y$ may be called the *combination class* of the two classes F and G, or the *couple conjunct* of these classes. An example is given by the class of all possible telephone connections that can be established between the subscribers in two cities. The class $F_x \vee G_y$ may be called the *couple disjunct* of the two classes F and G. It may be illustrated by the class of possible telephone connections that can be established between any subscriber in either of the two cities and another subscriber, including here all telephone subscribers in the world, i.e., by the class of all possible telephone connections for which at least one partner is registered in one of the two cities.

Although the couple disjunct is not often used in symbolic logic, it finds some important applications in the theory of probability.[1] In any case, if the combination class is defined, it will be logical to extend this sort of definition to include other kinds of propositional operations. In addition to the two classes defined in (1) and (2) we therefore can also define the classes:

$$x, y \in F_x \supset G_y =_{Df} (x \in F) \supset (y \in G) \qquad (3)$$
$$x, y \in F_x \equiv G_y =_{Df} (x \in F) \equiv (y \in G) \qquad (4)$$

We shall call these classes, respectively, the *couple implication class* and the *couple equivalence class*.

It should be noticed that the given definitions apply also if the classes F and G are identical. An illustration for a combination class of this sort, which we write in the form '$F_x.F_y$', is given by the class of telephone connections that can be established between the subscribers in one city. Similarly, we can use the class of all possible telephone connections for which at least one partner is registered in a certain city as an illustration of a class $F_x \vee F_y$. Furthermore, it is also possible to replace the class G by the complement $\bar{F}$ of F, and thus to consider classes of the form $F_x.\bar{F}_y$, etc.

Sometimes a narrower kind of couple class is used. We then introduce a one-one function $c(x, y)$ and consider only those couples which result from the coordination expressed in this function; to such couples we now apply the above definitions. Let us indicate the coordination of 'x' and 'y' by adding a common subscript; thus if x_i is a certain object, y_i will be the object coordinated to it by the function $c(x, y)$. We now define a *narrower couple conjunct*, or *narrower combination class*, by the condition

$$x_i, y_i \in F_{x_i}.G_{y_i} =_{Df} (x_i \in F).(y_i \in G).c(x_i, y_i) \qquad (5)$$

An example is given by the class of married couples for which one partner is a colored person and the other partner is a white person, i.e., by the class of mixed marriages between colored and white persons. Here the function $c(x_i, y_i)$ is interpreted as marriage.

---

[1] Cf. the author's *Wahrscheinlichkeitslehre*, A. W. Sijthoff, Leiden, 1935, § 7, § 9, § 10.

Similarly, we define the *narrower couple disjunct* by the condition

$$x_i, y_i \,\epsilon\, F_{x_i} \vee G_{y_i} =_{Df} [(x_i \,\epsilon\, F) \vee (y_i \,\epsilon\, G)].c(x_i, y_i) \qquad (6)$$

When we use the same interpretation as before, this class will be given by all married couples for which at least one partner is a colored or a white person. Another illustration is given by the kinship of two families F and G all of whose members are married but have no children. The narrower couple disjunct will then consist of all married couples for which at least one partner belongs to one of the two families.

It is easy to extend these definitions to the other propositional operations. Furthermore, it is clear that they apply also when the classes F and G are identical, or when one is the complement of the other. Similarly, an extension of the given definitions to the case of triplets is easily constructed.

VI. The Higher Calculus of Functions

§ 39. Functions of higher types

The extension of the logistic calculus to the inclusion of functions of higher types has its roots in a corresponding usage of conversational language. We often speak of *properties of properties*, and thus make a function the argument of another function. Consider, for instance, the statement 'economy is a useful property'. If we put 'f_1' for 'economy' and 'φ' for 'useful property', this statement is expressed by

$$\varphi(f_1) \tag{1}$$

We say that f_1 is a function of the *first type*, while φ is a function of the *second type*. Replacing 'f_1' by a variable 'f' denoting properties, we obtain a propositional function '$\varphi(\hat{f})$', which is also of the second type. Thus, if we put 'intelligence' for 'f', the resulting proposition is true, and if we put 'forgetfulness' for 'f', the resulting proposition is false. We can then construct statements which contain predicates as bound variables; thus the statement 'Peter has only useful properties' is represented by

$$(f)[f(x_1) \supset \varphi(f)] \tag{2}$$

There are functions of two or more arguments in which the arguments are of different types. They are called *heterogeneous* functions. For instance, in the statement 'Peter hates stinginess' the function 'hates' has in its first place an argument of the zero type, or thing type, and in its second place an argument of the first type. Such functions are considered as being one type higher than the highest type of their arguments; thus the function 'hates' in our example is of the second type. The statement 'there are useful qualities which are not developed by teachers' has the form

$$(\exists f)\{\varphi(f).(x)[t(x) \supset \overline{\psi(f, x)}]\} \tag{3}$$

where 't' stands for 'teacher' and 'ψ' for 'developed'.

In addition to statements of this kind we use statements which, although they do not contain functions of a higher type, refer to a totality of functions of the first type. Let us consider the statement 'Napoleon had all properties of a great general' (this example has been used by Russell). We can write it in the form

$$(x)[g_1(x) \supset f(x)] \supset f(x_1) \tag{4}$$

Here 'g_1' stands for 'great general' and 'x_1' for 'Napoleon'. 'f' is a free functional variable. The symbolization (4), therefore, can be read in the form: 'any property which all great generals have is a property of Napoleon'. In this form the statement belongs to the simple calculus of functions; only it is here an indefinite expression since it contains both a functional constant, namely, 'g_1', and a free functional variable, namely, 'f'. We can make it definite by binding the 'f', in the form:

$$(f)\{(x)[g_1(x) \supset f(x)] \supset f(x_1)\} \tag{5}$$

This form corresponds to the version above, 'Napoleon had *all* properties of a great general'. Although this statement does not contain a function of a higher type we shall regard it as belonging to the higher calculus of functions. We thus set up the rule that formulas containing a functional variable bound by an operator are to be classified as formulas of the higher calculus.

We now must consider an important difference between the totality of physical things and the totality of functions, or, more precisely, between the totality of physical things and any totality of functions of a certain type. The totality of things includes only *real* things, i.e., things which constitute objects of the physical world. A totality of functions includes not only properties of real objects but also properties which no physical thing possesses and which are constructed merely by the power of human imagination. Thus the property of being a sea serpent, or of being a perfectly virtuous man, is included in the totality of properties, although these properties do not apply to any object. We therefore say that any property which is *definable* is included in the totality of properties.

The reason for this conception of the totality of properties is that we wish to speak of properties both with and without corresponding objects. We regard the expression

$$(\exists x)f(x) \tag{6}$$

as *synthetic;* that is to say, we can speak also of properties f for which (6) is false. On the other hand, the expression

$$(\exists f)f(x) \tag{7}$$

is *analytic* and holds for every x. Statement (7), when asserted for the free variable 'x', says that everything has at least one property. This statement is true because, if there were a thing without any property, we could not even speak about it since 'being spoken of' would represent a predicate; such a thing, therefore, would not enter into any relations to other things and could not be regarded as existent.

Besides analytic statements of the form (7) there are also synthetic statements about properties which assert existence. They are verified by empirical observation like existential statements about physical objects. Of this sort is the statement 'there is a property in which Peter excels'. It states that a certain empirical object has a property of a certain kind, and it may be symbolized in the form

$$(\exists f)f(x_1).\varphi(f) \tag{8}$$

Empirical observation will always be necessary when the statement specifies the kind of property referred to and in addition includes a reference to physical objects. Similarly, the statement 'there is a kind of trustworthiness which Peter does not have' is also synthetic and empirical. But even if nobody should satisfy the standard of trustworthiness which we deny Peter, there still would exist such a property. Therefore a statement of the form

$$(\exists f)\varphi(f) \tag{9}$$

can be of a nonempirical nature; it then simply means that an f of this kind can be defined. We allow for the use of a proper name 'f_1' for such a property, and we may write as explained in (5, § 18)

$$(\exists f)\varphi(f).(f = f_1) \tag{10}$$

The existence of properties, therefore, is judged by the use of wider criteria than that of physical things and may be called a *fictitious* existence. [Cf. also the remarks added to formula (25, § 41), and § 49.]

The difference between the existential requirements expresses a characteristic asymmetry between variables of the thing type and of the functional type. It is because of this difference that the higher calculus of functions requires a treatment of its own, although, as to the form of the logistic technique, it follows closely the rules of the simple calculus. We shall present this technique in § 41.

Like functions, we shall also divide *classes* into classes of various types. To a function φ of the second type belongs a class Φ of the same type, and so on for higher types. We thus speak of a *class of classes;* a class can be a member of a higher class. This relation must be carefully distinguished from the relation of class inclusion. If the class F is included in the class G, every member of F is a member of $G;$ this is clear from the definition (21, § 35) of class inclusion. But if F is a member of the class Φ, the members of F are not members of Φ.[1] An example of a class of classes is furnished by the United Nations. The English nation, i.e., the class of Englishmen, is a member of the United Nations. But the relationship holds only for the English nation as a whole; an individual Englishman is not a member of the organization. The relation of class membership is not transitive[2]; this is the reason that the schema (17, § 36) furnishes no conclusion.

It should be noticed that even a class which has only one member must be distinguished from this member, as being of a higher type. This distinction can easily be made clear for a class of classes. Assume that a class Φ has only one class F as its member but that F has three members. Then, if we were to identify Φ and F, Φ would have three members, whereas according to its definition Φ has only one member. In order to avoid this contradiction we must distinguish Φ from F.[3]

---

[1] In § 40 we shall show that the precise formulation of the second clause would read: the expression 'the members of F are members of Φ' is meaningless.

[2] More precisely speaking, it is meaningless to say that the relation of class membership is transitive. Cf. § 40.

[3] This argument, which comes from Frege, is quoted by B. Russell and A. N. Whitehead in *Principia Mathematica*, second edition, Macmillan, New York, 1924, Vol. I, p. 340. Since it applies only to classes of classes, it leaves open the question of whether a class of one physical object must be distinguished from the object. The extension of the distinction to this case must be regarded as a convention made in accordance with the theory of types (cf. § 40).

§ 40. The antinomies and the theory of types

The division by types which we have carried through in the preceding section was introduced by Russell. The reason for this classification was his discovery that otherwise the system of logic would lead to contradictions. Let us now study these contradictions, or *antinomies*. In order to formulate them we must forget about our distinction of types previously introduced, and allow ourselves to follow a train of thought which, to the untrained mind, appears natural and logical.

In addition to classes such as used before, let us introduce classes that contain themselves as members. Using an example given in § 35, let us define for instance a class that contains the four members: the moon, my dog, Napoleon, and itself. This class, of course, is not the same as the class consisting of the moon, my dog, and Napoleon; and it is clear that classes containing themselves can be defined only by the use of *self-referent* definitions. But let us assume, for the present, that this is a permissible method of definition.

Now, when we regard all classes that can be defined in this world, we can divide them into classes which contain themselves as members, and classes which do not. This is an exhaustive classification; i.e., every class must be in one of the two categories. These categories, of course, will also represent classes. The first, let us call it A, contains all classes that contain themselves as members; and the second class, which we call B, contains all classes that do not contain themselves as members. The question arises: since this division is exhaustive, it must also apply to the classes A and B defined; into which of these two classes must we now incorporate the class B?

Let us assume that B is a member of A. Since A contains only classes that contain themselves as members, B then must also contain itself as a member. Now all members of B are classes that do not contain themselves as a member. It follows that B does not contain itself as a member. Therefore B must not be incorporated in A, as assumed, but in B.

Now let us assume that B is a member of B. Then B is a class that contains itself as a member. It follows that B must be incorporated in A, and not in B.

The contradiction at which we have arrived has the following structure. We define certain classes A and B and then derive the two implications:

$$B \,\epsilon\, A \supset B \,\epsilon\, B \tag{1}$$
$$B \,\epsilon\, B \supset B \,\epsilon\, A \tag{2}$$

Since, furthermore, the exhaustiveness of our classification leads to the statement

$$B \,\epsilon\, A \equiv \overline{B \,\epsilon\, B} \tag{3}$$

the contradiction assumes the form

$$B \,\epsilon\, B \supset \overline{B \,\epsilon\, B} \tag{4}$$
$$\overline{B \,\epsilon\, B} \supset B \,\epsilon\, B \tag{5}$$

This is, indeed, a vicious circle, or a vicious contradiction. As long as we only have a sentence which implies its own negation, i.e., a relation of the form

$$a \supset \bar{a} \tag{6}$$

we need not be afraid of the result. We then simply infer, by the use of the reductio ad absurdum (1g, § 8) that 'a' is false. But when we can prove, in addition, the relation

$$\bar{a} \supset a \tag{7}$$

we must also infer that '$\bar{a}$' is false. This result is an impermissible contradiction, i.e., an antinomy. Let us therefore understand by the latter term a case where both the relations (6) and (7) are derivable.

Let us now consider a second antinomy, also constructed by Russell. When we regard a property f we can ask whether this property has itself the property f. In general, this will not be the case. Thus the property *red* is not red. It is different with other properties; for instance, the property *imaginable* is imaginable, the property *determined* is determined, and the property *old* is old, since it certainly existed even in prehistoric times. Let us use the name *predicable* for properties of the second kind; the others are then called *impredicable*. As before, we have here an exhaustive classification; every property must be either predicable or impredicable. Where, then, must we classify the property *impredicable*?

Assume that *impredicable* is predicable; then it has the property which it represents, like *imaginable*, and therefore, *impredicable* is impredicable. Now assume that *impredicable* is impredicable. Our assumption states that *impredicable* has the property it represents; therefore, *impredicable* is predicable.

The two contradictions have the form, when we write '*pr*' for 'predicable' and '*ipr*' for 'impredicable':

$$pr(ipr) \supset ipr(ipr) \tag{8}$$

$$ipr(ipr) \supset pr(ipr) \tag{9}$$

Since we have, because of the exhaustiveness of the classification:

$$ipr(ipr) \equiv \overline{pr(ipr)} \tag{10}$$

we have two contradictions of the form (6) and (7), i.e., an antinomy.

Before we discuss a way to remove these contradictions, let us study some further examples. The given examples correspond in that they represent cases of *self-reference;* a class contains itself as a member, or a property has itself as an argument. There are other forms of antinomies in which the self-reference appears in a somewhat different form, namely, is stated with respect to linguistic expressions. Whereas the first group of antinomies usually is called *logical* antinomies, the second group is therefore known under the name of *semantical* antinomies. Some of the latter ones have the form of linguistic analogues of the logical antinomies. Let us study first a transcription of this sort for our second example, which has been constructed by K. Grelling, and then return to the first example.

Instead of asking whether a property applies to itself, let us ask whether the name of a property has the property under consideration. In general, this will not be the case. For instance, the name 'heavy' is not heavy; the name 'long' is not long. But the name 'short' is short, and the name 'old' is old. Let us comprise as *autological* all properties whose names have the property they denote; the others may be called *heterological*. This distinction is obviously exhaustive.

The antinomy results for the word 'heterological'. Where should we classify this word? Let us assume that the word 'heterological' is autological. Then it has the property it denotes, like the word 'short'; therefore the word 'heterological' is heterological. Now let us assume that the word 'heterological' is heterological. Our assumption has the same form as the statement "'short' is short" and therefore states that the word 'heterological' has the property it denotes; therefore, the word 'heterological' is autological. We have derived two contradictory statements of the form (6) and (7).

A similar transcription of the first antinomy can be constructed as follows. The catalogue of a library, according to the usual conception, names all books that belong to the library. Now the catalogue is itself a book which belongs to the library; at least, this holds for catalogues which are made up as bound volumes, but even the usual set of filed cards may be regarded as a book, in a wider sense. Should therefore the catalogue mention itself? If we follow the usual conception of a catalogue, it should; but we may also employ a somewhat different conception according to which a catalogue mentions all books in the library except itself. Let us therefore leave this decision to the librarians and assume that some catalogues do mention themselves, whereas others do not. Now assume that a man conceives the plan of compiling a catalogue of all catalogues that do not mention themselves. Should this catalogue mention itself? If it does, it is a catalogue which mentions itself, and therefore does not belong in the catalogues to be named in this one. If it does not, it is a catalogue which does not mention itself; according to the plan, it therefore should be mentioned in this catalogue.

In the semantical form, the two antinomies appear even more convincing. One may question whether it is permissible to say that a property has itself as a property; but that a word, i.e., a physical thing, may have the property it expresses seems to be a legitimate assumption. That, for instance, the word 'short' is short, seems to be unquestionable. Similarly, there seems to be no difficulty in assuming that a catalogue mentions itself. Thus we can easily write down, on a page of the catalogue of the University of California, the phrase 'the catalogue of the University of California', and add a number to it which we then write on the back of this catalogue. Such a procedure appears legitimate even when it is questioned whether we can speak of a class which contains itself.

The solution of the two groups of categories, in fact, must be sought along different lines. Let us first turn to the solution of the logical antinomies.

It is with respect to these antinomies that Russell has introduced the *theory of types*. This theory rules out certain expressions as *meaningless*. According to this theory, it is not permissible to say that a class contains itself as a member; such a combination of words has no meaning. A class must be regarded as being of a higher type than its members; then it cannot be said to contain itself. The emphasis is here on the rule that such expressions are to be regarded, not as false, but as meaningless. If they were false, their negations would be true. The antinomy then would remain derivable since, even if the class B of the above proof is empty, we can derive the relations (1)–(5) and thus prove that neither '$B \,\epsilon\, B$' nor '$\overline{B \,\epsilon\, B}$' is true. In introducing the category *meaningless* for such expressions, Russell recognized that the solution must be sought, not in a change of the *truth rules*, but in a revision of the *formation rules* of the language. In fact, the existence of the antinomies proves that the consistency of a language depends, to a great extent, on the formation rules. In our presentation of logic we have therefore taken care to give a precise formulation of the latter rules, separating, from the beginning, meaningless expressions from meaningful ones and carrying through the distinction of types.

Let us add here that the distinction of types refers only to functions, and is not carried over to the propositions in which these functions occur. All propositions are of the same type, whether or not they include functions of higher types. Therefore a proposition like (2, § 39) may be symbolized by the letter 'a' in the same way as propositions of the simple calculus of functions. The technique of the calculus of propositions, therefore, need not be given any supplementation with respect to higher functions.

Turning now to the consideration of the semantical antinomies, we see that they cannot be eliminated by the theory of types for the following reason. Words, or sentences, are physical things and cannot be said to be of a higher type; thus when a predicate is said to pertain to the name expressing it such a conception does not violate the rule of types. Since, however, this conception will also lead to contradictions, we must introduce rules for its exclusion.

These rules are given by the theory of levels of language, which we introduced at the beginning of our presentation (§ 3), and which may be regarded as an analogue of the theory of types constructed for linguistic levels. We set up the rule that within a language we cannot speak about this language, but must use for such purposes a metalanguage; in other words, within a language we can speak only about languages of lower levels. This leads to a reinterpretation of the above statements. When we say: "'short' is short', the second word 'short' belongs to a higher language than the first; therefore, the property *short* of the word 'short' of the object language is not the property denoted by this word, but an analogous property on a different linguistic level. The antinomy then disappears because in the statement "'heterological' is heterological' the second word 'heterological' is not the same as the first word, but belongs to a higher language. Similarly, we rule out as meaningless a catalogue which lists itself. A catalogue is a collection of names of books; if this collection were to include a name of itself we would have a linguistic expression speaking about itself, and thus a violation of the rules of levels of language.

Both the theory of types and the theory of levels of language include, therefore, a restriction concerning the construction of totalities. They require that the word 'all' be used with specific precautions. We can speak only of all things of the same type, or of all expressions of a certain language. The universal class V which we introduced in § 35 must therefore be regarded as the class including all things of the zero type, namely, all physical things. For properties of the first type we must construct another universal class, and so on for other types. Similarly, we cannot speak about all languages. The language in which we speak will always be one level higher than the highest language spoken about. Here, however, it is usually regarded as permissible to speak of all languages up to a given level, so that languages of different levels can be spoken about simultaneously, if only the language in which we speak is separated as being one level higher.

The question has been asked whether it is necessary to introduce these rules. One thing seems to be clear: the only reason we can give for such rules is that they exclude contradictions. There is no question of whether the theory of types, or the theory of levels of language, is true. These theories represent restrictions of formation rules and thus conventions introduced for the reason of making our language consistent. It can be asked, however, whether it is necessary to use these conventions or whether they can be replaced by other ones. Both theories appear rather rigorous, and we would prefer theories demanding weaker prohibitions. The construction of such weaker rules which reach the same effect, namely, the exclusion of antinomies, is still an open problem. It has been shown that at least in part a language may speak about itself, namely, as far as syntax is concerned, whereas the semantics must always be formulated in a different language.[1] We shall not go into these problems here.

The examples of antinomies given above constitute only a short selection of the antinomies which have been discussed in the literature of logic. All these further antinomies, however, are more or less similar to the given examples and can be incorporated either in the logical or the semantical group. All are eliminated, therefore, by the theories of types and of levels of language.

An antinomy of historical importance is known under the name of the antinomy of the *Cretan*. This paradox, which was formulated in ancient philosophy, is as follows. One Cretan says that all Cretans are liars; therefore he is a liar; therefore what he says is not true; therefore not all Cretans need be liars; therefore he need not be a liar; therefore what he says is possibly true — but then again he must be a liar. This antinomy is better stated in a somewhat different form in order to eliminate the complications resulting from the reference to all Cretans. We write:

$$\text{This sentence is false} \tag{11}$$

Here the word 'this' refers to the sentence in which it occurs. It is clear that, if the sentence is true, it is false; and that, if it is false, it is true. This antinomy is eliminated by the rule of levels of language. According to this rule the sentence is meaningless because it speaks about a linguistic expression in the language to which the expression belongs.

Some of the examples mentioned in the discussion, however, do not represent genuine antinomies and can be ruled out without resorting to the theory of types or the theory of levels of language. Thus a story is told about the barber of a company who is ordered by his captain to shave all men of the company who do not shave themselves, but no others. Since the barber himself belongs to the

---

[1] Cf. R. Carnap, *Logical Syntax of Language*, Harcourt, Brace, New York, 1937.

company, the poor man does not know whether or not he should shave himself. If he does, he should not; and if he does not, he should. This makes a good story, but it does not represent an antinomy. The barber is here defined, by the order of the captain, through two statements which contradict each other; so we have here a contradictory definition, and all we can infer is that there is no such thing as the barber of the company, in the sense defined.

The problem of the antinomies lies deeper. Its solution has introduced a period of criticism into logic; and the theories of types and of levels of language, as well as the analysis of the problem of consistency (cf. § 32), represent decisive steps in the new direction.

Historical remark. Most of the paradoxes known in ancient philosophy are not genuine antinomies. The antinomy of the Cretan is practically the only genuine one; the paradoxes of Zeno about the flying arrow and the race between Achilles and the tortoise are solved in the modern theory of the continuum in a way not involving the theory of types. Russell [1] constructed the antinomy of the class of all classes which do not contain themselves, and that of the notion *impredicable;* and he gave a list of further antinomies, including the Cretan and others developed in the theory of sets. For the solution he constructed the theory of types. Since this theory in its simple form does not solve the semantical antinomies, he extended it to the *ramified* theory of types which represents a rather involved theory. Since this theory appeared too rigorous in its prohibitions, he added the *axiom of reducibility,*[2] which, however, appeared scarcely acceptable. Although Russell, on another occasion,[3] suggested the extension of his theory to a theory of levels of language, this extension was actually carried through by Ramsey,[4] to whom we owe the distinction of logical and semantical antinomies, and by Tarski [5] and Carnap.[6] The ramified

[1] B. Russell and A. N. Whitehead, *Principia Mathematica*, Macmillan, New York, 1910, Vol. I, Introduction and chapter II.

[2] B. Russell and A. N. Whitehead, *op. cit.*, Vol. I, p. 161.

[3] In his introduction to L. Wittgenstein, *Tractatus Logico-Philosophicus*, Harcourt, Brace, New York, 1922, p. 23.

[4] F. P. Ramsey, *The Foundations of Mathematics*, Harcourt, Brace, New York, 1931, p. 20.

[5] A. Tarski, 'Der Wahrheitsbegriff in den formalisierten Sprachen,' *Studia Philosophica*, Leopoli, 1935.

[6] R. Carnap, *Logical Syntax of Language*, Harcourt, Brace, New York, 1937 (German edition, Vienna, 1934).

theory of types was then dropped.[1] An interesting attempt to replace the theory of types by weaker formation rules was constructed by Quine,[2] whose system employs certain ideas of von Neumann.

§ 41. The technique of the higher calculus

The technical rules of the higher calculus closely resemble those of the simple calculus. They are given by a transcription of the latter rules for functions and arguments of higher types, with the addition of an extension of the rule of substitution admitting the substitution of variables of higher types in the corresponding places of all formulas of the simple calculus. The definition of tautologies given in § 24, too, can be transcribed for higher functions. It is possible to define for higher functions also a rule of case analysis and thus to construct an initial stock of tautologies; but such a procedure can be dispensed with because all initial tautologies can be obtained equally well by the use of the mentioned extension of the rule of substitution.

Among the tautologies of the higher calculus there are, however, tautologies of a more general kind than those defined in § 24. The formula (7, § 39), for instance, cannot be said to hold for all matrices since it states only that for any selected x there is a matrix holding for it. We therefore introduce the following generalized definition of tautologies:

General definition of tautologies. A tautology is a true formula which contains no empirical constants, or a formula resulting from such a formula by specialization.

By an empirical constant we mean a constant denoting empirical objects, or properties: e.g., constants like 'Peter', 'copper', the verb 'see', and so on. We make this qualification because logical constants may be contained in a nonspecialized tautology, when we understand by logical constants terms of a logical nature, such as the propositional operations, or other terms [cf. the remarks following (6)].

It is clear that the tautologies occurring in the calculus of propositions and the simple calculus of functions satisfy the general definition; the definitions given in § 8 and § 24 are therefore in-

---

[1] Cf. B. Russell in his introduction to the second edition of the *Principles of Mathematics*, Norton, New York, 1938.

[2] W. Quine, *Mathematical Logic*, New York, 1940, Norton, § 24 and § 29.

cluded in the general definition. The difference between the general definition and the other ones is that the general definition includes formulas in which the functional variables are bound. Further examples of such formulas will be given below, for instance in (12), or in (37), where even all variables are bound. For formulas in which all variables are bound the addition concerning specialization cannot be used, since only free variables can be specialized. It follows that a formula containing constants can be tautologous only when, after the constants have been replaced by free variables, it is transformed into a formula which holds for all values of these free variables.

We now formulate the transcription of the rules of the simple calculus as follows:

Transcription of rules. All rules of the simple calculus can be applied in a corresponding form to functions and arguments of higher types.

The necessary extension of the rule of substitution is formulated as follows:

Extension of the rule of substitution. In a tautology of the calculus of propositions or the simple calculus of functions it is permissible to substitute for an elementary free propositional or functional variable a corresponding variable of a higher type; for the arguments of such functions, variables of a corresponding type must then be substituted.

An example of such a substitution is given by the various forms of the *tertium non datur* for higher functions:

$$(f)[\varphi(f) \lor \overline{\varphi(f)}] \tag{1}$$
$$(f)(x)[\varphi(f, x) \lor \overline{\varphi(f, x)}] \tag{2}$$

The first form follows from the corresponding formula (3, § 28) when we substitute 'φ' for 'f' and 'f' for 'x'. The second follows from the first form when we substitute the function '$\varphi(\hat{f}, x)$' for '$\varphi(\hat{f})$' and then apply the rule for free variables to 'x'.

Further examples are given by the formulas

$$(f)(x)\varphi(f, x) \equiv (x)(f)\varphi(f, x) \tag{3}$$
$$(\exists f)(\exists x)\varphi(f, x) \equiv (\exists x)(\exists f)\varphi(f, x) \tag{4}$$
$$(\exists f)(x)\varphi(f, x) \supset (x)(\exists f)\varphi(f, x) \tag{5}$$

Formulas (3) and (4) follow from the corresponding formulas (15a, 15b, § 25) by the substitution of the function '$\varphi(\hat{f}, \hat{x})$' for '$f(\hat{x}, \hat{y})$' and of '$x$' for '$y$'. Formula (5) follows from (16a, § 25) by the same substitution.

The transcription of rules is easily justified. Since the transcribed rules do not lead to the construction of expressions of higher types, but are used only for derivations within expressions of the same type, we can derive for each level the same metatheorems as established for the simple calculus. The justification of the transcribed rules, therefore, is accomplished in the same way as that of the original rules.

It is different with the extension of the rule of substitution. This extension, having no analogue in the simple calculus, requires a justification of another kind. We give it by proving the following theorem.

Metatheorem 17. (*Law of extended substitutions.*) When an extended substitution is made in a tautology, the resulting formula is a tautology.

The proof of this theorem involves some complications. We cannot assume that the manifoldness of matrices of higher functions is the same as that of lower functions. Thus we cannot establish a one-one correspondence between the matrices of a function $\varphi(\hat{x}, \hat{f})$ and those of a function $f(\hat{x}, \hat{y})$; therefore, we cannot infer that when '$\varphi(\hat{x}, \hat{f})$' is substituted for '$f(\hat{x}, \hat{y})$' the resulting formula will hold for all matrices if the original formula did so. The proof must be constructed along other lines, by reference to a homomorphism of derivations. This can be done as follows. For every formula 'p' demonstrable by case analysis in the simple calculus of functions we can construct a corresponding formula 'p'' in higher functions which is demonstrable by a procedure of case analysis corresponding to the procedure introduced in § 25, since the classification in terms of always true, always false, and mixed formulas can equally well be applied to higher functions. The tautologous character of 'p'' can therefore be directly shown. Furthermore, derivations made from 'p'' by the use of substitutions that do not change the type of the variables must lead to tautologies; this fact can be proved by analogy with the proof given for metatheorem 12, § 26.

On the basis of this result we can coordinate to every derivation in the simple calculus a corresponding derivation in the higher calculus. Thus if 'q' is a formula derived in the simple calculus from 'p', we shall have a formula 'q'' derivable from 'p'' in the higher calculus; thereby the relation between 'q'' and 'q' is of such a kind that 'q'' follows from 'q' by the same substitution of higher functions that leads from 'p' to 'p''. Metatheorem 17 is thus proved. We speak of a homomorphism, and not of an isomorphism, of derivations because, for a given derivation in the simple calculus, we can construct more than one corresponding derivation in the higher calculus; this follows because the substitutions can be of various kinds.

We must now turn to the consideration of a device that helps us to introduce a group of formulas which otherwise could not be constructed. Of this kind is formula (7, § 39). This formula, although belonging in the higher calculus, does not contain a higher function and therefore is not obtainable by the substitution of a higher functional variable in a tautology of the simple calculus. On the other hand, the binding of 'f' by an existential operator cannot be derived by one of the transcribed rules of the simple calculus.

In order to derive such formulas we use the following device. We regard an expression '$f(x)$' as a function both of 'f' and of 'x' and write it in the form '$\alpha[f, x]$'; i.e., we introduce the definition

$$\alpha[f, x] =_{Df} f(x) \qquad (6)$$

The function 'α' can be regarded as a function of the second type; it is not a variable, however, but a constant. It may be called a logical constant since it does not have any empirical content, such as empirical constants have (cf. also § 55). We call it a constant because of its technical properties: we can only substitute 'α' for other functions, but not other functions for 'α'. At the end of a derivation, 'α' can be eliminated by putting '$f(x)$' for it. The function 'α', therefore, has only an intermediate use; it appears neither in the beginning nor at the end of a derivation.

Let us illustrate the use of the α-method by an example. Starting from (14b, § 25), namely,

$$f(x) \supset (\exists y) f(y) \qquad (7)$$

we substitute the function '$\alpha[\hat{f}, x]$' for '$f(\hat{x})$', and 'g' for 'y':[1]

$$\alpha[f, x] \supset (\exists g)\alpha[g, x] \tag{8}$$

When we now reintroduce for 'α' its meaning from (6), we obtain

$$f(x) \supset (\exists g)g(x) \tag{9}$$

Here 'f' is a free variable, whereas 'g' is a bound variable. For 'f' we therefore can make substitutions while 'g' remains unchanged. Substituting '$\overline{f}$' for 'f', we obtain

$$\overline{f(x)} \supset (\exists g)g(x) \tag{10}$$

Using the schema (23, § 15) and putting 'f' for 'g' we derive

$$(\exists f)f(x) \tag{11}$$

Since this formula is derived from the tautologies (9) and (10), it is a tautology. We have thus derived formula (7, § 39). Applying the rule for free variables to 'x', we can also write the formula in the form

$$(x)(\exists f)f(x) \tag{12}$$

This example shows how the function 'α' is eliminated at the end of the derivation. The symbol 'α' has only a technical use; it makes it possible to give the necessary derivations without the introduction of further rules. The use of the α-symbol could be avoided by the introduction of specific rules which enable us to derive a relation like (11).

That a formula derived from tautologies by the α-method is a tautology can be shown as follows. As long as the expression '$\alpha[f, x]$' is employed, metatheorem 17 applies to the derived formula, which therefore is a tautology. When the α-symbol is eliminated by the transcription of the derived formula, the resulting formula must be tautologous because the meaning of the formula remains unchanged in the transcription. On the other hand, the α-method enables us to derive formulas, like (11) and (12), which cannot be obtained from a formula of the simple calculus by a direct substitution and which therefore have no analogues in the simple calculus. The use of the α-method thus leads also to derivations which possess no analogues in the simple calculus.

---

[1] This substitution can be more fully explained as follows: first we substitute a function '$f(\hat{x}, z)$' for '$f(\hat{x})$', obtaining '$f(x, z) \supset (\exists y)f(y, z)$'; then we put '$\alpha$' for '$f$', '$f$' for '$x$', '$g$' for '$y$', and '$x$' for '$z$'.

We can, of course, use the α-method also with respect to other functions. Thus we can define, giving a different value to the constant 'α':

$$\alpha[f, x] =_{Df} \overline{f(x)} \tag{13}$$

Putting this meaning of 'α' in (8), we obtain

$$\overline{f(x)} \supset (\exists g)\overline{g(x)} \tag{14}$$

Substituting in (14) '$\overline{f}$' for 'f', while 'g', which is a bound variable, remains unchanged, we obtain

$$f(x) \supset (\exists g)\overline{g(x)} \tag{15}$$

Using again schema (23, § 15) and putting 'f' for 'g', we arrive at the tautology

$$(\exists f)\overline{f(x)} \tag{16}$$

Applying the rule for free variables to 'x', (16) can also be written in the form

$$(x)(\exists f)\overline{f(x)} \tag{17}$$

We can derive even stronger formulas than (12) and (17). Using for '$\alpha[f]$' the definition

$$\alpha[f] =_{Df} (x)f(x) \tag{18}$$

and applying (8), we derive

$$(x)f(x) \supset (\exists g)(x)g(x) \tag{19}$$

Substituting for '$f(\hat{x})$' the function '$f(\hat{x}) \vee h(\hat{x})$', we obtain

$$(x)[f(x) \vee h(x)] \supset (\exists g)(x)g(x) \tag{20}$$

We now substitute '$\overline{f(\hat{x})}$' for '$h(\hat{x})$' and thus derive

$$(x)[f(x) \vee \overline{f(x)}] \supset (\exists g)(x)g(x) \tag{21}$$

Since the implicans is a tautology, we arrive at the relation, when we put 'f' for 'g',

$$(\exists f)(x)f(x) \tag{22}$$

This formula states the apparently far-reaching claim that there is one property which all things have. Actually, this does not mean very much since such a property can easily be defined. For instance, the property of being a thing is such a property; so is the property of being a member of the universal class V. The relation (22) may be regarded as a proof that the construction of the universal class is legitimate.

We can continue this proof as follows. Using the definition

$$\alpha[f] =_{Df} (x)\overline{f(x)} \tag{23}$$

we derive, applying (8),

$$(x)\overline{f(x)} \supset (\exists g)(x)\overline{g(x)} \tag{24}$$

Substituting '$\bar{f}$' for 'f', and then using the same substitutions as applied to (19), we arrive at the formula

$$(\exists f)(x)\overline{f(x)} \tag{25}$$

Formula (25) states that there is a property which no thing possesses. The formula, therefore, secures the existence of the null-class Λ in the same way as (22) does for the universal class. On the other hand, (25) makes clear that, as stated in § 39, the existential criteria used for functions are wider than those required for physical things. It shows that the technique of the calculus compels us to speak of existence also with respect to properties which no physical object possesses, and thus to use for properties a fictitious form of existence.

An important application of the α-method is found in the derivation of the relation between existential and all-operators holding for functions. We start from (13a, § 25):

$$\overline{(x)f(x)} \equiv (\exists x)\overline{f(x)} \tag{26}$$

Substituting for '$f(\hat{x})$' the function '$\alpha[\hat{f}, x]$' according to (6), we have

$$\overline{(f)\alpha[f, x]} \equiv (\exists f)\overline{\alpha[f, x]} \tag{27}$$

Replacing the α-function by its definiens, we arrive at the result

$$\overline{(f)f(x)} \equiv (\exists f)\overline{f(x)} \tag{28}$$

From this relation we can derive the relation, corresponding to (13b, § 25),

$$\overline{(\exists f)f(x)} \equiv (f)\overline{f(x)} \tag{29}$$

by putting '$\bar{f}$' for 'f' and taking the negation on both sides in (28).

Using (28) we can write (16) in the form

$$\overline{(f)f(x)} \tag{30}$$

Formula (30) states that no thing can have all properties. This result is clear because negative properties are included in the totality of properties, and no thing can have both the property f and $\bar{f}$. On the other hand, (16) and (30) show that (28) is restricted in its practical significance: it states only the equivalence of two tautologies. Similarly, (29) states the equivalence of two contradictions.

For the same reasons, the relations corresponding to (13c and 13d, § 25) are trivial. The implicates of these relations, written with 'f' instead of 'x' in the operator, are given, respectively, by (30) and (16) and thus are always true. Similarly, (11) makes the relation corresponding to (14b, § 25) trivial. The relation corresponding to (14a, § 25), with 'f' in the operator instead of 'x', is likewise trivial because the implicans, according to (30), is a contradiction.

We obtain, however, nontrivial relations when we replace the operand by a more complicated expression in which the function 'f' occurs several times. Thus the expression (5, § 39) represents a synthetic expression, and, when we take the negation of this expression, its equivalence to an existential statement has the same practical importance as the analogous relation in the simple calculus. Let us use the symbol

$$[f(x), f(y), \ldots] \tag{31}$$

to abbreviate compound expressions of this sort, which may also contain other functions in addition to 'f'. By the use of the α-method, namely, putting '$\alpha[f, x, y \ldots]$' for the whole expression (31), we then can derive the relations

$$\overline{(f)[f(x), f(y), \ldots]} \equiv (\exists f)\overline{[f(x), f(y), \ldots]} \tag{32}$$

$$\overline{(\exists f)[f(x), f(y), \ldots]} \equiv (f)\overline{[f(x), f(y), \ldots]} \tag{33}$$

In a similar form, the analogues of formulas (13c–d and 14a–b, § 25) can be written; and these relations will then also have the same practical significance as the relations from which they have been transcribed.

Another example for the use of the α-method is as follows. We start from the *tertium non datur*

$$(x)[f(x) \vee \overline{f(x)}] \tag{34}$$

Substituting here the α-function according to (6), we have

$$(x)\{\alpha[f, x] \vee \overline{\alpha[f, x]}\} \tag{35}$$

Applying the rule for free variables to the argument 'f', we write

$$(f)(x)\{\alpha[f, x] \vee \overline{\alpha[f, x]}\} \tag{36}$$

Substituting here for 'α' its value defined in (6), we arrive at:

$$(f)(x)[f(x) \vee \overline{f(x)}] \tag{37}$$

This is the *tertium non datur* written completely in bound variables. The derivation shows that we can apply the rule for free variables also to variables which do not occur as arguments and thus may go from (34) directly to (37). This rule, however, must be regarded as an extension of the rule for free variables since that rule, in the simple calculus as well as in a form transcribed for the higher calculus, originally applies only to variables occurring as arguments.

Formulas like (11), (12), and (37), which differ from those of the simple calculus only in that functions occur within operators while higher functions do not occur, may be said to belong in the *extended simple calculus of functions*. Under this name we subsume both the formulas of the latter kind and of the simple calculus.

Since all initial tautologies of the higher calculus are derivable from those of the simple calculus by substitutions, the higher calculus, in its usual form, does not require any additional axioms and is therefore reducible to the six axioms of the simple calculus mentioned at the end of § 28. It differs from the simple calculus only by the extension of the rules of derivation and formation. In contradistinction to the simple calculus, however, there are limitations to the method of derivation in the higher calculus, as was shown by K. Gödel. The simple calculus of functions is *complete*, i.e., every tautology belonging in it is derivable from the axioms. The higher calculus, on the other hand, is *incomplete*, i.e., it is not possible to give a set of axioms from which all tautologies are deducible. Some formulas cannot be derived within the system, although they can be shown to be true by other means, namely, by the use of metalinguistic considerations. If these formulas then are added as axioms, it can be shown that there are other underivable formulas, and so on. For all practical purposes, these limitations of derivability can be neglected since the underivable formulas do not possess a practical importance. But the fact that there is a limitation to the method of derivation is, of course, of great importance within the analysis of logic. A discussion of these results, however, lies beyond the scope of this book, and we refer the reader to other expositions.[1]

§ 42. The treatment of indefinite expressions

In the theory of derivation developed for the calculus of propositions and the simple calculus of functions we omitted the treatment of indefinite expressions containing free propositional or functional variables. Such expressions involve certain difficulties which, in general, can be overcome only by the use of the higher calculus of functions. We shall now enter into the analysis of this problem.

The distinction between definite and indefinite expressions was introduced in § 12, and the meanings of these terms were extended in § 21. We called an expression indefinite when it is synthetic but contains one or more free variables in addition to constants. The free variable may be an argument variable such as in the expression '$f_1(x)$', where 'f_1' is a constant; or it may be a propositional or functional variable, such as in the example (3, § 12) or (4, § 39). All the examples given belong to the simple calculus of functions; but it is obvious that similar examples can also be constructed within the higher calculus.

In order to study the use of such expressions more closely we must first show how to make derivations from indefinite expressions in which free propositional or functional variables occur. Now it is clear that the rule of substitution, which we have so far applied to functional or propositional variables only in tautologies, can equally be applied to such variables in indefinite expressions. In § 26 we formulated the rule of substitution (which in part δ includes the corresponding rule of the propositional calculus) in such a way that it includes the application to indefinite expressions, since the introductory clause refers to a true formula and is not restricted to tautologies. Similarly, metatheorem 12, § 26, applies also to indefinite expressions. Furthermore, it is clear that the rules of inference and of free variables also can be applied to indefinite expressions, even when they contain free propositional or functional variables. The same holds for formulas in the higher calculus of functions. Let us summarize this result in the following theorem.

[1] For instance, cf. D. Hilbert and P. Bernays, *Grundlagen der Mathematik*, Vol. II, J. Springer, Berlin 1939, § 5, 1.

Metatheorem 18. When a formula 'q' is derivable from a true formula 'p', the formula 'q' is true.

We do not state here any restrictions as to the formula 'p'; it may be a definite or an indefinite formula, and it may belong to the simple or to the higher calculus of functions.

Let us illustrate the method of derivation in an application to (4, § 39). The formula may be repeated:

$$(x)[g_1(x) \supset f(x)] \supset f(x_1) \tag{1}$$

Since 'f' is here a free variable we may substitute the constant 'g_1' for 'f'; we then obtain

$$(x)[g_1(x) \supset g_1(x)] \supset g_1(x_1) \tag{2}$$

Since the major implicans is here a tautology, the major implicate must be true. We thus derive

$$g_1(x_1) \tag{3}$$

that is, the statement 'Napoleon was a great general'. This statement is certainly one of the consequences of the statement considered.

This example shows that derivations from indefinite expressions are made in the same way as derivations from definite expressions, and do not meet with any obstacles. Difficulties arise, however, when the problem of tautological derivability is considered. We saw this in § 21 and § 27 where we considered indefinite expressions of the simple calculus in which only the argument variables were free. If 'p' is a formula of this kind, and 'q' is derivable from it, the formula '$p \supset q$' need not be tautologous. The same holds for indefinite expressions with free functional or propositional variables. Thus (3) is not tautologically implied by (1); i.e., the relation

$$\{(x)[g_1(x) \supset f(x)] \supset f(x_1)\} \supset g_1(x_1) \tag{4}$$

is not a tautology. It does not hold for all matrices of the functions 'g_1' and 'f'. Thus, if we put for '$g_1(\hat{x})$' the function '$\hat{x}$ is a Greek', and for '$f(\hat{x})$' the function '$\hat{x}$ is a woman', the expression '$(x)[g_1(x) \supset f(x)]$' is false, and thus the expression in the braces is true. But '$g_1(x_1)$' is then false when 'x_1' stands for 'Napoleon'; therefore (4) is false. This analysis shows that (4) is true only for special values of 'g_1' and 'f'.

In § 27 we showed that this difficulty can be overcome for indefinite expressions in which only the argument variables are free when these variables are bound by an all-operator. The same method can be used for free functional variables; thereby, however, we leave the simple calculus of functions, since the expression thus resulting belongs in the extended simple calculus. Thus we can write instead of (4):

$$(f)\{(x)[g_1(x) \supset f(x)] \supset f(x_1)\} \supset g_1(x_1) \qquad (5)$$

This expression is indeed a tautology. We can prove its tautological character as follows. When we choose for 'g_1' a function so that '$g_1(x_1)$' is false, we can always choose for 'f' a function so that the expression in the braces is false. This result is attained when we put 'g_1' for 'f'. Therefore the major implicans in (5), i.e., the term consisting of the '(f)' and the expression in the braces, must be false in this case, and thus (5) is true. On the other hand, when '$g_1(x_1)$' is true, (5) will also be true. This proof makes clear that the tautologous character of (5) originates from the fact that the major implicans in (5) has the same meaning as (1), whereas the major implicans of (4) has a different meaning, owing to the change of the scope of the free variable 'f'.

This change of meaning of the implicans can also be seen as follows. When we apply the rule of free variables and put the operator '(f)' before the whole formula (4), we can apply (11d, § 25); (4) then is transformed into the formula

$$(\exists f)\{(x)[g_1(x) \supset f(x)] \supset f(x_1)\} \supset g_1(x_1) \qquad (6)$$

It is obvious that the major implicans of this formula, consisting of the existential operator and the expression in the braces, does not have the meaning (1).

Our result can be stated in the following theorem.

Metatheorem 19. When a formula 'q' is derivable from an indefinite expression 'p', and when 'p'' is the corresponding closed expression, the formula '$p' \supset q$' will be a tautology.

The reason that we did not derive this theorem in § 27 is that the equipollent expression 'p'', so introduced, may belong in the higher calculus of functions, while 'p' is in the lower calculus.

When we wish to apply the same method in the calculus of propositions we must bind propositional variables. Let us illustrate this by an example. Consider the indefinite expression

$$a.p_1 \vee \bar{a}.p_2 \vee k_0 \tag{7}$$

where 'a' is a free variable and the other terms are constants. Substituting here 'p_2' for 'a', we derive

$$p_1.p_2 \vee k_0 \tag{8}$$

since the term '$\bar{p}_2.p_2$' drops out, being a contradiction. In spite of that, the relation

$$a.p_1 \vee \bar{a}.p_2 \vee k_0 \supset p_1.p_2 \vee k_0 \tag{9}$$

is not tautologous. It will be false, for instance, when we choose 'k_0' and 'p_2' as false while 'a' and 'p_1' are true. In order to have a tautology, we would have to write

$$(a)[a.p_1 \vee \bar{a}.p_2 \vee k_0] \supset p_1.p_2 \vee k_0 \tag{10}$$

But this is an unusual way of writing, and it may appear desirable to avoid bound propositional variables.

The reason is that situations, when expressed by bound variables in combination with existential operators, have the wider kind of existence that is used for properties. That it is impossible to restrict the existence of situations to physical existence is shown as follows. When a sentence 'a' is true, we write '$(\exists p)(p \equiv a)$'. We thus express the existence of a situation corresponding to a true sentence in a way similar to that used for thing objects; namely, for thing-objects we write '$(\exists x)(x = x_1)$', a relation which we shall derive in (19, § 43). Now it is obvious that the statement '$(p)(p \equiv a)$' cannot be true. If it were true, we could omit the all-operator and write '$p \equiv a$', where 'p' is a free variable and 'a' an uninterpreted constant that is true; but the formula '$p \equiv a$' can be made false by choosing for 'p' a false sentence. Therefore the formula '$\overline{(p)(p \equiv a)}$' must be true. This formula is easily transformed by means of (13a, § 25) and (7c, § 8) into '$(\exists p)(p \equiv \bar{a})$'. But the last formula states the existence of $\bar{a}$ in the same way as the analogous formula above states the existence of a; i.e., both a and $\bar{a}$ exist. This means that there exists also a situation corresponding to a false proposition; and we thus have a fictitious kind of existence. By the use of bound propositional variables, therefore, the calculus of propositions would be extended in such a way that it would assume the nature of the higher calculus of functions.

It can be shown, fortunately, that in order to carry through the principle of tautological derivability we need not resort to the expedient of binding propositional variables. We can devise a more convenient method of setting up a tautological implication between the original and the derived formula. When an indefinite expression 'p' in the calculus of propositions is given, it is always possible to construct an equipollent expression 'p_0' which does not contain a free propositional variable. When 'q' is derivable from 'p', the expression '$p_0 \supset q$' will then be a tautology. Instead of using the expression 'p'' which contains a bound propositional variable, we thus can use the expression 'p_0' which contains only constants.

The construction of 'p_0' is achieved as follows. Let a synthetic formula contain a free variable 'a', and let us assume that terms containing '$\bar{a}$' also occur. There may be other free variables; we disregard them for the present and eliminate them later by the same method as followed for 'a'. Now when we write the formula in the disjunctive normal form we can collect all terms containing 'a' in a term '$a.p_1$', and all terms containing '$\bar{a}$' in a term '$\bar{a}.p_2$'. The remaining terms may be comprised in a term 'k_0'. The formula will then have the form (7). The terms 'p_1', 'p_2', 'k_0' may include other free variables than 'a'.

Now it can be shown that (8) is equipollent to (7). We have derived (8) from (7); that conversely (7) is derivable from (8) follows because the relation

$$p_1.p_2 \vee k_0 \supset a.p_1 \vee \bar{a}.p_2 \vee k_0 \qquad (11)$$

is a tautology. This fact is easily verified: the implicans is true only when 'k_0' is true or when both '$p_1.p_2$' are true; but in both these cases the implicate will be true.

When we abbreviate the expression (7) by 'p', the expression (8) will therefore represent the equipollent expression 'p_0'. All that remains to show is that (7) includes as special cases all other forms in which a free variable 'a' can occur in a synthetic expression.

This fact is shown as follows. We derive these special cases by giving to the coefficients 'p_1', 'p_2', and 'k_0' the special values of a contradiction or a tautology. When 'p_1' is a contradiction the term in which it occurs as a factor can be canceled. When 'p_1' is a tautology it can be canceled as a factor while the rest of the term remains. These two cases correspond to the arithmetical cases of a factor which assumes the values 0 or 1. We thus derive the following special forms.

(a) If 'p_1' or 'p_2' is a contradiction, or, what is the same, if the term '$a.p_1$' or '$\bar{a}.p_2$' does not occur in (7), the relation (8) assumes the form

$$k_0 \tag{12}$$

(b) If 'p_1' is a tautology, (8) assumes the form

$$p_2 \lor k_0 \tag{13}$$

(c) If both 'p_1' and 'p_2' are tautologies, (8) is always true; the formula (7) then has the form '$a \lor \bar{a} \lor k_0$', which is tautologous.

(d) If both 'p_2' and 'k_0' are contradictions, formula (7) has the form '$p_1.a$'. This form is a contradiction, as is shown by the fact that then (8) also is a contradiction.

In all these forms, of course, (8) is equipollent to (7), as in the general form.

This result shows that in the calculus of propositions we can always eliminate indefinite expressions without using bound variables. For the calculus of functions, a similar theorem is not derivable. When we wish to eliminate indefinite expressions we must resort here to closing the expressions. We have seen that in general this leads into the higher calculus of functions. When we start from indefinite expressions in the simple calculus of functions, we shall be led only into the extended simple calculus which, however, represents also a part of the higher calculus.[1]

§ 43. The relation of identity

We have often used the relation of identity, denoted by the sign '$=$'. We considered it a primitive term whose meaning is known to us. It is possible, however, to define this relation by the use of other primitive terms, as has been shown by Russell. He defines

$$(x = y) =_{Df} (f)[f(x) \equiv f(y)] \tag{1}$$

---

[1] There are, however, special forms of expressions, containing free functional variables and belonging in the simple calculus of functions, in which these free variables can be eliminated instead of being bound. We shall not specify such forms here.

This formula can be paraphrased as follows: two things are identical if they correspond in all properties. The formulation, however, is rather awkward since, in a predication of identity, we cannot speak of two things. A correct interpretation requires the use of the meta-language: two symbols denote the same thing if any two corresponding sentences, which contain these symbols in corresponding places, have equal truth-values.

The significance of this definition should not be underestimated. It reduces *identity* to *equality*, namely, equality of certain properties. Thus in order to apply the definition we must be able to know whether the signs used in the corresponding sentences are equal to each other, in the sense of the similarity relation between tokens (cf. § 2); furthermore, we must know what equality of truth-value is. The objection might be raised that this is not sufficient; that we could not apply the definition if the meaning of identity were not implicitly assumed. For instance, we must be able to see that the '*x*' at the beginning of relation (1) is only one sign, and does not split into two signs. That is true; but this fact does not render definition (1) superfluous. Although we may be compelled to *apply* the concept of identity before we understand the definition, we need not *say* it. What the definition does is *formulate* a concept, thus making explicit what so far has been expressed only through an attitude. It does so in introducing a new symbol and showing how this symbol is reducible to other symbols previously introduced.

Moreover, the reach of the definition goes far beyond an interpretation of the primitive identity of concrete objects. The definition is meant to hold for every type level and thus formulates also the identity of abstract objects, like properties or physical states. In this extension, it has sweeping consequences, the analysis of which constitutes a separate chapter in the history of philosophy. We may refer here to the historical remarks made in § 2 with respect to the verifiability theory of meaning; it is clear that the definition of identity belongs in the framework of this theory since it constitutes the basis of the second of the two principles in which we formulated the definition of meaning.

A closer analysis shows that the definition (1) is capable of several interpretations and thus leads to different conceptions of identity. The way it is written it uses an adjunctive equivalence, and the identity so defined may be called an adjunctive identity. We can, however, add to (1) the requirement that the definiens be a tautology; the identity so defined may be called a tautological identity. Furthermore, we can use, instead, a wider form of connective operation including a synthetic equivalence of a nature such as defined in chapter VIII; we then shall speak of a connective identity. Instead of asking which of all these definitions is the correct one, we should use all of them, always specifying the meaning to which we refer within a particular consideration.

Finally, we must mention here the definition (8, § 35) given for the identity of classes. We said at the end of § 35 that this definition, too, may be subdivided into an adjunctive and a connective identity. On the other hand, since we can coordinate to every function a corresponding class, namely, its extension, there arises the question of the relation between the identity of functions and that of classes. When two classes are identical, will the corresponding functions be identical in the sense of (1)? This question has the following meaning. Transcribing (1) for the next higher level, we obtain

$$(\varphi)[\varphi(f) \equiv \varphi(g)] \qquad (2a)$$

On the other hand, transcribing (8, § 35) for functions, we obtain

$$(x)[f(x) \equiv g(x)] \qquad (2b)$$

Now when (2a) holds, (2b) must also hold, since, if there is an x for which f holds while g does not, we can construe this fact as a property φ of f which g does not have. But will the inverse relation hold? Can we infer from (2b) that (2a) must hold? Russell [1] has voiced the opinion that this inverse relation can be assumed at least for mathematical functions, whereas there may be other functions for which it cannot be maintained. It seems that the problem should be investigated anew in consideration of the distinction between adjunctive and connective identity. It may be advisable to restrict functions of functions in such a way that, when the extensions are connectively identical, the functions must be, likewise. We do not wish to anticipate further analysis of this problem and

[1] B. Russell and A. N. Whitehead, *Principia Mathematica*, Macmillan, New York, 1924, second edition, Vol. I, p. 187.

content ourselves with stating the necessary distinctions on which such an inquiry should be based.

Let us put aside all these questions and turn to the technical discussion of formula (1), assuming that the formula uses the equivalence sign merely in the adjunctive sense. When we interchange 'x' and 'y', the definiens does not change its meaning. This fact proves the symmetry of the identity relation

$$(x = y) \equiv (y = x) \qquad (3)$$

Similarly, the transitivity of equivalence, applied in the form (46, § 28) with the operator '(f)' instead of the operator '(x)', leads to the transitivity of identity

$$(x = y).(y = z) \supset (x = z) \qquad (4)$$

Furthermore, definition (1) shows immediately the reflexivity of identity, which is formulated in relation (8a), to be introduced presently.

It is sometimes useful to regard also the relation of nonidentity. We use here the sign '$\neq$' introduced in § 18, which is defined by the relation

$$x \neq y =_{Df} \overline{x = y} \qquad (5)$$

From (3) we infer

$$(x \neq y) \equiv (y \neq x) \qquad (6)$$

This relation, therefore, is also symmetrical. Of course it is not transitive; instead we infer from (4), using contraposition,

$$(x \neq z) \supset (x \neq y) \vee (z \neq y) \qquad (7)$$

This relation states that, when two things x and z are not identical, they cannot both be identical with a third thing y.

Formula (1) belongs in the extended simple calculus of functions. It cannot be written in the simple calculus because, when we drop the operator '(f)' and write 'f' as a free variable, the meaning of (1) changes. This fact becomes evident when we treat the sign '$=_{Df}$' like an equivalence sign; then (11e, § 25) shows that an operator '(f)' before the whole formula does not mean the same as the operator '(f)' placed on the right-hand side, as in (1). It is possible, however, to write two important consequences of the definition (1) in the simple calculus, the second one by the use of a free variable 'f'. These are the relations:

$$x = x \tag{8a}$$
$$(x = y) \supset [f(x) \equiv f(y)] \tag{8b}$$

Because of (11c, § 25), (8b) means the same as

$$(x = y) \supset (f)[f(x) \equiv f(y)] \tag{9}$$

A number of applications of the identity relation can be derived from relations (8) alone, without the use of (1), and are therefore derivable in the simple calculus.

First, it can be shown that the symmetry and transitivity of identity are derivable from relations (8) even when we replace the equivalence in (8b) by an implication. As to the proof, we refer the reader to the literature.[1]

We shall give an example of a derivation in the simple calculus by deriving relation (5, § 18) from relations (8). Putting 'x_1' for 'y' in (8b) we have

$$(x = x_1) \supset [f(x) \equiv f(x_1)] \tag{10}$$

Using the implication from right to left instead of the equivalence, and applying (6d, § 8), we write

$$f(x_1) \supset [(x = x_1) \supset f(x)] \tag{11}$$

Using the rule for free variables and applying (11c, § 25), we derive

$$f(x_1) \supset (x)[(x = x_1) \supset f(x)] \tag{12}$$

This relation holds also when the major implication is reversed. We show this by applying (14a, § 25) to the major implicate:

$$(x)[(x = x_1) \supset f(x)] \supset [(x_1 = x_1) \supset f(x_1)] \tag{13}$$

Transposing by the use of (6d, § 8), we have

$$(x_1 = x_1) \supset \{(x)[(x = x_1) \supset f(x)] \supset f(x_1)\} \tag{14}$$

Since '$x_1 = x_1$' is always true, according to (8a), the expression in the braces must always be true; but this expression is the reversion of (12). We therefore have the equivalence

$$f(x_1) \equiv (x)[(x = x_1) \supset f(x)] \tag{15}$$

[1] Cf. D. Hilbert and P. Bernays, *Grundlagen der Mathematik*, Springer, Berlin, 1934, Vol. I, p. 167.

When we negate both sides and then substitute '$\bar{f}$' for 'f', we obtain

$$f(x_1) \equiv (\exists x)f(x).(x = x_1) \tag{16}$$

This is relation $(5, \S\ 18)$.

We can derive a further relation. Using in (16) only the implication from left to right, and applying $(16, \S\ 28)$, we derive by the transitivity of implication

$$f(x_1) \supset (\exists x)(x = x_1) \tag{17}$$

Applying the rule for free variables to 'f', and then using $(11d, \S\ 25)$, we derive

$$(\exists f)f(x_1) \supset (\exists x)(x = x_1) \tag{18}$$

According to $(11, \S\ 41)$ the implicans is always true. Therefore we have

$$(\exists x)(x = x_1) \tag{19}$$

Although this relation has been derived from relations (8) alone, we have used in the derivation the higher calculus, since (18) is in the extended simple calculus. Formulas (16) and (19), of course, remain correct when 'the symbol 'y' is put in the place of 'x_1'. This fact shows that our argument-symbols are so used that they always refer to existing objects.

Another application of the identity relation is given by the formula

$$(x)(y)[(x = y) \supset f(x, y)] \equiv (x)f(x, x) \tag{20}$$

We prove this formula as follows. Substituting in (8b) the function '$f(x, \hat{x})$' for '$f(\hat{x})$', we have

$$(x = y) \supset [f(x, x) \equiv f(x, y)] \tag{21}$$

Using the implication instead of the equivalence, and transposing according to $(6d, \S\ 8)$, we have

$$f(x, x) \supset [(x = y) \supset f(x, y)] \tag{22}$$

Applying the rule for free variables and using $(9d, \S\ 25)$, we derive

$$(x)f(x, x) \supset (x)(y)[(x = y) \supset f(x, y)] \tag{23}$$

since the operator '(y)' on the left-hand side drops out. We thus have derived the implication from right to left in (20). To prove the implication from left to right also we proceed as follows. Applying the tautology

$$a \supset (b \equiv c) \equiv [a \supset b \equiv a \supset c] \tag{24}$$

to (21) we first derive the intermediary relation

$$(x = y) \supset f(x, x) \equiv (x = y) \supset f(x, y) \tag{25}$$

From the expression

$$(x = y) \supset f(x, x) \tag{26}$$

we now can derive

$$f(x, x) \tag{27}$$

The derivation is carried out by substituting 'x' for 'y' in (26); the implicans then is always true, according to (8a), and drops out. It follows that (27) is also derivable from the right-hand side of (25). We thus can put an implication sign between the expressions when they are closed in the argument variables:

$$(x)(y)[(x = y) \supset f(x, y)] \supset (x)f(x, x) \tag{28}$$

This formula is the implication from left to right in (20).

In all these derivations we have used only formulas (8), but not the definition (1) itself. Let us now study a problem where we need (1). We write (1) in the form of an equivalence:

$$(x = y) \equiv (f)[f(x) \equiv f(y)] \tag{29}$$

We now take the negation on both sides, applying (32, § 41) and (7c, § 8):

$$(x \neq y) \equiv (\exists f)[f(x) \equiv \overline{f(y)}] \tag{30}$$

Using (7b, § 8) and (10b, § 25), we can write this

$$(x \neq y) \equiv (\exists f)f(x).\overline{f(y)} \vee (\exists f)\overline{f(x)}.f(y) \tag{31}$$

We now use the α-method and define

$$\alpha[f, x, y] =_{Df} f(x).\overline{f(y)} \tag{32}$$

Substituting the function '$\alpha[f, x, y]$' for '$f(\hat{x})$' in (14b, § 25), and putting 'g' for 'y', we have

$$\alpha[f, x, y] \supset (\exists g)\alpha[g, x, y] \tag{33}$$
$$f(x).\overline{f(y)} \supset (\exists g)g(x).\overline{g(y)} \tag{34}$$

Substituting '$\overline{f}$' for 'f' in (34), we have

$$\overline{f(x)}.f(y) \supset (\exists g)g(x).\overline{g(y)} \tag{35}$$

Applying the rule for free variables to 'f' and using (11d, § 25), we derive, when we put 'f' for 'g',

$$(\exists f)\overline{f(x)}.f(y) \supset (\exists f)f(x).\overline{f(y)} \tag{36}$$

We now apply the schema (25, § 15), using as first premise the right-hand side of (31), and (36) as second premise. We thus derive

$$(x \neq y) \supset (\exists f) f(x) . \overline{f(y)} \qquad (37)$$

This result states that when x and y are not identical there is at least one property f which x has but y does not have; i.e., there is at least one distinguishing property for the two things.

§ 44. The definition of number

The relation of identity allows us to construct a logistic expression for numbers.

When we wish to say that a certain property f belongs to one and only one object, we can formulate this statement as follows:

$$(\exists x) f(x) . (y)[f(y) \supset (y = x)] \qquad (1)$$

The first part of this formula, '$(\exists x) f(x)$', states that there is *at least* one x having the property f; and the following part states that there is *at most* one such x.

The statement that there are exactly two things x having the property f is written

$$(\exists x)(\exists y) f(x) . f(y) . (x \neq y) . (z)[f(z) \supset (z = x) \vee (z = y)] \qquad (2)$$

Here we have added, in the first part, the condition '$(x \neq y)$'; then this part states that there are at least two things having the property f. The remainder, as before, states the limitation that there are no more such things.

To state that the property f belongs to exactly three things, we write

$$(\exists x)(\exists y)(\exists z) f(x) . f(y) . f(z) . (x \neq y) . (y \neq z) . (x \neq z)$$
$$.(u)[f(u) \supset (u = x) \vee (u = y) \vee (u = z)] \qquad (3)$$

Using class notation, we shall say that, if (1) is satisfied, the class F has exactly one member. If (2) is valid, F has exactly two members; and, if (3) holds, F has exactly three members.

We see that numbers appear here as properties of *classes*, or *functions*, but not as properties of individual objects. Thus the property expressed through the number twelve belongs to the class of apostles but not to any of the individual apostles. The number property must therefore be expressed through a function, or class, of the second type. This class can be determined as follows. Since, for instance, having the number 3 is a common property of all classes F which have three members, we can apply Russell's principle of abstraction (§ 37) and conceive the number 3 as the class of all classes F which have three members. Similarly, any other integer can be conceived as a class of classes. We therefore define individual integers in the following form:

$$F \epsilon 1 =_{Df} (\exists x)(x \epsilon F).(y)[(y \epsilon F) \supset (x = y)] \tag{4}$$
$$F \epsilon 2 =_{Df} (\exists x)(\exists y)(x \epsilon F).(y \epsilon F).(x \neq y).(z)[(z \epsilon F) \supset (z = x) \vee (z = y)] \tag{5}$$
$$F \epsilon 3 =_{Df} (\exists x)(\exists y)(\exists z)(x \epsilon F).(y \epsilon F).(z \epsilon F).(x \neq y).(y \neq z).(x \neq z)$$
$$.(u)[(u \epsilon F) \supset (u = x) \vee (u = y) \vee (u = z)] \tag{6}$$

This method of defining can be continued for any positive integer. We define here the classes 1, 2, 3, . . . by defining the expressions '$F \epsilon 1$', '$F \epsilon 2$', '$F \epsilon 3$' in the same way as we defined in § 35 the class F by defining the expression '$x \epsilon F$'.

The definitions of integers thus constructed, which were given by Russell, are called *logical* definitions of numbers. They express the individual integers in terms of merely logical concepts, which comprise the existential and all-operator and the propositional operations. Since the relation of identity referred to in these definitions is also reducible to these concepts, as shown in (1, § 43), there are only elementary logical concepts presupposed in the number definitions. Considering in particular the number 1, we see that this concept is reduced to the concept 'at least one', which is expressed in the existential operator. This is no circularity; rather we must say that the concept 'at least one' is logically prior to the concept 'one'. The fact that the phrase 'at least one' is written in three words which include the word 'one' is of course irrelevant for this logical relation. Russell's definition shows that we need not regard 'one' as a primitive term; instead, we can use 'at least one' as primitive.

Russell has shown that we can give another definition of number, which is not of the logical kind. We can reduce the concept of *number* to that of *the same number* by a method like the one developed in § 37. We explained there that the concept *the same weight* is logically prior to the concept *weight* because empirical methods determine directly the same weight of two bodies and lead to the weight of a body only in an indirect way. Similarly, we can find out that two classes have the same number without knowing their number. This is possible when we can establish a one-one-corre-spondence between the two classes. Thus in a monogamous society we know that there are as many husbands as there are wives, when we exclude widows and widowers. We know here an equality of two numbers without knowing the numbers themselves. Using this method, we can define the number 3, for instance, by pointing to a trio of objects, say the men Brown, Jones, and Robinson, and then adding: the class of all classes having the same number as this class. Such a definition of number may be called an *ostensive* definition since it is based on pointing out an empirical object of the kind con-sidered. Compared with this method, the logical definition has the advantage that it does not refer to empirical objects. The two definitions are, of course, identical when the class chosen as standard in the ostensive definition of a number satisfies the logical definition of that number. The logical definition tells us the conditions under which a class has the number three; that the class given by Brown, Jones, and Robinson has this property is ascertainable only by empirical observation, which informs us that this class satisfies the conditions (6).

The objection has been raised against the logical definition that it makes an implicit use of empirical objects occurring in the respec-tive number, namely, in the form of signs. Thus definition (6) of the number 3 contains three existential operators; and it is clear that every definition of this form will contain as many existential operators as correspond to the number defined. We do not think that this objection is valid, however. Definition (6), although it contains three existential operators, does not *refer* to these operators; it uses these signs but does not speak about them. It would be different were we to give, say, a definition of 'green' by writing the word 'green' in green ink and then saying 'green is the color of this sign'. This would be an ostensive definition given by self-referent symbols. The logical definition of number, however, is not of this kind. The correspondence between the number defined and the number of existential operators used in its definition must be re-garded as accidental. It originates from a certain isomorphism between the logistic symbolism and the number classes expressed

through that symbolism, and it would disappear for another form
of the symbolism.

Let us add a remark about the general definition of the number n.
It cannot be written in a form corresponding to (6) because we then
would have to use the number n on the right-hand side: we would
have to indicate, for instance by the use of subscripts, that there
must be n existential operators in the definiens. In that case, in
fact, the above objection that we used in the definiens the number
to be defined would be applicable. For the general definition we
therefore must use another form. We define the number $n + 1$ in
terms of the number n; i.e., we define a number in terms of the
number preceding it. This is achieved by the definition:

$$F' \; \epsilon \; n + 1 = _{Df} (\exists F)(F \; \epsilon \; n).(\exists x)(x \; \epsilon \; F').\overline{(x \; \epsilon \; F)}$$
$$.(y)[(y \; \epsilon \; F').\overline{(y \; \epsilon \; F)} \supset (y = x)] \quad (7)$$

This definition says that a class F' has the number $n + 1$ if it has
a subclass F which has the number n and if one and only one element of F' does not belong to F. We thus arrive at a recursive
definition of numbers.

The method of logical definition can be extended to numbers
other than positive integers, such as negative numbers and frac-
tions. Such numbers are defined in a complicated way by reference
to the positive integers, also called *natural numbers*. We do not go
into this chapter of logic because it would exceed the scope of this
book, leading into the logical analysis of mathematics. Likewise,
we do not give here the definition of the general concept *natural
number*. It is clear that this concept will be defined as the class of
all individual natural numbers, and will therefore be of the third
type. But the delimitation of the word 'all' used in this definition
involves particular problems referring to the principle of mathematical induction, which we cannot discuss in this book.

VII. Analysis of Conversational Language

§ 45. The deficiencies of traditional grammar

In the present chapter we shall attempt to apply the methods of symbolic logic to an analysis of conversational language. Let us begin by comparing our analysis of the inner structure of propositions with the analysis given by traditional grammar.

It is a serious deficiency of grammar that it has no knowledge of the concept of propositional functions. Grammar presents us, instead, in its arsenal of word categories, with three different kinds of terms, all of which we must incorporate into propositional functions. These are nouns, adjectives, and verbs. The separation of nouns from the two others can be justified in so far as the noun can be interpreted to indicate the class for which the predicate holds, or the extension. The division into adjectives and verbs, however, is of questionable significance; we shall show later (cf. § 51) that it is better to put the adjective on a par with a tense of a verb.

The important division of functions according to the number of variables appears in grammar in the distinction between transitive and intransitive verbs but is not extended to nouns and adjectives. It is interesting to see that we have functions of one, two, and three variables in each of the three grammatical categories, although unequally distributed among them. One-place functions are mainly nouns or adjectives, as '$\hat{x}$ is a house', or '$\hat{x}$ is red'; however, there are also verbs (the intransitive verbs) which represent one-place functions, like '$\hat{x}$ sleeps'. Among the two-place functions verbs dominate, as all transitive verbs with one object fall into this category, like '$\hat{x}$ sees $\hat{y}$'; but we include in the two-place functions also verbs connected with a second argument by a preposition, as '$\hat{x}$ speaks to $\hat{y}$', '$\hat{x}$ differs from $\hat{y}$', which grammarians classify as intransitive. It may appear advisable, though, to regard the compound expressions 'speaks to', 'differs from', as the propositional functions. Adjectives in the comparative degree belong to this group, such as used in the function '$\hat{x}$ is taller than $\hat{y}$'; however, we also find here adjectives in the positive degree, such as 'x is similar to y'. Among nouns belonging to this group are the family-relation terms, such as 'x is the mother of y'; another noun example of this group is 'x is an antagonist of y'. Three-place functions are mostly verbs, namely, those with two objects, such as occur in 'Peter sends a telegram to Paul', where 'Peter', 'a telegram', and 'Paul' are the arguments, and 'sends' is the function. An example with a noun as three-place function is the statement 'this book is a present from John to Mary', which has the form 'x is a present from y to z'. Among these functions we find also the word 'between', which grammar awkwardly classifies as a preposition; for 'x is between y and z' is a three-place function.

In many-place functions the order of the variables is relevant to the meaning of the expression. Thus 'x loves y' is different from 'y loves x'. In symbolic logic these two cases are distinguished as '$f(x, y)$' and '$f(y, x)$'. Language has developed in the declension of nouns and pronouns a system of inflections, suffixes and prepositions which makes us, at least partially, independent of the order of words; the place of the variable is thus characterized by a mark attached to the argument. Thus 'he loves her' can be transformed into 'her he loves' or even 'her loves he'; although this order of words is unusual the meaning remains clearly the same. Similarly the sentence 'he gave the book to John' can be transformed into 'to John he gave the book' or 'to John the book he gave'. The deviations from the usual order of words thus made possible are used for emphasis, or for other purposes of style; languages rich in suffixes, like Latin, make wide use of this liberty in the arrangement of words. In logic this means of expression plays no part; and marks indicating the place of the variable can be dispensed with because the order of the signs is rather rigorously determined.

Whereas these considerations show that the structure of language is obviously adapted to the use of many-place predicates, traditional grammar does not recognize such functions. It conceives every sentence as being written in the subject-predicate form, i.e., in our terminology, as being derived from a one-place predicate. Thus in 'Peter is taller than Paul' the word 'Peter' is conceived as the subject, and the phrase 'is taller than Paul' is construed as the predicate. Such an interpretation, however, does violence to the structure of the sentence, in which both the terms 'Peter' and 'Paul' occupy the same type of logical position; they are both subjects in the sense that something is said about them, formulated in the phrase 'is taller than'.

The idea that grammatical subject and object are of different logical rank and that the object constitutes a part of the predicate leads to the further difficulty that the logical nature of the *converse function* cannot be understood. We defined the converse in (9, § 22); it is clear that this definition can be given only for two-place predicates. Thus 'shorter' is the converse of the two-place function 'taller'. Therefore, if the above sentence is conceived as a one-place functional '$g(x)$', where 'g' means 'is taller than Paul', the converse cannot be defined. The sentence 'Paul is shorter than Peter' would have to be written as '$h(y)$', where 'h' means 'is shorter than Peter'; this function 'h', however, is not the converse of 'g', since 'h' contains the term 'Peter' and 'g' contains the term 'Paul'. This shows once more that language has been misconstrued by traditional grammar, since this grammar has no logical place for the existing linguistic forms of the converse. For adjectives and nouns as functions, we have correlative terms, such as 'taller' and 'shorter', 'parent' and 'child'; for verbs, language has developed the passive voice. Thus 'x sees y' can be transformed into 'y is seen by x'.

We do not want to say that it is false to conceive a combination of a predicate and an argument as a new predicate. It is permissible to consider the phrase 'being taller than Paul' as a one-place predicate 'f', namely as a complex predicate which has absorbed one of the arguments, such as defined in (20, § 22). Using this terminology we can say that the grammarians do not recognize simple predicates of more than one argument, but always regard such predicates as complex predicates of one argument. Furthermore, it is not seen that a similar contraction into a one-place function is equally possible with respect to the second argument as subject. Thus our sentence can also be construed as saying of Paul that Peter is taller than he. Here 'Peter is taller than' is the complex function having absorbed the argument 'Peter' and taking 'Paul' as the remaining argument, or subject.

Similarly, in the sentence 'John loves Mary' we can regard 'Mary' as the subject and 'John loves' as the predicate; in this interpretation, the sentence says of Mary that John loves her.

It is a further consequence of the disregard of propositional functions that grammar does not distinguish the structure of propositions having bound variables from that of propositions resulting by specialization. Consider the three statements:

Socrates is mortal	(1)
All men are mortal	(2)
All men have fathers	(3)

Their logistic symbolization is given, in the same order, by the formulas

$$mt(x_1) \tag{4}$$
$$(x)[m(x) \supset mt(x)] \tag{5}$$
$$(x)(\exists y)[m(x) \supset f(y, x)] \tag{6}$$

Here 'mt' means 'mortal', 'm' means 'men', 'f' means 'father'. The three statements are considered in traditional grammar as being of the same form, having a subject and a predicate, the only qualifications being that in (2) and (3) the indefinite adjective (!) 'all' is added and in (3) the predicate has an object. We see that (1) and (2) are of essentially different structures, since (2) includes a bound variable, indicated by the word 'all'. Giving to (2) a linguistic form corresponding more closely to (5), we can formulate this statement: 'if something is a man, it is mortal'. Here the indefinite pronoun 'something' represents the bound variable. In class terminology, (1) represents the relation of class membership, whereas (2) states the relation of class inclusion. Turning to (3) we see from (6) that the difference from (2) consists not only in the appearance of a so-called object, i.e., of a second argument, but also in the presence of a second operator, namely, an existential operator, which binds the second variable as explained in the discussion of this example in (16, § 20).

The existence of forms pertaining to many-place functions, like the inflections of nouns and the passive voice, of expressions for operators and variables, and of other devices for the indication of structure, reveals that the instrument of language as it has been developed in the course of human civilization is superior to the theory of the instrument constructed by logicians. Traditional grammar reflects the primitive stage in which logic remained up to the beginning of logistic. We should not be astonished when the instruction in syntax, in grammar schools and colleges, meets with antagonism, in particular on the part of intelligent students. Our present grammar as it is taught, with its artificial classifications and gratuitous constructions, is based on obvious misunderstandings of the structure of language. We should like to hope that the results of symbolic logic will some day, in the form of a modernized grammar, find their way into elementary schools.

It seems to us that the deficiencies of traditional grammar are equally visible in the science of language in its present condition. The high level of historical and psychological analysis in philology is not matched by a similar level in the understanding of the logical side of language. If philologists would try to make use of a modernized grammar for linguistic purposes they might discover new means of elucidating the nature of language. In the hope that our appeal will be heard outside the camp of the logicians and will be taken up by the few linguists who are aware that a science of language cannot be constructed without a scientific logic, we present, in the following sections, the present status of logistic analysis of conversational language and indicate the outlines of a logistic grammar.

§ 46. Proper Names

A logical classification of the parts of speech of the object language begins with a division of words into three main categories. In the first we have terms used as *arguments;* in the second, terms used as *functions;* in the third, *logical* terms, i.e., terms like 'is' and 'or', which indicate the logical structure of the sentence. In §§ 46–51 we deal with the first category; in §§ 52–54, with the second; and in §§ 55–57, with the third.

The simplest kind of argument is a *proper name.* A proper name is a symbol coordinated by definition to an individual thing. We use proper names for persons, commercial firms, some animals, cities, countries, ships, stars, etc. If a term is to be called a proper name it is necessary that there is a corresponding thing. Mythical names, like 'Zeus', are therefore not proper names, but only *names,* i.e., words used like proper names, namely, as arguments of functions. Sentences asserting properties about such mythical individuals are therefore not objectively true but express a fictitious form of existence with which we shall deal later (§ 49).

That true statements containing proper names assert existence of respective individuals is expressed in relation (5, § 18). Moreover, the statement of existence is derivable even if the sentence containing the proper name is not true. This fact is shown as follows. If '$f(x_1)$' is false, the sentence '$\overline{f(x_1)}$' will be true. Conceiving the function '$\bar{f}$' as representing a certain function 'g' we then can write the true sentence '$g(x_1)$'; and with the use of (5, § 18) we derive

$$(\exists x)g(x).(x = x_1) \tag{1}$$

Replacing 'g' by '$\bar{f}$' we thus arrive at the statement

$$(\exists x)\overline{f(x)}.(x = x_1) \tag{2}$$

We see that the syntax of proper names is so constructed that a statement containing a proper name involves existence of a respective individual, whether or not the statement is true. This fact is also expressed in relation (19, § 43) which asserts existence independently of any true statement '$f(x)$' or '$\overline{f(x)}$'. The knowledge thus conveyed by the occurrence of a proper name, however, is not knowledge concerning the physical world; it refers only to the language we use. All that a proper name tells us is that there is a thing corresponding to it; but such knowledge supplies an information only about language since it merely classifies the proper name as a term that denotes something. This is the same kind of knowledge as we acquire from the occurrence of other symbols; in fact, a classification into symbols denoting things, symbols denoting functions, symbols for propositional operations, etc., must be known if we wish to employ symbols successfully.

§ 47. Descriptions

The number of proper names is small compared with the number of individual things in which we are interested. Language has therefore developed another method of characterizing an individual thing, the method of *description*. One kind of description is given by means of a predicate, or a combination of predicates, chosen in such a way that there is only one thing satisfying it. Thus we can describe a man as 'the first man who saw a living human retina'. It happens that in this case the description determines a man having a well-known proper name, namely, 'Hermann von Helmholtz', since the physicist carrying this name, by means of the ophthalmoscope invented by him, was the first man to have the experience mentioned. The description 'the first man who made a fire' also determines a definite person, but we do not know his name, and perhaps he had none. Considering a combination of predicates as one predicate, we may say that this kind of description is given by a predicate whose extension is one individual.

A description of a second kind differs from the first in that it contains, in addition to a predicate, a proper name of an individual by reference to which another individual is determined. The description 'Napoleon's mother' contains the two-place predicate 'mother' and the proper name 'Napoleon'; the person in question is determined as the individual that has the relation *mother* to Napoleon. Of the same kind are descriptions like 'John's house', or 'the ship in which Columbus discovered America'. This is the sort of description mostly used. If we consider such expressions to be descriptions given by complex predicates, the two kinds of descriptions assume the same form, and we can therefore subsume all descriptions under the definition: a description is given by a predicate whose extension is one individual.

When we use a description we must make sure whether there is an individual satisfying that description. Thus the description 'king of France', meant to refer to the present time, does not determine an individual, because there is at present no king of France, whereas the description 'king of England' does determine a certain person. We indicate the existence of one, and only one, corresponding thing by putting the definite article before the description. Thus we say 'the king of England', and employ this expression like a proper name as an argument in sentences, for instance in the sentence 'the king of England was crowned in Westminster Abbey'. We must not combine the definite article with the phrase 'king of France'; sentences thus constructed, like 'the king of France is a Frenchman', are false because they include the assertion that there is such a person. The use of the word 'the' distinguishes the sentence from the mere implication 'if there were a king of France he would be a Frenchman', a sentence which is presumably true. On the other hand, the article 'the' construed with a singular noun excludes also the case that there is more than one individual satisfying the description.[1] Thus it is false to say 'the minister of England made a speech'; the correct form would be 'a minister of England made a speech'. The indefinite article indicates that there is more than one individual satisfying the predicate.

With Peano and Russell we use for descriptions the notation

$$(\imath x) f(x) \tag{1}$$

meaning 'the thing x having the property f'. The operator in this expression is an inverted Greek 'ι', and is therefore called the iota-operator; 'x', in this expression, is a bound variable. The object described will be called the *descriptum*.

Languages which have no definite article, like Latin or Turkish, use for a description the noun in the singular form, as in 'Rex Francorum', meaning 'the king of the French'. Since Latin has no indefinite article either, the distinction between the descriptional use of the noun and the use expressing class membership is not indicated here but must be known from the context. The Turkish language has an indefinite article; therefore the descriptional form constructed without an article is here unambiguous. In exceptional cases, a similar usage is followed in English: the definite article can sometimes be omitted, as in 'George VI is king of England', or in appositive phrases. On the other hand, the definite article is sometimes used even when the description is ambiguous, as in sentences like 'the train will arrive at 7 P.M.' The necessary addition then is understood. It usually consists in a reference to a preceding utterance; for instance, it may be assumed in the form 'the train of which we spoke'. In such cases the definite article assumes the function of what is called in grammar 'a demonstrative adjective', with whose meaning it is historically connected. In fact, the word 'the' is an old demonstrative, related to 'that'. Similarly, the French articles 'le' and 'la' are derived from the Latin demonstrative adjectives 'ille' and 'illa', meaning 'that'. We shall

---

[1] For the interpretation of the definite article construed with a plural noun cf. p. 314.

analyze the meaning of the demonstrative terms, which represent a special kind of description and certainly are no adjectives, in § 50.

Statements containing a proper name, and statements containing a description, are alike in so far as they tell us that there is a corresponding thing. However, they differ in that a proper name is introduced by definition as a sign of the thing, whereas if we apply a description to a thing we must ask whether the application is correct. Thus the statement 'London is called 'London'' tells us only that there is a thing having that name; the rest is tautologous by definition. Contrarily, the statement 'London is the capital of England' not only tells us that there is a thing named 'London' but also adds synthetic information about the thing. The word 'is' in this example expresses the relation of identity (cf. § 43). Thus the sentence is symbolized by

$$x_1 = (\imath x)c(x, z_1) \tag{2}$$

where 'x_1' means 'London', 'z_1' means England, and 'c' means 'capital'.

We have so far mentioned only physical things and persons as argument-objects of functions. A second class of argument is given in space-time determinations. The sentence 'John met Jeanne in Hollywood on Tuesday at 8 P.M.' has as arguments the two proper names 'John' and 'Jeanne', the space indication 'Hollywood', and the time indication 'Tuesday 8 P.M.'; it therefore represents a four-place function '$f(x, y, s, t)$'. As we have only a few proper names for individual space-time indications (such as 'Trafalgar Square', 'Diluvium', etc.), language has developed for this purpose a system of descriptions which corresponds to mathematical coordinates. For time a zero point is defined as the time point simultaneous with a certain event, for instance, the birth of Jesus Christ; we thus have a description by means of a complex function. The time indication 'December 5, 1940' means 'the day which is 1940 years, 11 months, and 5 days after the zero point'. Similarly, space indications are given by means of distances from a zero point. The numbers introduced by these descriptions take the place of the names used in the more primitive characterization of arguments. The superiority of this method consists in presenting the argument-objects in an ordered form; this not only simplifies the finding of a thus-described object but also makes possible the introduction of mathematical methods. Time and space indication by words like 'now' and 'here' and by the tenses of verbs will be dealt with later (§§ 50–51).

The use of space-time indications is an important means of making descriptions of things unambiguous. Thus papers of identification, like passports, show place and time of birth of persons, in addition to the name. This kind of description is used even for the introduction of proper names of persons. In the birth registers we find statements of the form 'the child born to Mrs. N on June 17, 1917, receives the name 'Isabelle''. This statement is symbolized by

$$x_1 = _{Df} (\imath x)b(x, z_1, t_1) \tag{3}$$

where 'b' means 'born'. Here the sign '$=_{Df}$' stands between arguments, not between propositions. Of course, it is not always necessary to use space-time indications of births in descriptions used for the introduction of proper names. Many proper names are originally descriptions, the meaning of which later has been forgotten, such as 'Leonardo da Vinci', 'Edward the Confessor', 'Smith'. The line of demarcation between proper names and descriptions can therefore not always be clearly drawn.

Statements using the identity relation, like (2), are not the only kind of statements containing descriptions. We can introduce descriptions as arguments in every propositional function. Let us start from the example 'George VI was crowned at Westminster Abbey', which we symbolize by

$$c(x_1, y_1) \tag{4}$$

where 'c' means 'crowned', 'x_1' means 'George VI', and 'y_1' means 'Westminster Abbey'. For 'George VI' we can put 'the king of England'; i.e., we have the statement

$$x_1 = (\imath x)k(x, z_1) \tag{5}$$

where 'k' means 'king' and 'z_1' means 'England'. Putting (5) in (4) we have

$$c[(\imath x)k(x, z_1), y_1] \tag{6}$$

for the statement 'the king of England was crowned at Westminster Abbey'. We see that a sentence with a description as argument has a structure similar to that of a sentence with a proper name as argument, with the difference, however, that the argument itself has an inner structure; it is a compound term including a propositional function, other arguments, and a bound variable.

When we are asked to give a definition of the iota-operator we must say that we cannot define a term like (1), but that we can give a definition for every sentence in which the iota-operator occurs. That is, we have only a *definition in use* for the iota-operator. Thus we define the sentence (6) by the sentence

$$(\exists x)\{k(x, z_1).c(x, y_1).(u)[k(u, z_1) \supset (u = x)]\} \tag{7}$$

We see that a sentence containing a description can be replaced by an existential sentence which includes a qualification that there is only one argument satisfying the function; this qualification is expressed in the last term of (7) stating that all things u satisfying the function k are identical with x (cf. 1, § 44). The notation (6) corresponds exactly to the form of speaking used in conversational language; (7) is a transcription in terms of an existential operator.

In order to simplify our notation we shall introduce an abbreviation for a condition stating that there is one and only one individual having a certain property f. We write

$$(\exists x)f^{(1)}(x) =_{Df} (\exists x)f(x).(u)[f(u) \supset (u = x)] \tag{8}$$

We then can write the general definition of the iota-operator in the form

$$g[(\imath x)f(x)] =_{Df} (\exists x)f^{(1)}(x).g(x) \tag{9}$$

The syntax of the iota-operator requires some additional remarks. This syntax is very simple when the operator is *properly used*, i.e., when there is one and only one object having the property f. In this case definition (9) can be applied with respect to every expression having the iota-expression as an argument. This may be shown as follows.

Let us consider the expression

$$\overline{g[(\imath x)f(x)]} \tag{10}$$

We have two ways of eliminating the iota-operator. First we can apply definition (9) to the function 'g', and then add the negation bar over the whole resulting expression. We then have

$$\overline{(\exists x)f^{(1)}(x).g(x)} \tag{11}$$

The second way is to regard the function occurring in (10) as a function '$\bar{g}$' so that we have to apply (9) with the substitution of '$\bar{g}$' for 'g'; we thus arrive at

$$(\exists x)f^{(1)}(x).\overline{g(x)} \tag{12}$$

On the condition that there is one and only one object having the property f, these two expressions are equivalent, since we have the tautological relation

$$(\exists x)f^{(1)}(x) \supset [\overline{(\exists x)f^{(1)}(x).g(x)} \equiv (\exists x)f^{(1)}(x).\overline{g(x)}] \tag{13}$$

In a proper use of the iota-operator, i.e., when a descriptum exists, we therefore need not distinguish between (11) and (12). Similar considerations hold for all sorts of compound expressions containing the iota-argument. If the operator is properly used it is therefore irrelevant in which way the operator is eliminated; the resulting expressions will always be equivalent. The iota-argument has here the syntactical properties of a proper name.

It is different for an *improper use* of the iota-operator, i.e., when there is no object, or more than one object, satisfying '$f(x)$'. In such cases the iota-notation must be supplemented by the use of a symbol indicating the *scope* of the iota-operator, i.e., a symbol indicating that part of the formula to which definition (9) is meant to apply. We shall use half-brackets '$\ulcorner \urcorner$' for the indication of this scope.[1] The expression (10) then admits of two interpretations:

$$\ulcorner g[(\imath x)f(x)]\urcorner \tag{14}$$
$$\ulcorner \overline{g[(\imath x)f(x)]}\urcorner \tag{15}$$

In the interpretation (14) the expression leads to the meaning (11), whereas in the interpretation (15) we arrive at the meaning (12). The two meanings are not equivalent when no descriptum exists. An example for (14) is the statement: it is false that the present king of France is forty years old; an example for (15) is the statement: the present king of France is not forty years old. The first statement is true; the second is false.

---

[1] A scope indication has been introduced by Russell in combination with his treatment of descriptions. Our scope symbol, however, differs from the symbol used by him (B. Russell and A. N. Whitehead, *Principia Mathematica*, Macmillan, New York, 1924, Vol. I, second edition, p. 173, 14.01).

Another example is given by the two statements:

$$\ulcorner g[(\imath x)f(x)] \vee \overline{g[(\imath x)f(x)]}\urcorner \tag{16}$$

$$\ulcorner g[(\imath x)f(x)]\urcorner \vee \ulcorner \overline{g[(\imath x)f(x)]}\urcorner \tag{17}$$

They are translatable, respectively, into the two statements

$$(\exists x)f^{(1)}(x).[g(x) \vee \overline{g(x)}] \tag{18}$$

$$(\exists x)f^{(1)}(x).g(x) \vee \overline{(\exists x)f^{(1)}(x).g(x)} \tag{19}$$

With an improper use of the iota-operator (18) is false, whereas (19) is true even in this case. Thus it is false to say, using (18): the present king of France is forty years old or not forty years old; but it is true to say: the present king of France is forty years old or it is false that he is forty years old. (16) shows through its transcription into (18) that in an improper use the iota-argument does not behave like a proper name and therefore does not permit the assertion of the tertium non datur.

If the iota-operator is properly used, the scope indication can be dropped. Conversely, we shall interpret formulas containing the iota-operator without a scope indication as representing a proper use of the operator, i.e., as asserting the statement '$(\exists x)f^{(1)}(x)$' in addition to their direct meaning. The introduction of the scope symbol for an improper use of the operator has the advantage that such statements as 'the present king of France is forty years old' need not be regarded as meaningless, but are simply false, and that they can even be made true by the addition of a negation outside the scope.

Although descriptions are widely used as arguments in conversational language, the definite article does not always indicate a description. Language has a tendency to *equalization*, to express logically complicated structures by the linguistic form of simple structures. Thus general statements are often expressed in a form analogous to specialized statements. We say 'the lion is a ferocious animal', and mean by this: 'all lions are ferocious animals'. The definite article is used here, not as indicating a description, but in order to express generality. In the logical interpretation of such statements we therefore must be careful; the correct formalization of the statement 'the lion is a ferocious animal' would be

$$(x)[l(x) \supset f(x).a(x)] \tag{20}$$

where '*l*' means 'lion', '*f*' means 'ferocious', and '*a*' means 'animal'. The misinterpretation of such statements is the root of the medieval controversy of realists and nominalists; medieval realism, in interpreting general terms as referring to entities having an existence of their own, must be considered a victim of the equalization tendency of language.

We shall call all terms determining individuals, i.e., proper names and proper descriptions of individuals, including space-time determinations, *individual-terms*. The operation of specialization (cf. § 17) is then defined as the replacement of a free variable by an individual-term. A specialized statement is a statement in which at least one argument is expressed by an individual-term; a fully specialized statement, or *particular* statement, is a statement in which all arguments are expressed by individual-terms. As these individual-terms may consist of descriptions, a particular statement can include bound variables.

Statements in which no argument is an individual-term, and which cannot be written in such a form, are called *universal* statements. This qualification is necessary because individual-descriptions can be eliminated by giving the statement a form like (7). If the description is given without the use of other individual-terms, as in the examples above, the transformed statement then will contain no individual-terms. According to our definition, however, such a statement is not a universal statement because it can be written in a tautologically equivalent form which contains an individual-term.

In addition to the descriptions so far considered, there exists a second kind of description. If it is necessary to distinguish the two kinds, the descriptions of the first kind are called *definite descriptions* since they determine one and only one individual. In the second kind of description the condition of uniqueness is omitted; they are therefore called *indefinite descriptions*. Whereas the definite descriptions of conversational language are constructed by means of the definite article 'the', the indefinite descriptions are constructed in terms of the indefinite article 'a'.

Consider the sentence 'a man answered Peter'. The phrase 'a man' may be regarded as an indefinite description, i.e., a description of an indefinite individual; it asserts existence but does not

state uniqueness. For the indefinite description the symbol
'$(\eta x)f(x)$' is used [1]; the sentence may then be written in the form

$$g[(\eta x)f(x), y_1] \qquad (21)$$

where 'f' stands for 'man', 'g' for 'answered', and 'y_1' for 'Peter'.
The definition of the η-symbol is given by the rule that (1) has the
meaning

$$(\exists x)f(x).g(x, y_1) \qquad (22)$$

This definition differs from (7) by the omission of a condition such
as stated in the brackets of (7), i.e., a condition stating that there
is *only* one x of the property f.

While in the symbolic representation the form (22) appears just
as convenient as (21), the form (21) has the advantage that it
corresponds more closely to the usage of conversational language.
We are inclined to interpret the sentence as stating a relation between a man and Peter, so that these two terms should occupy
analogous positions in the symbolic formulation.

From a technical viewpoint, however, the use of indefinite
descriptions is connected with certain disadvantages. The relation
(13) does not hold when the superscript in '$f^{(1)}$' is omitted; consequently, indefinite descriptions must be equipped with specific
scope rules even in proper usage, i.e., when there is an x of the
property f. We therefore set up the rule that the scope of the
η-operator is always the narrowest scope possible. Thus the expression

$$\overline{g[(\eta x)f(x), y_1]} \qquad (23)$$

is to mean

$$\overline{(\exists x)f(x).g(x, y_1)} \qquad (24)$$

but (23) does not mean

$$(\exists x)f(x).\overline{g(x, y_1)} \qquad (25)$$

This scope rule corresponds to the usage of language; thus when
we say 'it is not true that a man answered Peter' we mean (24),
not (25), since (25) will be true even if a man answered Peter.

The scope rule for indefinite descriptions is connected with a
rule concerning the use of indefinite descriptions in definitions. We
might attempt to introduce a function h by the definition

$$h[(\eta x)f(x), y_1] =_{Df} \overline{g[(\eta x)f(x), y_1]} \qquad (26)$$

[1] D. Hilbert and P. Bernays, *Grundlagen der Mathematik*, Vol. II, Berlin, 1939,
Springer, p. 10. The Greek letter 'η' is pronounced 'eta'.

When we now apply the definition of the η-operator to the left-hand side of (26), we arrive at the meaning (25). The reason is that the rule 'use the narrowest scope possible' leads to a different result when the function 'h' is used. We therefore set up the rule that indefinite descriptions must not be used in definitions. This rule follows from rule 4, p. 123, since the indefinite description represents an existential operator in the definiendum.

Because of these complications indefinite descriptions are seldom employed in symbolic logic. They are necessary, however, for a logistic grammar because of their application in conversational language.

Like the definite article, the indefinite article is sometimes applied for the expression, not of existence, but of generality; in this meaning it does not represent a description but has the meaning of 'any'. We refer to the examples at the end of § 21. [Ex.]

§ 48. The problem of individuals

We must now turn to a closer analysis of the term 'individual'. We may define it as something occupying a continuous and limited part of space and time. Whereas this definition may be correct as far as space or time points, or areas, are concerned, it leads into difficulties with respect to things. The definition would exclude things that consist of spatially separate parts; and in fact, we would, for instance, say that the furniture of a certain house is not an individual, but a class of individuals. Only the chairs, tables, etc., of which the furniture is composed are regarded as individuals. Physicists tell us, however, that the individual chair is composed of atoms in the same way as the furniture is composed of chairs and tables; matter is not continuous, and the interstices between its parts are relatively much greater than those between the chairs in a house. Perhaps we can consider as individuals the ultimate particles discovered by the physicists, such as electrons, protons, positrons; but who knows whether they are ultimate? What we know, in any case, is that their individuality is of a dubious nature since they do not fit the space-time-causality order of our environmental world.

There is only one way to get out of this difficulty: to drop the condition of a physical connection of the parts and to consider the determination of the individual as a matter of convention. It is not necessary, for the purposes of daily life, to regard atoms as the only individuals; it is permissible to consider as individuals pieces of macroscopic matter. However, we must then also admit a terminology for which the furniture of a house is an individual, and another one in which a chair is composed of individual parts, such as legs, a seat, a back. Whereas the latter terminology may be useful for the joiner, the first may sometimes be preferred by the men in the moving business. We see that in these interpretations the condition of continuous occupation of space is abandoned, and that, on the other hand, the existence of a physical connection does not prevent us from dividing a thing into several individuals.

We discussed the arbitrariness in the definition of the individual when we analyzed usage predicates (cf. § 17). We say that when the same thing is used in different ways we sometimes speak of different things. Thus the turban and the sling are regarded as different things although they are forms of the same scarf. When we use a teapot as a coffeepot, should we say that there are two different things? Here we would hesitate. But it seems obvious that, when we unravel a woolen sweater and knit mittens of the same wool thread, the mittens will not be the same thing as the sweater. Here language follows unwritten rules whose conventional nature is visible every time we are confronted by a borderline case and do not know how to decide.

There is a further arbitrariness which we must now explain. Apart from space-time points and areas, the individuals so far considered are all of the *thing type;* they are, like the human body, aggregates of matter keeping together for a certain time. In the theory of relativity modern physics has introduced individuals of another kind, which are of the *event type,* i.e., which are space-time coincidences and do not endure. A thing is then considered a class of events. For physics, events are more fundamental units than things.

An analysis shows that the distinction between these two types of individuals is also made in conversational language, and that language has developed forms of speech for both kinds of arguments. The reason is that sometimes events are important units also for the purposes of daily life. A coronation, an assassination, an earthquake, an automobile accident, are events, not things; but language contains designations of such events and uses them as arguments of sentences. Such designations are mostly individual-descriptions, not proper names. Thus we say: 'the coronation of George VI took place at Westminster Abbey', or 'the earthquake was followed by the explosion of the factory'. The latter sentence formulates a two-term relation between two events; the former, a two-term relation between an event and a thing.

Now it is frequently possible to eliminate event arguments. Our first sentence, for instance, can be stated in the equivalent form: 'George VI was crowned at Westminster Abbey'. Similarly, the second sentence can be transformed into: 'the earth shook at a time t_1 and the factory exploded at a time t_2, and $t_2 > t_1$'. We see that only in the first example event-arguments have disappeared; in the second, although the original event arguments are eliminated, we have new event arguments in the symbols 't_2' and 't_1', since time points are events (or rather, classes of simultaneous events). There are some relations which can be formulated only as relations between events, such as time sequence; therefore event designations cannot be entirely eliminated in such sentences. Another important example of a relation between events is causality.

We have used the term *situation* to designate the object corresponding to a proposition. By describing a situation in a proposition composed of function and argument we split the situation into argument-object and predicate-object (or property). Our preceding discussion shows that there are two ways of splitting a situation; we distinguish these ways as *thing-splitting* and *event-splitting*. Between the corresponding propositions we then have a tautological equivalence (as to the accent cf. § 9):

$$f(x_1) \triangleq g(v_1) \tag{1}$$

where 'v_1' denotes the event, and 'g' the event property.

This equivalence may be used to define an event and its property in terms of a thing and its property. It is more convenient to express this idea in the metalanguage, as a relation between terms. We then say that an event-argument and its predicate can be defined as a function of a thing-argument and its predicate. Thus, if '$f(x_1)$' means 'George VI is crowned', 'g' is the predicate 'coronation of George VI', which is a function of both the predicate 'is crowned' and the argument 'George VI'. We shall use an asterisk for the indication of the transition to event-splitting and write the function 'g' in the form '$[f(x_1)]^*$'. Then the expression '$g(v_1)$' can be replaced by '$[f(x_1)]^*(v_1)$'. The argument 'v_1' used here is the name of the event which has the property $[f(x_1)]^*$ and which is determined if both the predicate 'is crowned' and the argument 'George VI' are given. Usually v_1 is denoted, not by a proper name, but by a description using the function '$[f(x_1)]^*$'; therefore the event-argument sign 'v_1' can be written in the form

$$(\imath v)[f(x_1)]^*(v) \tag{2}$$

The event is here indicated by a bound variable 'v'. This mode of expression, prevalent in conversational language, leads to the use of such predicates as 'takes place' and 'occurs', which merely express existence. Thus we say 'the coronation of George VI took place'. In symbolic language the last sentence is represented by a bound variable and an existential operator, in the form

$$(\exists v)[f(x_1)]^*(v) \tag{3}$$

Synonymously with the word *event* we shall use the word *fact*. The objective function $[f(x_1)]^*$ will be called a situational *fact-function;* the corresponding linguistic function '$[f(x_1)]^*$' will then be named a propositional fact-function. In contradistinction to this kind of function, the functions f or 'f' will be called, respectively, situational or propositional *thing-functions*.

The transformation from thing-argument to event-argument, expressed in (1), may be called a *holistic transformation* since only the expressions '$f(x_1)$' and '$g(v_1)$' as *wholes* are equivalent to each other, whereas there is no direct correspondence between the parts 'f' and 'x_1' on the one side, and the parts 'g' and 'v_1' on the other side. Such transformations possess an analogue in the holistic transformations used in mathematics. The usual transformation of a mathematical function by a transformation of coordinates is not of this sort; it is a *point transformation*, i.e., a transformation in which each point of one system of coordinates corresponds to one point in the other system. However, other transformations used in mathematics have a holistic character. Of this sort are the *contact transformations*. Like point transformations, these transformations coordinate to a curve or mathematical function $\varphi(x, y) = 0$ in the x, y-plane, a curve $\psi(u, v) = 0$ in the u, v-plane. But here a given point x, y does not determine a point u, v; the coordinated point u, v will depend, not only on x, y, but also on the shape of the curve $\varphi(x, y) = 0$ (namely, on the direction of the line element in the latter point). Therefore only the curves as wholes correspond to each other. Another mathematical example of holistic transformations is provided by the integral transformations represented by Fourier expansions.[1] We must also conceive as holistic transformations certain transformations which have been called, by F. Klein,[2] *transformations with a change of the space element*, as for instance, transformations in which points as coordinates are replaced by straight lines as coórdinates. Thus we can define a triangle as a set of three straight lines x_1, x_2, x_3, which have the property f that each two of them have a point in common, but that there is no point in common for all of them; or we can define a triangle as a set of three points x_1', x_2', x_3', which have the property f' that there is no straight line on which all of them are situated. We then have an equivalence

$$f(x_1, x_2, x_3) \triangleq f'(x_1', x_2', x_3') \tag{4}$$

corresponding to (1); the arguments of (4) on the left-hand side are straight lines, on the right-hand side, points.

If the thing-function 'f' has several arguments, fact-functions can be constructed in different ways, according as we include all arguments or only a part of them in the fact-function. Thus the sentence 'Amundsen flew to the North Pole in May 1926', symbolized in thing-splitting by

$$f(x_1, y_1, t_1) \tag{5}$$

can be transformed into event-splitting in various ways. One is to use the fact-function 'Amundsen's flight to the North Pole in May 1926', symbolized by '$[f(x_1, y_1, t_1)]^*$'; we then write

$$(\exists v)[f(x_1, y_1, t_1)]^*(v) \tag{6}$$

In words this reads: 'a flight by Amundsen to the North Pole in

---

[1] Cf. the author's *Philosophic Foundations of Quantum Mechanics*, University of California Press, Berkeley, 1944, § 12.

[2] Cf. Felix Klein, *Elementarmathematik vom höheren Standpunkte aus*, Springer, Berlin, 1925, Vol. II, pp. 63, 117.

May 1926 took place'. Another form obtains when we use the
fact-function '$[f(\hat{x}_1, y_1)]^*$', and write:[1]

$$(\exists v)[f(x_1, y_1)]^*(v, t_1) \tag{7}$$

This can be read as 'a flight by Amundsen to the North Pole took
place in May 1926'. A third form is given by the use of the fact-
function '$[f(x_1)]^*$'; we then have

$$(\exists v)[f(x_1)]^*(v, y_1, t_1) \tag{8}$$

In words: 'one of Amundsen's flights took place at the North Pole
in May 1926'.

When the thing-function contains bound variables, these are
always included in the fact-function. Consider, for instance, the
thing-function '$\hat{x}$ is married', which has the form '$(\exists y)m(\hat{x}, y)$',
where 'm' is 'married'. We can abbreviate this function by '$f(\hat{x})$';
then (2) expresses the fact denoted by the expression 'the being
married of x'. If no abbreviation is used, the fact-function will be
expressed in the form '$[(\exists y)m(x, y)]^*$'.

When the sentence given in thing-splitting is indicated, not by
a function, but by a propositional variable 'a', we shall also use the
asterisk to indicate a fact-function. Thus we write

$$a \equiv (\exists v)a^*(v) \tag{9}$$

The event, or fact, corresponding to 'a' is then written in the form

$$(\imath v)a^*(v) \tag{10}$$

It should be noticed that the asterisk is not commutative with
the negation. The fact-functions '$\bar{a}^*$' and '$\overline{a^*}$' are not identical.
Thus, when 'a' is true, we have in general not only a fact v corresponding to 'a', as stated in (9), but also facts which do not correspond to 'a'; this existence is expressed by the relation

$$(\exists v)\overline{a^*(v)} \tag{11}$$

However, we then do not have a fact corresponding to '$\bar{a}$'; i.e., the
relation

$$(\exists v)\bar{a}^*(v) \tag{12}$$

is not true when 'a' is true. Thus, when Peter has a car, there is
the fact of his having a car, and besides, there are many facts which
are not his having a car; but when he has no car, and only then,

[1] The function '$f(x_1, y_1)$' can be regarded as defined in the form '$(\exists t)f(x_1, y_1, t)$'.

is there the fact of his having no car. Facts, therefore, have the physical existence of things, and not the fictitious existence of situations, or properties; the derivation given on page 238, which shows that there exists a situation corresponding to a false proposition, cannot be constructed for fact-functions because of the noncommutativity of the asterisk and the negation. We can therefore regard facts, or events, as being of the same type as things.

A relation concerning the interchange of asterisk and negation line is given by the following formula:

$$\overline{(\exists v)a^*(v)} \equiv (\exists v)\bar{a}^*(v) \tag{13}$$

This equivalence follows because both sides are equivalent to '$\bar{a}$'; for the left-hand side this is seen when we negate both sides of (9), for the right-hand side it follows when we substitute '$\bar{a}$' for 'a' in (9). When we shorten the negation line in (13), we arrive at the relation:

$$(v)\overline{a^*(v)} \equiv (\exists v)\bar{a}^*(v) \tag{14}$$

This formula says that the negation line can be contracted over an asterisk when the functional is preceded by an all-operator, the all-operator then being reversed into an existential operator. A corresponding rule for a negated functional preceded by an existential operator, however, is not derivable.

Language uses the particle 'that' to indicate a transition from thing-splitting to event-splitting. We say, for instance, 'Peter saw that the cat jumped down'; here the particle 'that' indicates that Peter saw the event. This is not the same as 'Peter saw the cat' since we can see things without seeing the motion of these things, for example, the hands of a watch. The word 'that', therefore, represents an iota-operator introducing an event, used when the sentence is given in thing-splitting.[1] The fact-function is frequently indicated by the use of the suffixes 'ing' or 'ion', added to the verb-root. Thus the above sentence can be transformed into 'Peter saw the jumping of the cat'. Since the fact-function is here explicitly stated, the particle 'that' is no longer necessary and can be replaced by the iota-operator 'the'. Furthermore, the impersonal use of 'it' usually indicates reference to an event argument, as in 'it was my tenth birthday'.

---

[1] As to a different usage of 'that' cf. p. 279.

The parallelism between events and things is shown also by the following consideration. Like things, events are carriers of various properties; therefore different sentences can be asserted concerning the same event. Thus we say 'the earthquake was terrific' and 'the earthquake was caused by a volcanic eruption'; or 'the jumping of the cat was graceful' and 'the appearance of a dog caused the jumping of the cat'. Although we usually do not give proper names to events, we could do so; perhaps terms like 'Renaissance', 'Restoration', which have practically lost their descriptional character, can be regarded as proper names of events.

Summarizing our remarks about arguments we find that the possibilities in the choice of the argument show a broad variety. First, in the thing sphere, we may select as arguments smaller and smaller units, like cells of organic matter, atoms, or protons and electrons. Second, we may introduce events as argument-objects. Third, we may combine events in different ways to form things, by considering, different sequences of events as things. Thus the different states of a water particle may be conceived as forming the thing called 'water particle'; or the different states of different water particles may be united in the thing called 'water wave'. A water wave is not a thing in the sense of a permanent complex of matter; but those complexes of matter are by no means 'better' things. We know that many material things differ from water waves only in that the exchange of matter is slower in them; the human body, for instance, is said to change its matter within seven years.

Disregarding the arbitrariness in the choice of the argument-object has led to an unfortunate absolutism in certain philosophical systems. Thus materialism seems to be guilty of an absolutism of thing-arguments; other philosophical systems are on the search for absolute argument-things in the construction of substances beyond material things. On the other hand, insight into the arbitrariness of the argument has led to the mistake of denying the existence of things. When Heraclitus says 'you cannot step twice into the same river', he is right if he intends to say that a river is not a thing in the sense of an enduring substance but an event sequence in which the matter does not remain the same; he is wrong, however, if he wants to infer that it is not permissible to consider a river as a thing. The meaning of the phrase 'the same river' is a

matter of definition; and with the definition of the word 'river' as
denoting a thing that consists in an event sequence it *is* possible
to step twice into the same river. We therefore do not agree with
philosophical systems that want to abolish things; instead, we consider the definition of the argument-object a matter of volitional
decision. [Ex.]

§ 49. Fictitious existence

The existential operator introduced by us represents *physical
existence*, i.e., the sort of existence applying to the concrete objects
of our daily environment as well as to the objects discovered by the
methods of science. It is sometimes convenient, in addition, to
speak of existence in a fictitious sense and thus to introduce classes
of fictitious objects which facilitate the description of the world.

The first objects of this kind to be mentioned here are the objects
of perception. We speak of seeing an object not only if the object
is physically present; we say that we see certain objects also when
dreaming, or when looking at physical objects of a different sort —
thus the black and white patches on the screen of a movie theater
are seen by us as persons and physical objects like those of our daily
environment. Such objects are fictitious; but it is convenient to
deal with them as though they were real objects. We shall call
them *subjective things*. The name *immediate things* will be used by
us to include both objective things which are perceived, and subjective things; thus if a thing is immediate it is left open whether
it is at the same time objective. We shall use the word 'perceive'
in order to indicate the occurrence of immediate things and reserve
the word 'observe' for objective, or physical, existence. The word
'see' is used ambiguously; sometimes the expression 'I see a dog' is
used also for dreams, sometimes only in the sense of 'I observe a dog'.

The sentence 'x_1 perceives a dog' will be symbolized by

$$(\exists y)_{x_1} d(y).pc(x_1, y) \tag{1}$$

We speak here of *immediate existence;* the subscript 'x_1' added to
the existential operator is read as 'there exists immediately for x_1
an entity y'. We use 'd' for 'dog', 'pc' for 'perceives'. The corresponding sentence 'x_1 observes a dog', which implies physical
existence, is written

$$(\exists y)d(y).obs(x_1, y) \tag{2}$$

where the symbol '*obs*' stands for 'observes'. According to our convention about the use of the immediate-existence operator, (1) does not exclude that simultaneously (2) is true; it leaves this open. Thus (1) holds for a dream or a movie show as well as for a real dog; (2), only for a real dog. A comparison of (1) and (2) shows the analogy in the structure of the two statements.

When we consider subjective things *fictitious* we follow the conception of *logical empiricism* as well as the interpretation of *common sense*. According to this conception, what can be asserted as physically existent during a dream is only a physiological process in the human body of such a kind as would take place if there were a real physical object of the same sort. This statement can be expressed for the example of the perceived dog, when we use the abbreviation '*bst*' for 'bodily state', in the form

$$(\exists z)\{bst(x_1, z)[(\exists y)obs(x_1, y).d(y) \ni bst(x_1, z)]\} \tag{3}$$

This means: there is a z which is a bodily state of x_1 and of such a kind that, when x_1 observes a dog y, then x_1 is in this bodily state z.[1] Physical existence of the dog is not asserted here because the respective sentence occurs in the implicans of an implication. This implication is written as a connective implication in order to avoid the paradoxes of the adjunctive implication; it is to be understood as a synthetic connective implication such as defined in chapter VIII.

The equivalence of expressions (1) and (3) constitutes the thesis of logical empiricism. This equivalence shows that expressions stating immediate existence can be eliminated. On the other hand, it shows that even the empiricist is allowed to use such expressions if they serve special purposes.[2] They represent an indirect mode of speech which is translatable into physical existence. Metaphysicians who hold the opinion that there are so-called 'mental' objects,

---

[1] For a more precise formulation a time argument t must be introduced, in the form
$$(\exists z)bst(x_1, z, t_1).(t)[(\exists y)obs(x_1, y, t).d(y) \ni bst(x_1, z, t)] \tag{3a}$$
For (1) we then write
$$(\exists y)_{x_1} d(y).pc(x_1, y, t_1) \tag{1a}$$
This means: x_1 perceives a dog at the time t_1.

[2] Thus the author has used the term 'immediate existence' in a sense translatable into physical existence in his book *Experience and Prediction*, University of Chicago Press, Chicago, 1938, § 24, also p. 266. We refer to this book for a further explanation of the above notation.

existing in a sphere inaccessible to empiricist conceptions, are seriously mistaken. We see rather that the 'unphysical' existence of such objects derives from the fact that in the empiricist translation (3) the existential operator stands in the implicans of an implication, and that therefore the existential term is not asserted. We have here an instructive example of the clarifying power of logical analysis: such analysis reveals that the occurrence of an existential term in an implicans has been misconstrued so as to give rise to the metaphysical conception of different realms of existence.

A second sort of fictitious existence is sometimes employed which extends the domain of existence much farther than does immediate existence. Following the second conception, we speak of existence whenever the assumption of physical existence is not contradictory. We thus introduce a category of *logical existence*, and formulate it by the symbol

$$(\exists x)_l f(x) \tag{4}$$

According to our definition, (4) means the same as

the statement '$(\exists x)f(x)$' is not contradictory   (5)

(4) is a statement in the object language, (5) in the metalanguage. Therefore we cannot put an equivalence sign between the two; we have, instead, a relation of *equipollence*. The advantage of using logical existence is that it allows us to express in the object language statements that, strictly speaking, belong in the metalanguage; however, such usage leads to what were called above (§ 5; cf. also § 62) *improper object statements*. We use a *shifting of the level of language*.

Logical existence is used, for instance, in the sign theory of Charles Morris,[1] who distinguishes between *designation* and *denotation*. Not every sign *denotes;* but every sign *designates* if the corresponding existential statement is not contradictory. Thus the word 'sea serpent' designates; its designatum, the sea serpent, has, a logical existence. Another use of logical existence is made in such expressions as 'logically possible things', 'logically possible facts'. A somewhat different meaning is given by the phrase 'physically possible facts', which we shall define in § 65. This sort of possibility

[1] Cf. footnote, p. 4.

can also be used for the definition of a fictitious existence; it is applied, for instance, when we say that a physical theory holds for all possible facts.

A third sort of fictitious existence appears in statements which refer to what Russell [1] has called *propositional attitudes*, i.e., attitudes towards propositions. Of this form are statements as 'x_1 believes that . . .', 'x_1 doubts that . . .', 'x_1 denies that . . .', 'x_1 says that . . .', and 'x_1 knows that . . .'.

Consider the statement: 'Peter believes that John was killed'. Interpreting the particle 'that' as indicating event-splitting (cf. p. 272), we can write this statement in the form 'Peter believes the being killed of John'. In saying so we do not wish to say that there exists physically a fact such as described by the clause, i.e., that John was killed; nor do we wish to assert the contrary. We rather regard the fact as existing in a fictitious sense, and then say Peter believes it. Since this sort of existence is similar to the one previously used we shall speak here, as before, of logical existence; but we shall distinguish it from the other logical existence by using a capital 'L' as subscript.

In order to write the statement symbolically let us denote by 'p' the clause 'John was killed'; the fact-function 'John's being killed' will then be designated by 'p^*' (cf. 9, § 48). We now write the whole statement in the form

$$(\exists v)_L \; bl(x_1, v).p^*(v) \tag{6}$$

the sign 'bl' standing for 'believes'.

In order to make the form of (6) clearer, let us now consider the statement 'Peter knows that John was killed'. When we use the word 'knows' we indicate that the event considered actually happened; and therefore we can formulate this statement by the use of the unqualified existential operator, in the form

$$(\exists v) kn(x_1, v).p^*(v) \tag{7}$$

The sign 'kn' stands for 'knows'.

The structure of (7) shows the origin of the form of (6). A statement about belief is construed as analogous to a statement about

[1] Bertrand Russell, *Inquiry into Meaning and Truth*, Norton, New York, 1940, p. 22, pp. 336–342.

knowledge. It is this *tendency to analogy* which leads conversational language to the introduction of logical existence.

We must now give a transcription which translates statements of this kind into statements containing objective existence. As before, this can be done only by the use of the metalanguage. We shall then conceive, as the object of belief, not the fact, but the sentence 'p', or '$(\exists v)p^*(v)$', asserting the existence of the fact, and write

$$bl(x_1, \text{'}p\text{'}) \tag{8}$$

or when we use a form analogous to (6):

$$(\exists s)bl(x_1, s).eqs(s, \text{'}p\text{'}) \tag{9}$$

Here 's' is a sentence name variable, i.e., a variable whose special values are names of sentences (cf. p. 13). The symbol 'eqs' means 'equisignificant', the relation dealt with in § 2 and § 9; it can be regarded as a relation in the metalanguage holding when the corresponding sentences are equivalent in a connective sense.[1]

The rule of transcription to be used for logical existence of this sort is made clear by the transition from (6) to (9). It requires the replacement of the fact variable v by a sentence name variable, and of the expression '$p^*(v)$' by the expression '$eqs(s, \text{'}p\text{'})$'. The interpretation (8) or (9), which was introduced by Carnap, uses only objective existence because sentences exist physically as ink marks on paper or sound waves produced in human throats, even if there is no corresponding fact, i.e., if they are false.

The form (8) requires some qualification. The meaning of quotation marks depends on the sort of equisignificance relation between tokens assumed for the definition of the quotes (cf. § 3). In the usual interpretation of quotes, (8) would state that x_1 believes the sentence 'p' in the special wording given. Now we do not wish to restrict (8) to this narrow meaning; we wish to leave it open whether Peter believes a statement of the form 'John was killed', or 'John died, but not a natural death', or a corresponding statement in a foreign language. Therefore, the equisignificance relation assumed for (8) must be understood in the widest sense, as including all possible versions of the statement 'p'. The form (9)

---

[1] Cf. chapter VIII. The connective equivalence to be used here can be regarded either as tautological or as synthetic, depending on the meaning assumed for the equisignificance.

does not need this qualification for the quotes because the equi-significance relation is here explicitly stated and may be assumed in any convenient sense. The quotes in (9) therefore may have the usual narrow meaning.

It is true that the interpretations (8) and (9) do not strictly correspond to the wording of sentences about belief since such sentences do not contain the clause 'p' in quotes, but state it in the object language. Therefore (6) symbolizes the actual wording of such sentences, which employs logical existence. When we wish to make the interpretation (9) plausible we may relate it to actual language as follows: the clause following the particle 'that' describes a possible fact, and the sentence s believed by x_1 is a certain sentence which asserts this fact; only the form of this sentence is unknown to us. But this explanation, once more, employs logical existence by the use of the term 'possible fact'. When we wish to avoid logical existence we must understand the particle 'that' as representing, not event-splitting, but quotation marks, characterized by the widest use of the equisignificance relation; this interpretation leads directly to the forms (8) and (9).

A similar interpretation can be given to statements about other propositional attitudes. Only for the function 'knows' do we not need a transcription, since this function, as apparent in (7), can be directly expressed without the use of logical existence. It is convenient, however, to construe this function analogously to the other functions of this group, and to regard the object of knowledge also as consisting in a sentence. We shall then replace (7) by either one of the two formulas

$$kn(x_1, \text{'}p\text{'}) \tag{10}$$
$$(\exists s)kn(x_1, s).eqs(s, \text{'}p\text{'}) \tag{11}$$

The objection has been made to the interpretations (8)–(11) that a man who believes, or knows, something need not formulate his belief, or knowledge, in a sentence, at least as far as his own thinking is concerned. We may answer that if a man thinks there will be images, or psychological processes of some sort, which will have the nature of a language and may be included in the range of the sentence name variable s; they have sentence character because they will be made either true or false by respective facts. These psychological processes are sometimes called the *interpretant* of a sentence, and are thus distinguished from language proper. Actually they have the nature of an 'inner language', which differs from language proper only by the fact that its syntax is not clearly defined.

Note on Russell's conceptions. Russell regards, as the object of belief, not the fact indicated by the bound variable v, but the whole situation coordinated to 'p'. He then writes

$$bl(x_1, p) \tag{12}$$

This way of writing has the disadvantage that principle (5, § 18) must be dropped; although (12) is objectively true, we cannot coordinate to it a sentence stating objective existence, but rather must write

$$(\exists q)_L \, bl(x_1, q).(q \overset{\cdot}{\equiv} p) \tag{13}$$

where 'q' is a propositional variable, or, what is the same, where q is a situational variable. The accent on the equivalence sign represents the connective equivalence introduced in (2, § 9), i.e., a tautological equivalence, which takes the place of the identity relation in (5, § 18).

The disadvantages connected with the use of bound propositional variables were discussed in § 42. All such syntactical difficulties are avoided by the use of the fact-function. This interpretation eliminates also the problem of extensionality discussed by Russell with respect to the form (12).

Another form of fictitious existence refers to what may be called *intentional objects*. We say, 'he desires to live in New York', 'he plans to become an actor', 'he attempts to write a novel', etc.; in these sentences, the living in New York, the becoming an actor, and the writing of a novel constitute intentional objects, about whose real existence nothing is said since we do not know whether the intention will ever be realized. When we conceive terms like 'desire', 'plan', 'attempt', as functions, we therefore are compelled to interpret principle (5, § 18) as referring to a fictitious existence. This existence is comparable to the subjective, or immediate, existence used above, with the difference, however, that the intentional object is not given by a perception. It may be associated with some more or less vague images; but it is not these images which are intended, since the intention is directed toward the realization of these images. Using the symbol '$(\exists x)_{in}$' for intentional

objects, we can write the sentence 'Peter desires to live in New
York' in the form:

$$(\exists v)_{in}\,[f(x_1,\,y_1)]^*(v).ds(x_1,\,v) \qquad (14)$$

Here '$f(x_1,\,y_1)$' means 'Peter lives in New York', and 'ds' means
'desires'. The particle 'to', in this interpretation, is regarded as
introducing event-splitting.

The translation of such expressions into sentences using only
physical existence is rather involved. We cannot use sentences as
objects of the intention since it is not the sentence 'Peter lives in
New York' which Peter desires. Nor can we say that Peter is in
a bodily state such as would result if he were living in New York:
this would be false since when he lives in New York the desire is
fulfilled and Peter, therefore, is not in the state of a desire. We
rather must say that Peter is in a bodily state for which a life in
New York would be fulfillment. That is, we define intentional
objects in terms of the psychological notion of *fulfillment*. For
instance, we define the sentence (14) by the following sentence
which uses only physical existence:

$$(\exists z)bst(x_1,\,z).[f(x_1,\,y_1)\,\ni\,ff(z)] \qquad (15)$$

Here 'bst' means 'bodily state', as in (3), and 'ff' means 'fulfillment'. This sentence does not assert that Peter will live in New
York, since the term '$f(x_1,\,y_1)$' stands in the implicans. As before,
the fictitious existence is eliminated because the sentence '$f(x_1,\,y_1)$',
or its corresponding event argument 'v', is used, not as the argument of a function, but as a term connected by a propositional
operation with other terms. We use the accent implication for the
reasons explained with respect to (3).

In a similar way, other intentional terms can be defined. The
difference between desiring, planning, expecting, can be reduced to
different kinds of fulfillment. Terms like 'planning' and 'attempting' include a further statement that the person commits certain
actions which, at least, will imply the aim with a certain probability.
For the complete definition of such terms we therefore must use the
probability implication. We cannot include a presentation of the
probability implication in this book but refer the reader to another
publication by the author.[1]

[1] H. Reichenbach, *Wahrscheinlichkeitslehre*, A. W. Sijthoff, Leiden, 1935, § 9.

There are further forms of fictitious existence. Thus novels deal with fictitious objects whose existence is assumed when sentences concerning such objects are stated in a book. The fictitious existence of these objects, which may be called *literary existence*, is therefore translatable into the physical existence of sentences in a book. Using the subscript '*li*' for literary existence we write

$$(\exists x)_{li} f(x) \tag{16}$$

and interpret this expression as equipollent to the statement of the metalanguage

$$\text{the sentence } `(\exists x)f(x)\text{' is stated in a book} \tag{17}$$

Some philosophers believe that this interpretation belittles the intrinsic value of literary creation, contending that literature deals with a higher sort of truth. Such criticism is based on a misunderstanding. Only the factual statements of works of literature have the sort of truth indicated by our subscript '*li*'.[1] The behavior of the fictitious persons in their fictitious environment should be so presented that it satisfies the laws of psychology holding for actual persons; in other words, the *laws* assumed for the behavior of the fictitious persons should be *objectively true*. The satisfaction of this requirement of objective truth is one of the criteria by which the quality of fiction is judged. A further requirement is that the laws expressed by the behavior of the fictitious persons play an important role in our own lives and therefore help us to understand human behavior in general.

The examples given show that fictitious existence plays a useful part in conversational language since it simplifies language considerably. It may be compared to the fictitious existence of mathematical objects, such as represented by the infinitely distant point in which two parallels intersect. There is no danger in employing such expressions when the translatability into expressions using physical existence is kept in mind. Likewise, the syntax of such expressions includes no difficulties, as is shown by our examples.

In particular, the relation (5, § 18) between truth and existence need not be abandoned. If a sentence is physically true, it can al-

---

[1] It is also possible to translate literary existence into the existence of images and emotions in the reader. This interpretation will lead to a fictitious existence similar to immediate existence.

ways be translated into physical existence. Only sentences which
are true in a modified sense require fictitious existence in the application of (5, § 18). Thus if a person sees an automobile accident in
his dream, the sentence 'two cars collide' will be only subjectively
true; the application of (5, § 18) then requires the use of immediate
existence for the statement 'there is an automobile accident'. For
logical existence of the first sort the property of being noncontradictory plays the role of truth and is translatable into logical
existence. For literary existence, the substitute truth is defined as
the property of being asserted in a book.

On the other hand, sentences stating fictitious existence, such as
(1) or (6), are objectively true. They are translatable into objectively true sentences concerning other objects which have physical
existence; but as long as the fictitious object is retained the qualified
existential operator cannot be eliminated. In order to carry through
this principle we introduce the rule that fictitious objects cannot be
given proper names. They can only be described and therefore are
expressed by means of variables bound by qualified existential
operators. The word 'Hamlet', therefore, is not a proper name, but
an abbreviation standing for the description of a fictitious personal-
ity. Names of this sort may be indicated by the addition of a
superscript indicating the sort of existence used. Thus the statement 'Hamlet killed his stepfather' can be symbolized, when '$y_1{}^{li}$'
stands for 'Hamlet', 'st' for stepfather, and 'k' for 'killed', in this
form:
$$(\exists x)_{li}\, st(x, y_1{}^{li}).k(y_1{}^{li}, x) \tag{18}$$

When we apply here (5, § 18) the argument '$y_1{}^{li}$' will be translated
into a second operator of literary existence.

Finally, we may mention here the existence of properties as a
form of fictitious existence. We saw in § 39 that this existence is
based on criteria wider than those used for physical existence, and
that a property which is definable is regarded as existent in the sense
that it is given a proper name. The attempt might be made to
reduce this kind of existence to physical existence by a procedure
analogous to that used for the fictitious existence of objects of the
thing level. Such a plan, however, leads into complications since, as
we saw in (25, § 41), the technique of the higher calculus of functions leads automatically into a fictitious existence of properties.

This result makes understandable the desire of some logicians to eliminate the higher calculus altogether. So far, however, a suitable substitute has not been devised. [Ex.]

§ 50. Token-reflexive words[1]

We saw that most individual-descriptions are constructed by reference to other individuals. Among these there is a class of descriptions in which the individual referred to is the act of speaking. We have special words to indicate this reference; such words are 'I', 'you', 'here', 'now', 'this'. Of the same sort are the tenses of verbs, since they determine time by reference to the time when the words are uttered. To understand the function of these words we have to make use of the distinction between *token* and *symbol*, 'token' meaning the individual sign, and 'symbol' meaning the class of similar tokens (cf. § 2). Words and sentences are symbols. The words under consideration are words which refer to the corresponding token used in an individual act of speech, or writing; they may therefore be called *token-reflexive* words.

It is easily seen that all these words can be defined in terms of the phrase 'this token'. The word 'I', for instance, means the same as 'the person who utters this token'; 'now' means the same as 'the time at which this token is uttered'; 'this table' means the same as 'the table pointed to by a gesture accompanying this token'. We therefore need inquire only into the meaning of the phrase 'this token'.

We can interpret the function of the phrase 'this token' as an operation similar to the operation of the quotes explained in § 3; the new operation will be symbolized by *token quotes*. Whereas the ordinary-quotes operation leads from a word to the name of that word, the token-quotes operation leads from a token to a token denoting that token. Let us use little arrows for the token quotes; then the sign

$$\searrow a \nearrow \tag{1}$$

represents, not a name for the token 'a' in (1),[2] but a token for it.

---

[1] The ideas presented in this section were developed after Professor Bertrand Russell had kindly given me the opportunity of reading a manuscript of his, subsequently published in his book, *An Inquiry into Meaning and Truth*, Norton, New York, 1940, chapter VII, 'Egocentric Particulars'.

[2] This phrase is an abbreviation for 'the token of (1), similar to the token $\searrow a \nearrow$'.

(1) is not a name because the token (1) is a reflexive token and cannot be repeated; thus

$$\searrow a \nearrow \tag{2}$$

not only is a token different from (1), but also refers to a different token.

If we want to introduce a name, i.e., a symbol and thus a class of tokens, for the token (1), we can do so by using nonreflexive tokens. For a better understanding let us consider first the operation of introducing new symbols standing for other symbols. If we write the definition

$$p =_{Df} a \vee b \tag{3}$$

or the corresponding statement using names of these symbols:

$$\text{'}p\text{' has the same meaning as '}a \vee b\text{'} \tag{4}$$

this statement means, strictly speaking:

$$\text{every token similar to } \searrow p \nearrow \text{ is equisignificant to}$$
$$\text{every token similar to } \searrow a \vee b \nearrow \tag{5}$$

We see that the introduction of a synonymous symbol involves application of the relation *equisignificant* between tokens, and of the token-quotes operation.

In a similar way we can introduce nonreflexive tokens equisignificant to a reflexive token. For instance, we can write

$$\text{every token similar to } \searrow w \nearrow \text{ is equisignificant to}$$
$$\text{the token } \searrow\searrow b \nearrow\nearrow \tag{6}$$

This statement means that 'w' is the name of the token 'b' in (6). In the same way we could introduce a name for the token 'a' in (1); we then would have to write a phrase corresponding to (6) on the place on the paper immediately before the place where the token (1) stands, adding a second set of token quotes to the token (1). Incidentally, we have already used such a name for (1) by using the symbol '(1)'; writing this number-symbol at the side of a reflexive token is to be interpreted in the sense of (6).[1]

[1] Our use of the number-symbol '(1)' as name of a token differs from the ordinary use of number-symbols for formulas because usually the number-symbols are names of symbols, i.e., refer to any token similar to the one at the side of which they are written. Number-symbols of reflexive tokens are always used in the way that our symbol '(1)' is used.

We now understand the function of the symbol 'this token'. For instance, the sentence

$$\text{the piece of paper covered by this token} \qquad (7)$$

could be replaced by a sentence consisting of the token (7) in token quotes and the words 'the piece of paper covered by' written before it. Or, using the token name '(7)', we could write on any sheet of paper the sentence 'the piece of paper covered by the token (7)'. In this form the token-reflexivity has been eliminated.[1]

We see, furthermore, that the symbol 'this token' is not a phrase, or set of words, in the ordinary meaning of 'phrase'. The different tokens, similar to one another, constituting the symbol 'this token', are not equisignificant to one another. We shall therefore call the symbol 'this token' a pseudo-phrase. The symbol 'this token' is used to indicate an operation; the meaning of this operation cannot be formulated in the language itself but only in its metalanguage. All token-reflexive words are pseudo-words; they can be replaced by proper words in combination with an operation, namely, the operation 'this token'.

To have a convenient notation, we shall introduce the symbol 'Θ' as the name of a specific token similar to the one in which 'Θ' is used. Assume, for instance, that a man says, 'this boy is tall'; then the token used by the man (i.e., the whole sentence uttered by him) will be named 'Θ'. The sentence uttered by the man can now be symbolized in the form:

$$t[(\imath x)b(x).rf(x,\Theta)] \qquad (8)$$

Here 't' means 'tall', 'b' means 'boy', and 'rf' means 'referred to'. The reference usually is given by spatio-temporal proximity. The token denoted by 'Θ' is not the token used for the above formulation of the sentence (8), but another token of the same sentence, namely, the one uttered by the man mentioned. The symbol 'Θ', therefore, is not token-reflexive, but reflexive in another sense: it refers to a similar token, to be specified separately. It is, therefore, a proper

---

[1] There is a slight difference in the usage of the phrase 'this token' and the usage of the arrow quotes. The token including the arrow quotes refers to the same token with the exclusion of the arrow quotes, whereas the phrase 'this token' refers to the latter token including the token of the word 'this'. The tokens of the word 'this' are therefore reflexive, whereas the arrow quotes are not. This difference, however, is irrelevant.

name of a particular nature, used in the metalanguage. In fact, the word 'this', when used in quotations of direct speech, has the meaning of 'Θ', not of the token-reflexive 'this' of the original speech.

By means of this notation we can write sentences which originally were uttered in a token-reflexive form, in such a way that the symbol used is not token-reflexive. To have a further example, let us symbolize the sentence 'here I stand', uttered by Luther on the *Reichstag* at Worms in 1521. The utterance by Luther we denote by 'Θ'. The word 'I' then can be given in the form 'the x that spoke Θ'; the word 'here' will be expressed by the phrase 'the place z where Θ was spoken'. We thus have, using the function '$\hat{x}$ speaks $\hat{y}$ at $\hat{z}$', with 'sp' for 'speak' and 'st' for 'stand':

$$st[(\imath x)(\exists z)sp(x,\Theta,z), (\imath z)(\exists x)sp(x,\Theta,z)] \qquad (9)$$

It is clear that the words 'here' and 'I', used in a quotation of Luther's words, do not have the token-reflexive meaning of the original words but the meaning stated in terms of 'Θ' and used in (9).

As a substitute for arrow quotes and in order to have a notation that portrays closely the wording of conversational language, we shall employ the symbol 'Θ*' in the meaning of the original 'this', namely, as a name for the whole token within which the token of the symbol 'Θ*' occurs. Thus the sentence 'the page on which this sentence is written has the number 287' can be symbolized, with 'p' for 'page', 'wr' for 'is written on', and 'hn' for 'has the number', in the form

$$hn[(\imath x)p(x).wr(\Theta^*, x), 287] \qquad (10)$$

Although the sentence (10) is true for every copy of this book its meaning is different for every copy; it refers to the page of the individual copy in which it is printed. The token of the symbol 'Θ*' occurring in (10) denotes the individual token of the sentence (10) in which it occurs.[1] (Ex.)

§ 51. The tenses of verbs

A particularly important form of token-reflexive symbol is found in the tenses of verbs. The tenses determine time with reference to

---

[1] If we want to define the token 'θ^*' used in (10) in terms of the token quotes we can proceed as follows. We first introduce a name 'w' for the token (10) by analogy with (6). We then define: the token 'θ^*' used in w is equisignificant with w.

the time point of the act of speech, i.e., of the token uttered. A closer analysis reveals that the time indication given by the tenses is of a rather complex structure.

Let us call the time point of the token the *point of speech*. Then the three indications, 'before the point of speech', 'simultaneous with the point of speech', and 'after the point of speech', furnish only three tenses; since the number of verb tenses is obviously greater, we need a more complex interpretation. From a sentence like 'Peter had gone' we see that the time order expressed in the tense does not concern one event, but two events, whose positions are determined with respect to the point of speech. We shall call these time points the *point of the event* and the *point of reference*. In the example the point of the event is the time when Peter went; the point of reference is a time between this point and the point of speech. In an individual sentence like the one given it is not clear which time point is used as the point of reference. This determination is rather given by the context of speech. In a story, for instance, the series of events recounted determines the point of reference which in this case is in the past, seen from the point of speech; some individual events lying outside this point are then referred, not directly to the point of speech, but to this point of reference determined by the story. The following example, taken from W. Somerset Maugham's *Of Human Bondage*, may make these time relations clear:

But Philip ceased to think of her a moment after he had settled down in his carriage. He thought only of the future. He had written to Mrs. Otter, the *massière* to whom Hayward had given him an introduction, and had in his pocket an invitation to tea on the following day.

The series of events recounted here in the simple past determine the point of reference as lying before the point of speech. Some individual events, like the settling down in the carriage, the writing of the letter, and the giving of the introduction, precede the point of reference and are therefore related in the past perfect.

Another illustration for these time relations may be given by a historical narrative, a quotation from Macaulay:

In 1678 the whole face of things had changed . . . eighteen years of misgovernment had made the . . . majority desirous to obtain security for their liberties at any risk. The fury of their returning loyalty had spent itself in its first outbreak. In a very few months they had hanged and half-hanged, quartered and emboweled, enough to satisfy them. The Roundhead party seemed to be not merely overcome, but too much broken and scattered ever to rally again. Then commenced the reflux of public opinion. The nation began to find out to what a man it had intrusted without conditions all its dearest interests, on what a man it had lavished all its fondest affection.

The point of reference is here the year 1678. Events of this year are related in the simple past, such as the commencing of the reflux of public opinion, and the beginning of the discovery concerning the character of the king. The events preceding this time point are given in the past perfect, such as the change in the face of things, the outbreaks of cruelty, the nation's trust in the king.

In some tenses, two of the three points are simultaneous. Thus, in the simple past, the point of the event and the point of reference are simultaneous, and both are before the point of speech; the use of the simple past in the above quotation shows this clearly. This distinguishes the simple past from the present perfect. In the statement 'I have seen Charles' the event is also before the point of speech, but it is referred to a point simultaneous with the point of speech; i.e., the points of speech and reference coincide. This meaning of the present perfect may be illustrated by the following quotation from Keats:

> Much have I traveled in the realms of gold,
> And many goodly states and kingdoms seen;
> Round many western islands have I been
> Which bards in fealty to Apollo hold.

Comparing this with the above quotations we notice that here obviously the past events are seen, not from a reference point situated also in the past, but from a point of reference which coincides with the point of speech. This is the reason that the words of Keats are not of a narrative type but affect us with the immediacy of a direct report to the reader. We see that we need three time points even for the distinction of tenses which, in a superficial considera-tion, seem to concern only two time points. The difficulties which grammar books have in explaining the meanings of the different tenses originate from the fact that they do not recognize the three-place structure of the time determination given in the tenses.[1]

We thus come to the following tables, in which the initials 'E', 'R', and 'S' stand, respectively, for 'point of the event', 'point of reference', and 'point of speech', and in which the direction of time is represented as the direction of the line from left to right:

Past Perfect	Simple Past	Present Perfect
I had seen John	I saw John	I have seen John
$E \quad R \quad S \longrightarrow$	$R,E \quad S \longrightarrow$	$E \qquad S,R \longrightarrow$

Present	Simple Future	Future Perfect
I see John	I shall see John	I shall have seen John
$\longrightarrow$ S,R,E	$S,R \qquad E \longrightarrow$	$S \quad E \quad R \longrightarrow$

In some tenses, an additional indication is given concerning the time extension of the event. The English language uses the present participle to indicate that the event covers a certain stretch of time. We thus arrive at the following tables:

Past Perfect, Extended	Simple Past, Extended	Present Perfect, Extended
I had been seeing John	I was seeing John	I have been seeing John
$\boxed{} \quad E \quad R \quad S \longrightarrow$	$R,E \qquad S \longrightarrow$	$\boxed{} \quad E \qquad S,R \longrightarrow$

Present, Extended	Simple Future, Extended	Future Perfect, Extended
I am seeing John	I shall be seeing John	I shall have been seeing John
$\qquad E$ $\longrightarrow \boxed{}$ S,R	$S,R \qquad E \longrightarrow$	$S \quad E \quad R \longrightarrow$

The extended tenses are sometimes used to indicate, not duration of the event, but repetition. Thus we say 'women are wearing larger

---

[1] In J. O. H. Jespersen's excellent analysis of grammar (*The Philosophy of Grammar*, H. Holt, New York, 1924) I find the three-point structure indicated for such tenses as the past perfect and the future perfect (p. 256), but not applied to the interpretation of the other tenses. This explains the difficulties which even Jespersen has in distinguishing the present perfect from the simple past (p. 269). He sees correctly the close connection between the present tense and the present perfect, recognizable in such sentences as 'now I have eaten enough'. But he gives a rather vague definition of the present perfect and calls it 'a retrospective variety of the present'.

hats this year' and mean that this is true for a great many instances. Whereas English expresses the extended tense by the use
of the present participle, other languages have developed special
suffixes for this tense. Thus the Turkish language possesses a
tense of this kind, called *muzari*, which indicates repetition or
duration, with the emphasis on repetition, including past and
future cases. This tense is represented by the diagram

Turkish Muzari

görürüm

$$\underset{S,R}{\underrightarrow{E \;\; E \;\; E \;\; E \;\; E \;\; E}}$$

An example of this tense is the Turkish word 'görürüm', translatable as 'I usually see'. The syllable 'gör' is the root meaning
'see', 'ür' is the suffix expressing the muzari, and the 'üm' is the
suffix expressing the first person 'I'.[1] The sentence 'I see' would be
in Turkish 'görüyorum'; the only difference from the preceding
example is given by the inflection 'üyor' in the middle of the word,
expressing the present tense. The Greek language uses the *aorist*
to express repetition or customary occurrence in the present tense.
The aorist, however, is originally a nonextended past tense, and
has assumed the second usage by a shift of meaning; in the sense
of the extended tense it is called *gnomic aorist*.[2]

German and French do not possess extended tenses, but express
such meanings by special words, such as the equivalents of 'always',
'habitually', and so on. An exception is the French simple past.
The French language possesses here two different tenses, the
imparfait and the *passé défini*. They differ in so far as the *imparfait*
is an extended tense, whereas the *passé défini* is not. Thus we have

Imparfait	*Passé défini*
je voyais Jean	je vis Jean

$$\underset{R,E \qquad\qquad S}{\underrightarrow{\rule{2cm}{0.4pt}\qquad\qquad\quad}} \qquad\qquad \underset{R,E \qquad\qquad S}{\underrightarrow{\rule{2cm}{0.4pt}\qquad\qquad\quad}}$$

---

[1] Turkish vowels with two dots are pronounced like the German vowels 'ö' and 'ü'.

[2] This shift of meaning is explainable as follows: One typical case of the past is stated,
and to the listener is left the inductive inference that under similar conditions the same
will be repeated in the future. A similar shift of meaning is given in the English 'Faint
heart never won fair lady'. Cf. W. W. Goodwin, *Greek Grammar*, Ginn, Boston,
1930, p. 275.

We find the same distinction in Greek, the Greek imperfect corresponding to the French imparfait, and the Greek aorist, in its original meaning as a past tense, corresponding to the French passé défini. Languages which do not have a passé défini sometimes use another tense in this meaning; thus Latin uses the present perfect in this sense (historical perfect).

We may add here the remark that the adjective is of the same logical nature as the present participle of a verb. It indicates an extended tense. If we put the word 'hungry', for instance, in the place of the word 'seeing' in our tables of extended tenses, we obtain the same extended tenses. A slight difference in the usage is that adjectives are preferred if the duration of the event is long; therefore adjectives can often be interpreted as describing permanent properties of things. The transition to the extended tense, and from there to the permanent tense, is seen in the examples 'he produces', 'he is producing', 'he is productive'.

When we wish to express, not repetition or duration, but validity at all times, we use the present tense. Thus we say 'two times two is four'. There the present tense expressed in the copula 'is' indicates that the time argument is used as a free variable; i.e., the sentence has the meaning 'two times two is four at any time'. This usage represents a second temporal function of the present tense.

Actual language does not always keep to the schemas given in our tables. Thus the English language uses sometimes the simple past where our schema would demand the present perfect. The English present perfect is often used in the sense of the corresponding extended tense, with the additional qualification that the duration of the event reaches up to the point of speech. Thus we have here the schema

English Present Perfect, Second Usage

I have seen him

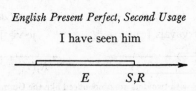

$$E \qquad S,R$$

In the sense of this schema we say, for instance, 'I have known him for ten years'. If duration of the event is not meant, the English language then uses the simple past instead of the present perfect, as in 'I saw him ten years ago'. German and French would use the present perfect here.

When several sentences are combined to form a compound sentence, the tenses of the various clauses are adjusted to one another by certain rules which the grammarians call the rules for the *sequence of tenses*. We can interpret these rules as the principle that, although the events referred to in the clauses may occupy different time points, the reference point should be the same for all clauses — a principle which, we shall say, demands *the permanence of the reference point*. Thus, the tenses of the sentence, 'I had mailed the letter when John came and told me the news', may be diagramed as follows:

$$
\begin{array}{lll}
\text{1st clause:} & E_1 - R_1 & - S \\
\text{2nd clause:} & R_2, E_2 - S & \\
\text{3rd clause:} & R_3, E_3 - S &
\end{array}
\tag{1}
$$

Here the three reference points coincide. It would be incorrect to say, 'I had mailed the letter when John has come'; in such a combination the reference point would have been changed. As another example, consider the compound sentence, 'I have not decided which train I shall take'. That this sentence satisfies the rule of the permanence of the reference point is seen from the following diagram:

$$
\begin{array}{ll}
\text{1st clause:} & E_1 - S, R_1 \\
\text{2nd clause:} & S, R_2 - E_2
\end{array}
\tag{2}
$$

Here it would be incorrect to say: 'I did not decide which train I shall take'.

When the reference point is in the past, but the event coincides with the point of speech, a tense $R - S, E$ is required. In this sense, the form 'he would do' is used, which can be regarded as derived from the simple future 'he will do' by a back-shift of the two points R and E. We say, for instance, 'I did not know that you would be here'; this sentence represents the diagram:

$$
\begin{array}{ll}
\text{1st clause:} & R_1, E_1 - S \\
\text{2nd clause:} & R_2 \quad - S, E_2
\end{array}
\tag{3}
$$

The form 'I did not know that you were here' has a somewhat different meaning; it is used correctly only if the event of the man's being here extends to include the past time for which the 'I did not know' is stated, i.e., if the man was already here when I did not know it. Incidentally, in these sentences the forms 'would be' and 'were' do not have a modal function expressing irreality (cf. § 57); i.e., they do not represent a conditional or a subjunctive, since the event referred to is not questioned. The nonmodal function is illustrated by the sentence 'I did not know that he was here', for which the form 'that he were here' appears incorrect.

When a time determination is added, such as is given by words like 'now' or 'yesterday', or by a nonreflexive symbol like 'November 7, 1944', it is referred, not to the event, but to the reference point of the sentence. We say, 'I met him yesterday'; that the word 'yesterday' refers here to the event obtains only because the points of reference and of event coincide. When we say, 'I had met him yesterday', what was yesterday is the reference point, and the meeting may have occurred the day before yesterday. We shall speak, therefore, of the *positional use of the reference point;* the reference point is used here as the carrier of the time position. Such usage, at least, is followed by the English language. Similarly, when time points are compared by means of words like 'when', 'before', or 'after', it is the reference points to which the comparison refers directly, not the events. Thus in the above example (1) the time points stated as identical by the word 'when' are the reference points of the three clauses, whereas the event of the first clause precedes that of the second and the third. Or consider the sentence, 'How unfortunate! Now that John tells me this I have mailed the letter'. The time stated here as identical with the time of John's telling the news is not the mailing of the letter but the reference point of the second clause, which is identical with the point of speech; and we have here the schema:

$$\text{1st clause:} \quad S, R_1, E_1$$
$$\text{2nd clause:} \quad E_2 - S, R_2 \tag{4}$$

For this reason it would be incorrect to say, 'Now that John tells me this I mailed the letter'.

If the time relation of the reference points compared is not identity, but time sequence, i.e., if one is said to be before the other, the rule of the permanence of the reference point can thus no longer be maintained. In 'he telephoned before he came' R_1 is said to be before R_2; but, at least, the tenses used have the same structure. It is different with the example, 'he was healthier when I saw him than he is now'. Here we have the structure:

$$\text{1st clause: } R_1, E_1 - S$$
$$\text{2nd clause: } R_2, E_2 - S \qquad (5)$$
$$\text{3rd clause: } \qquad S, R_3, E_3$$

In such cases, the rule of the permanence of the reference point is replaced by the more general rule of the *positional use of the reference point*. The first rule, therefore, must be regarded as representing the special case where the time relation between the reference points compared is identity.

Incidentally, the English usage of the simple past where other languages use the present perfect may be a result of the strict adherence to the principle of the positional use of the reference point. When we say, 'this is the man who drove the car', we use the simple past in the second clause because the positional principle would compel us to do so as soon as we add a time determination, as in 'this is the man who drove the car at the time of the accident'. The German uses here the present perfect, and the above sentence would be translated into 'dies ist der Mann, der den Wagen gefahren hat'. Though this appears more satisfactory than the English version, it leads to a disadvantage when a time determination is added. The German is then compelled to refer the time determination, not to the reference point, but to the event, as in 'dies ist der Mann, der den Wagen zur Zeit des Unglücksfalles gefahren hat'. In such cases, a language can satisfy either the principle of the permanence of the reference point or that of the positional use of the reference point, but not both.

The use of the future tenses is sometimes combined with certain deviations from the original meaning of the tenses. In the sentence 'Now I shall go' the simple future has the meaning $S, R - E$; this follows from the principle of the positional use of the reference point. However, in the sentence 'I shall go tomorrow' the same principle compels us to interpret the future tense in the form $S - R, E$. The simple future, then, is capable of two interpretations, and since there is no prevalent usage of the one or the other we

cannot regard one interpretation as the correct one.[1] Further deviations occur in tense sequences. Consider the sentence: 'I shall take
your photograph when you come'. The form 'when you will come'
would be more correct; but we prefer to use here the present tense
instead of the future. This usage may be interpreted as follows.
First, the future tense is used in the first clause in the meaning
$S — R, E$; second, in the second clause the point of speech is
neglected. The neglect is possible because the word 'when' refers
the reference point of the second clause clearly to a future event.
A similar anomaly is found in the sentence, 'We shall hear the
record when we have dined', where the present perfect is used instead of the future perfect 'when we shall have dined'.[2]

Turning to the general problem of the time order of the three
points, we see from our tables that the possibilities of ordering the
three time points are not exhausted. There are on the whole 13 possibilities, but the number of recognized grammatical tenses in English is only 6. If we wish to systematize the possible tenses we can
proceed as follows. We choose the point of speech as the starting
point; relative to it the point of reference can be in the past, at the
same time, or in the future. This furnishes three possibilities. Next
we consider the point of the event; it can be before, simultaneous
with, or after the reference point. We thus arrive at $3 \cdot 3 = 9$ possible
forms, which we call *fundamental forms*. Further differences of
form result only when the position of the event relative to the point
of speech is considered; this position, however, is usually irrelevant.
Thus the form $S — E — R$ can be distinguished from the form $S,
E — R$; with respect to relations between S and R on the one hand
and between R and E on the other hand, however, these two forms
do not differ, and we therefore regard them as representing the
same fundamental form. Consequently, we need not deal with all
the 13 possible forms and may restrict ourselves to the 9 fundamental forms.

---

[1] The distinction between the French future forms *je vais voir* and *je verrai* may
perhaps be regarded as representing the distinction between the order $S, R — E$ and
the order $S — R, E$.

[2] In some books on grammar we find the remark that the transition from direct to
indirect discourse is accompanied by a shift of the tense from the present to the past.
This shift, however, must not be regarded as a change in the meaning of the tense; it
follows from the change in the point of speech. Thus 'I *am* cold' has a point of speech
lying before that of 'I said that I *was* cold'.

For the 9 fundamental forms we suggest the following terminology. The position of R relative to S is indicated by the words
'past', 'present', and 'future'. The position of E relative to R is
indicated by the words 'anterior', 'simple', and 'posterior', the
word 'simple' being used for the coincidence of R and E. We thus
arrive at the following names:

Structure	New Name	Traditional Name
$E - R - S$	Anterior past	Past perfect
$E, R - S$	Simple past	Simple past
$R - E - S$ ⎫ $R - S, E$ ⎬	Posterior past	—
$R - S - E$ ⎭		
$E - S, R$	Anterior present	Present perfect
S, R, E	Simple present	Present
$S, R - E$	Posterior present	Simple future
$S - E - R$ ⎫ $S, E - R$ ⎬	Anterior future	Future perfect
$E - S - R$ ⎭		
$S - R, E$	Simple future	Simple future
$S - R - E$	Posterior future	—

We see that more than one structure obtains only for the two
retrogressive tenses, the posterior past and the anterior future, in
which the direction $S - R$ is opposite to the direction $R - E$. If
we wish to distinguish among the individual structures we refer to
them as the first, second, and third posterior past or anterior future.

The tenses for which a language has no established forms are
expressed by transcriptions. We say, for instance, 'I shall be going
to see him' and thus express the posterior future $S - R - E$ by
speaking, not directly of the event E, but of the act of preparation
for it; in this way we can at least express the time order for events
which closely succeed the point of reference. Languages which have
a future participle have direct forms for the posterior future. Thus
the Latin 'abiturus ero' represents this tense, meaning verbally
'I shall be one of those who will leave'. For the posterior past
$R - E - S$ the form 'he would do' is used, for instance in 'I did not
expect that he would win the race'. We met with this form in an
above example where we interpreted it as the structure $R - S, E$;
but this structure belongs to the same fundamental form as
$R - E - S$ and may therefore be denoted by the same name. In-

stead of the form 'he would do', which grammar does not officially recognize as a tense,[1] transcriptions are frequently used. Thus we say, 'I did not expect that he was going to win the race', or, in formal writing, 'the king lavished his favor on the man who was to kill him'. In the last example, the order $R — E — S$ is expressed by the form 'was to kill', which conceives the event E, at the time R, as not yet realized, but as a destination.

Incidentally, the historical origin of many tenses is to be found in similar transcriptions. Thus 'I shall go' meant originally 'I am obliged to go'; the future-tense meaning developed because what I am obliged to do will be done by me at a later time.[2] The French future tense is of the same origin; thus the form 'je donnerai', meaning 'I shall give', is derived from 'je donner ai', which means 'I have to give'. This form of writing was actually used in Old French.[3] The double function of 'have', as expressing possession and a past tense, is derived from the idea that what I possess is acquired in the past; thus 'I have seen' meant originally 'I possess now the results of seeing', and then was interpreted as a reference to a past event.[4] The history of language shows that logical categories were not clearly seen in the beginnings of language but were the results of long developments; we therefore should not be astonished if actual language does not always fit the schema which we try to construct in symbolic logic. A mathematical language can be coordinated to actual language only in the sense of an approximation. [Ex.]

[1] It is sometimes classified as a tense of the conditional mood, corresponding to the French conditional. In the examples considered above, however, it is not a conditional but a tense in the indicative mood.

[2] In Old English no future tense existed, and the present tense was used both for the expression of the present and the future. The word 'shall' was used only in the meaning of obligation. In Middle English the word 'shall' gradually assumed the function of expressing the future tense. Cf. *The New English Dictionary*, Oxford, Vol. VIII, Pt. 2, S–Sh, 1914, p. 609, col. 3.

[3] This mode of expressing the future tense was preceded by a similar development of the Latin language, originating in vulgar Latin. Thus instead of the form 'dabo', meaning the future tense 'I shall give', the form 'dare habeo' was used, which means 'I have to give'. Cf. Ferdinand Brunot, *Précis de grammaire historique de la langue française*, Masson et Cie., Paris, 1899, p. 434.

[4] This is even more apparent when a two-place function is used. Thus 'I have finished my work' means originally 'I have my work finished', i.e., 'I possess my work as a finished one'. Cf. *The New English Dictionary*, Oxford, 1901, Vol. V, Pt. I, H, p. 127, col. 1–2. The German still uses the original word order, as in 'Ich habe meine Arbeit beendet'.

§ 52. Classification of functions

The grammatical classification of functions into nouns, adjectives, and verbs, criticized by us in § 45, has only a technical significance. The verb carries the tense and personal suffixes and is thus indicated as a propositional function. Nouns and adjectives are used as functions only in combination with the auxiliary verb 'to be'. A logical difference is combined with this linguistic distinction only so far as nouns can be regarded as terms denoting classes, whereas the adjective has the nature of a tense and is therefore better included with the verbs (cf. § 51).

A logically relevant classification of functions can be constructed by the use of the concepts developed in § 22; furthermore, the distinctions explained in § 39 and § 48 must be considered. We thus arrive at the following classification.

1. We begin by collecting in one group several forms of divisions, namely, all divisions of functions with respect to *place properties*. Here we have the division concerning the number of variables (where space-time arguments are usually not counted), concerning univocality, concerning the structure of the field (divided field, nondivided field, uniform functions, interconnective functions), concerning symmetry, transitivity, reflexivity. Each function has its category in each of these divisions; this rule holds, correspondingly, also for the following divisions. The reflexive verbs are, in our terminology, mostly mesoreflexive functions; for instance, 'I struck myself' may be true, but need not be true. The grammatical term 'reflexive' means the logical 'not irreflexive'. Incidentally, the English language frequently drops the reflexive pronoun and then construes reflexive verbs as one-place functions, as in 'the enemy surrendered'. Other languages do not have this usage; thus the sentence quoted would read in German: 'der Feind ergab sich', in French: 'l'ennemi se rendit'.

2. The second form of classification is given by the division concerning the *inner structure* of a function. Here we have the distinction of simple and complex functions. We shall regard a function as simple if it is expressed by an independent word not indicating a derivation from other functions. Thus we shall conceive, for instance, the function 'brother' always as a simple function. It is true that there is no logical necessity to do so. Thus we could

regard the function '$\hat{x}$ is the brother of $\hat{y}$' as a complex function of the form 'there is a u and a v so that u is the father of $\hat{x}$ and $\hat{y}$ and v is the mother of $\hat{x}$ and $\hat{y}$, and $\hat{x}$ and $\hat{y}$ are not identical, and $\hat{x}$ is male'. We prefer, however, not to make a formal use of this definition. What we call a simple function, therefore, depends on the language used; when this language possesses a special term for the function it will appear convenient to regard it as a simple function. In expressions like 'a brother of William', on the other hand, the linguistic composition indicates the structure of a complex function. It is easily divided into simple parts, namely, the function 'brother' and the argument 'William'. Sometimes we are compelled to speak of complex functions even when a division into linguistic parts is not possible, as for instance when the word through its relation to other words indicates the character of a contracted symbol. Certain tenses of verbs are of this form. For instance, the indication of the past tense in the strong conjugation is not given by a suffix, but by a change of the vowel of the verbal root; the resulting word then expresses simultaneously the function and its time argument. Thus 'broke' is a contracted term expressing in one word the function 'break' and a time argument in the past tense. We therefore cannot separate here a term expressing the function from a term expressing the time argument, as can be done in the past tense of a weak verb, like 'killed'. But here the complex character of the word is clear from its relation to the root 'break'. Another example is the word 'presidential', which in such combinations as 'the presidential yacht' means 'belonging to the president', and therefore includes a function, namely, 'belonging', and its specialized argument.

3. The third form of classification is given by the division with respect to the *nature of the argument*, namely, the division into thing-functions and fact-functions (cf. § 48). Most verbal roots expressing functions are thing-functions; and the transition to an event-function is usually achieved by means of a suffix, as in 'the beating'. There are, however, also other forms of event-functions that are not constructed by means of a suffix, such as 'the sale'. Historically, of course, such words may be reducible to thing-functions; thus 'sale' is derived from the thing-function 'to sell'. In words like 'coronation' the origin from a verb, i.e., a thing-function,

is visible only in the Latin root of the word, the verb 'coronare'. The English verb 'to crown', although also derived from this root, differs from it by a phonetic development to which the noun was not subjected, and therefore does not represent the logical root of the corresponding fact-function.

4. Finally, we shall use the classification with respect to *type*, resulting in the distinction between lower and higher functions. This distinction requires a special analysis which we shall present in the following section.

§ 53. Functions of higher types

The necessity for the use of functions of higher types was made clear in § 39. In a logistic grammar, we need functions of higher types in particular for the interpretation of *functional modifiers*. These terms are known in traditional grammar by the name of *adverbs*, or *adverbial complements*. To understand the nature of these modifiers, let us compare the statement 'John drives slowly' with the statement 'Royce Hall is a red building'. The word 'slowly' is a modifier of the function 'drives'; the word 'red', however, is not a modifier of the function 'building', but is an independent function. The independence is evident when our second sentence is written in the form 'Royce Hall is a building and Royce Hall is red'; for the first sentence such a transformation is not possible. Traditional grammar is seriously mistaken in interpreting all adjectives as modifiers of nouns; most adjectives are independent functions, like nouns and verbs. The symmetrical nature of adjective and noun in sentences like the given one was seen by Leibniz, who conceived such propositions as conjunctions of two propositions.

However, not all adjectives are independent functions. Let us write, instead of the first sentence, 'John is a slow driver'. This sentence differs from the first only in so far as it asserts a constant property of John and therefore represents an extended tense; it is equivalent to the sentence 'John always drives slowly'. Therefore the word 'slow' is here also a modifier of the function 'driver'. This is clear, too, because we cannot divide the sentence into two sentences 'John is slow and John is a driver'. What is said is not that John is slow in general but only that John is slow in his driving; thus the word 'slow', as before, operates as a modifier of 'drive'. It would be logical, therefore, to say 'John is a slowly driver'. It is the tendency to equalization which prevents the English language (and other languages as well) from distinguishing two different usages of the adjective in combination with a noun, although English clearly indicates the modifier character in such words if they are used in combination with a verb, namely, by adding the suffix 'ly'. The German language, strangely enough, makes the opposite mistake. In general it drops the distinction between adjective and adverb when the adverb modifies a verb; but for some adjectives used as modifiers of nouns it indicates the modifier character by writing adjective and noun in one word. Thus the two forms of our sentence would be in German 'Johann fährt langsam' and 'Johann ist ein Langsamfahrer', although in the second version also the form 'Johann ist ein langsamer Fahrer' could be used in German. Usually only some established combinations are used in German in this way, such as 'Schnellfahrer', 'Schwerverwundeter', 'Schwerarbeiter', etc. This is an instance where a logically clear distinction has not found its full expression in conversational languages.

To understand the nature of modifiers we must realize that most predicates are used in such a way that they leave open some space for further specification. The function 'x moves', for instance, leaves open a specification of the speed, telling us only that the speed is not zero. The addition of an adverb like 'slowly' includes the speed in a narrower interval. We may consider the individual motion of x, determined as to speed, direction, and so on, as a property f, which we call a *specific property;* the various properties f of this kind then may be included in a class of motion-properties represented by a function $\mu(f)$. The statement 'x_1 moves' tells us that x_1 has one of the motion-properties; in symbols, with '$m(x)$' for 'x moves':

$$m(x_1) =_{Df} (\exists f) f(x_1) . \mu(f) \tag{1}$$

We see that '$m(x)$' is a contracted symbol standing for a complex function which includes a predicate as a bound variable. m is the property every x has that has a property f belonging to μ; being a property of an object of the zero type, m is of the first type. In class terminology, the class M coordinated to m is the disjunct of all the classes F coordinated to functions f having the property μ.

We can now introduce the modifier 'slowly'. Writing '$msl(x)$' for 'x moves slowly', and '$\sigma(f)$' for 'f is slow', we obtain

$$msl(x_1) =_{Df} (\exists f)f(x_1).\mu(f).\sigma(f) \qquad (2)$$

This analysis shows why adverbs can be constructed from adjectives by the addition of a suffix; they are predicates, like adjectives, not denoting properties of *things*, however, but of *properties*. The suffix 'ly' indicates this usage.[1]

The functions μ and σ used in (2) differ in one important point. The properties f comprehended in μ are *mutually exclusive;* i.e., if a thing x_1 has one of these properties, it cannot have one of the others. Thus if x_1 moves at 30 miles an hour, it cannot move simultaneously at 40 miles an hour. The properties f comprised in σ, however, are *nonexclusive.* Thus x_1 may move slowly and at the same time eat slowly, or think slowly.

Usually a first-type property is introduced in the form (1) only for higher functions pertaining to mutually exclusive properties, such as μ. If the higher function relates to nonexclusive properties, like the function σ, the corresponding first-type property is usually defined by the requirement that a great number of specific properties of x_1 belong to σ, or that those specific properties of x_1 which belong to a certain function κ are contained in σ. Here κ may mean, for instance, properties involving a change in time, or at least a group of *relevant* properties of this sort. The first-type property then is defined in the form

$$sl(x_1) =_{Df} (f)[\kappa(f).f(x_1) \supset \sigma(f)] \qquad (3)$$

This, for instance, is the way in which the adjective 'slow' is defined. The statement 'x_1 is slow' is given by the expression '$sl(x_1)$' defined in (3). The rather complicated relation between $sl(x_1)$ and $\sigma(f)$, shown by (3), explains why the meanings of the first-type predicate 'sl', i.e., the adjective 'slow', and of the second-type predicate σ, i.e., the word root 'slow' used in the adverb 'slowly', do not strictly correspond to each other.

Most adverbs represent functions σ of the nonexclusive type. Thus adverbs like 'very', 'greatly', 'extremely', are of this sort.

---

[1] It is impossible to symbolize the sentence 'x_1 moves slowly' in the form '$m(x_1).\sigma(m)$' because the property *moving* in general is not slow; only the specific sort of moving which refers to x_1 is slow. This is the reason that the specific property f must be introduced, as shown in (2).

A man may have many specific properties to which the modifier 'very' applies; he may be very intelligent, very strong, very helpful. Not always does language possess a word for the first-type property defined in (3). Thus we do not say 'x is very', although this expression might be defined according to (3). Perhaps the term 'x_1 is outstanding' may be conceived as derived from the adverb 'very' by (3), when in this formula 'σ' is interpreted as 'very'.

Now let us compare an adverbial modification with a statement asserting two independent properties. The statement cited above, 'Royce Hall is a red building' can be written in the form

$$bl(x_1).r(x_1) \tag{4}$$

where 'bl' stands for 'building' and 'r' for 'red'. As in the preceding instance it is permissible to regard each of these properties as including a number of specific properties; thus a building is capable of many forms and sizes, and the color red is divisible into many different shades of red. Therefore we can conceive these first-type properties as derived from second-type properties:

$$bl(x_1) =_{Df} (\exists f)f(x_1).\lambda(f) \qquad r(x_1) =_{Df} (\exists f)f(x_1).\rho(f) \tag{5}$$

Here λ corresponds to the class of specific sorts of buildings, and ρ to the class of shades of red. Inserting (5) in (4) we arrive at the statement

$$[(\exists f)f(x_1).\lambda(f)].[(\exists f)f(x_1).\rho(f)] \tag{6}$$

Comparing this with (2) we see a remarkable difference as to structure: (6) says that there is a specific property f of x_1 belonging to λ, and likewise such a property belonging to ρ; (2) says that it is *the same specific property f* which belongs to the functions μ and σ. The identity of specific properties, expressed by the adverbial suffix 'ly', may be regarded as formulating the difference between an adverbial qualification of a function and the use of two independent functions.

The relation between the two forms (2) and (6) is as follows. A form corresponding to (6), namely,

$$[(\exists f)f(x_1).\mu(f_1)].[(\exists f)f(x_1).\sigma(f)] \tag{7}$$

can be derived from (2), although the reverse relation does not hold; (2) says more than (7). We therefore can derive from an adverbial qualification a statement asserting two independent properties. But only the first of these two properties is the same as that used in the adverbial statement; this is the property derived from the exclusive function μ, in our example the first bracket of (7), which according to (1) means 'x_1 moves'. The second independent property, derived from the nonexclusive function σ, does not represent the adjective corresponding to the adverb. In our example, the second bracket of (7) defines a first-type property for which there is no name in conversational language, since the adjective 'slow' is defined, not by this bracket, but by (3). This is the reason that the statement 'x_1 moves and x_1 is slow' is not derivable from the statement 'x_1 moves slowly'.

Similar considerations hold where other forms for the definition of the adjective are used. Thus the derivation of the adjective 'beautiful' from a corresponding function of the second type does not follow (3), but cannot be represented by the second bracket in (7) either. Consider the statement 'Annette dances beautifully'; it has the form

$$(\exists f) f(x_1) \delta(f) . \beta(f) \tag{8}$$

where δ represents the function comprehending properties of dancing, and β the function comprising properties of beauty. The first-type statement 'Annette dances' is definable as

$$d(x_1) =_{Df} (\exists f) f(x_1) . \delta(f) \tag{9}$$

But the first-type predicate 'beautiful' is not defined in the same way. When we say 'Annette is beautiful' we mean that a particular property of Annette, namely, her bodily form, is beautiful. This leads to the following definition of the adjective 'beautiful':

$$bt(x_1) =_{Df} (\exists f) f(x_1) . \alpha(f) . \beta(f) \tag{10}$$

where α is the function comprehending all individual forms of human bodies such that each f belonging to it specifies the shape of the whole body. We see that this definition corresponds to (2), not to (1); it can be read in the form 'Annette is beautifully shaped'. This shows that, although a form corresponding to (7) is derivable from (8), statement (8) does not imply a statement of the form 'Annette dances and is beautiful'.

This analysis explains why, although expressions like (2) and (8) are symmetrical in the two functions of higher order, verb and adverb do not occupy symmetrical positions in conversational language. Only if we were to use a definition of the adjective 'beautiful' corresponding to (9), namely, in the form

$$bt(x_1) =_{Df} (\exists f) f(x_1).\beta(f) \tag{11}$$

could we give to (8) a symmetrical interpretation; namely, we then could read (8) also in the form 'Annette is beautiful dancingly'. But although such a terminology might not appear incorrect, it certainly does not correspond to actual usage.

Let us add here a remark about the necessity of introducing higher functions for the interpretation of adverbs. It might be suggested that such functions be avoided by the application of the relation of class inclusion. Instead of the function δ, we then would use the function d occurring in (9), or the corresponding class $D;$ similarly, we would use the function bt defined in (11), or the corresponding class Bt. These classes then would not be defined by the definitions (9) and (11), but would be regarded as primitive terms, like δ and β in the other interpretation. Applying the relation of class inclusion instead of class membership, we then would write the sentence 'Annette dances beautifully' in the form:

$$(\exists F)(x_1 \epsilon F).(F \subset D).(F \subset Bt) \tag{12}$$

This expression would replace (8). It can be shown, however, that such an interpretation cannot be carried through. Formula (12) is equivalent to the expression

$$(x_1 \epsilon D).(x_1 \epsilon Bt) \tag{13}$$

since when x_1 belongs to the common class of D and Bt, there is always a subclass F of both classes to which x_1 belongs. We therefore cannot distinguish here between the adverbial modifier and the genuine adjective. Whereas the use of higher functions enables us to construct, for this distinction, the two different forms (7) and (2), the two corresponding forms (12) and (13) are identical. Incidentally, this interpretation would not rid us of the higher calculus of functions since the occurrence of a class F as a bound variable, as in (12), represents an operation within this calculus.

On the other hand, we do not wish to say that functions of a higher type are always clearly distinguishable, in conversational language, from functions of the first type. The logistic interpretation of conversational language cannot be given without certain arbitrary restrictions of meanings. General predicates like 'color' and 'motion' are not always used in strictly the same sense. Thus when we say 'red is a color', we use 'color' as a second-type predicate; in the sentence 'a red thing is a colored thing' we use it as a first-type predicate.

Let us now turn to another interpretation of adverbs which, although it also applies higher functions, leads to the advantage that the adverb is reduced to a function of the first type. The adverb is construed, in this conception, not as a predicate of the argument of the sentence, but of the *fact* coordinated to it. Following this interpretation we write the sentence 'Annette dances beautifully' in the form

$$(\exists f)f(x_1).\delta(f).bt\{(\imath v)[f(x_1)]^*(v)\} \tag{14}$$

where δ, as before, comprises all specific properties of dancing, while 'bt' is employed here as a primitive term not derived from a higher function. Formula (14) can be read in the form: 'Annette dances and her dancing is beautiful'. The phrase 'her dancing' is regarded here as the description of the event, and not as a description of the specific property. This interpretation offers the advantage of a very simple conception of adverbs, namely, as adjectives referred to the event indicated by the sentence.

It is necessary, however, to add a qualification. We wish to say that there is a connection between the way of dancing and the property of being beautiful; that is, the property of the event is meant to qualify the specific property f in such a way that it must have a certain specific character. This connection is not expressed in (14), since (14) will hold also when no such connection exists. For instance, the sentence 'x_1 was murdered and the murder of x_1 led to a war' can also be expressed in the form (14), when we interpret the symbol 'δ' as 'was murdered' and 'bt' as meaning 'led to a war'. However, the fact that the event led to a war does not specify the manner in which the murder was committed; it only states a consequence which originated, not from the way the murder was committed, but from the political circumstances in which it occurred. This example shows that (14) does not convey the full meaning of the adverbial sentence about the dancing, and therefore must be supplemented by a qualification.

The addition to be made to (14) is given by the statement of a connection between f and bt, which can be expressed in the form

$$f(x_1) \mathbin{\vartheta} bt\{(v)[f(x_1)]^*(v)\} \tag{15}$$

This expression must be so added to (14) that the scope of the existential operator includes (15). The addition then states that the property f which x_1 has connectively implies the property bt of the event; in this way a specification of f is attained. We use here a connective implication of the synthetic kind (cf. chapter VIII) in order to exclude the paradoxes of the adjunctive implication. The application of an adjunctive general implication would not suffice. When adverbial expressions are interpreted in the form (14), a qualification in the form (15) must therefore be understood.

Grammar places in the category of adverbs a great many terms which should not be regarded as adverbs, i.e., as modifiers of functions. The word 'not', for instance, is classified by many grammarians as an adverb; but it is a logical term. Neither are the so-called adverbs of time and place *adverbs;* they are terms specifying the time and space argument to which the sentence as a whole refers. There are also higher functions which are not adverbs. They occur, for instance, when classes of classes are referred to, as in the term 'United Nations', or in the definition of numbers. Language frequently uses numerical qualifications, such as the words 'frequently', 'seldom', and 'usually'; these terms, which grammar classifies as adverbs, are not modifiers of functions, but express frequency relations concerning the occurrence of objects or events. They therefore constitute higher functions used as *numerical qualifiers*. Thus, in the sentence 'chemical reactions are frequently accompanied by a production of heat', the word 'frequently' does not modify the function 'accompany' since it does not delimit a specific form of accompaniment; it states rather that the class of events described by the sentence as a whole is large.

On the other hand, there are certain terms that are incorrectly classified as adjectives. To call a word like 'all' an adjective when it precedes a noun is a logical absurdity. It is a logical term, and not a function. It is incorrect, furthermore, to regard the first term in a noun compound as an adjective. Thus, in the compound 'door bell', the word 'door' is not used as an adjective, since the sentence 'x is a door bell' cannot be split into the two sentences 'x is a door' and 'x is a bell'. But the word 'door' is not used as an adverb either. It constitutes, rather, a part of a complex function, as explained in (23, § 22). If a suitable terminology is desired, we may call the first term, in our example the word 'door', the *indirect part* of the compound; the second term, the word 'bell' in our example, may be called the *direct part*, it being the function which applies directly to the argument of the sentence. There are other forms of complex functions which do not follow the pattern (23, § 22). Thus the function 'the last Goths' is a complex function in which 'last' does not have the character of an adjective, since 'x is a last Goth' cannot be split into the two statements 'x is last' and 'x is a Goth'; that is, x is last only in so far as it is a Goth. The word 'last' stands here as an abbreviation for a qualification saying that there were no later Goths; the qualification could be easily symbolized with the help of logical terms. The word 'last', therefore, cannot be construed as an independent function but must be regarded as an incomplete symbol which has a meaning only in combination with functions. The distinction between direct and indirect part may be applied also to complex functions of this kind. Such functions, incidentally, can be constructed for any type. 'The last integer below 100', for instance, is a description given by means of a complex function of a higher type.

Let us add a remark about the use of predicative adjectives after verbs other than the copula. We say, for instance, 'the wine seems good', using the adjective 'good' as qualifying the wine and not as modifying the property of seeming. This appears correct; in fact, the sentence means 'it seems that the wine is good'; or, in a more elaborate version, 'the wine has a certain appearance which makes the speaker believe that the wine is good'. Here the word 'good' occurs as a predicate of the argument, 'the wine'.

There are other forms which, although at first glance appearing to be of a similar construction, require a different explanation. We say, 'oranges taste good'; but we mean, obviously, that the taste is good. The word 'good', therefore, has here the character of an adverb, and it would appear preferable to say: 'oranges taste well'. The adjectival form presumably originates from a shift in the meaning of the word 'taste'. It was used originally in the transitive form 'I taste the orange'. From a transitive verb we can always derive an intransitive one by means of binding one variable by an existential operator. Thus we have the two forms 'I taste' in the sense of 'there is something that I taste', and 'the orange is tasted' in the sense of 'there is somebody who tastes the orange'. But the verb 'taste' as well as some others, like 'look' and 'feel', underwent an unusual development: the transition to the intransitive form was combined with a change in the place of the argument variables, and we say now 'the orange tastes'. The adverb 'well' was originally combined only with the transitive meaning; in the sentence 'I tasted the orange well' it modifies, not the taste of the orange, but the abilities of the taster. In the reversed intransitive meaning, the adjective 'good' was used; 'the orange tastes good' stands as an abbreviation for 'I tasted the orange and found it good to my taste'. But since the verb 'taste' has now fully assumed the reversed intransitive meaning, it would be logical to combine it also in this meaning with the adverb 'well'.

In phrases like 'looks good' we may distinguish two different meanings. When we say 'this meal looks good', we mean that the meal looks as though it were a good meal; here the adjectival form is correct. In speaking of a good-looking girl, however, we wish to say, not that the girl is presumably good, but that the way she looks is good; here 'good' is used as an adverb, and it would appear preferable to speak of a well-looking girl. What makes such usage inopportune is that the adverb 'well' has acquired, in relation to persons, the meaning of the adjective 'healthy'; to say that a girl looks well means that she looks as though she were healthy. The historical explanation for the use of the adjective may be found in the transition to the reversed intransitive form, as explained above.

Because of its reference to functions of a higher type, the study of the adverb represents a rather intricate chapter of grammar. We should not be astonished that this part of conversational language does not always stand up to the gauge of logic. It is amazing enough that conversational language possesses means to delimit a category of terms that belong in the higher calculus of functions. [Ex.]

§ 54. Descriptional functions

From propositional functions we turn now to a second kind of function, which has developed out of descriptions, and which we call *descriptional functions*. Such functions do not represent a separate class of words but are constructed from propositional functions by the aid of grammatical devices. We therefore did not mention them in the classification given in § 52.

Consider a statement in functional notation

$$f(y_1, x_1) \tag{1}$$

for instance, the statement 'y_1 is the father of x_1'. If the functional

$$f(y, x) \tag{2}$$

contains a one-one or one-many function, we can write the statement (1), in descriptional notation,

$$y_1 = (\imath y)f(y, x_1) \tag{3}$$

Introducing the abbreviation

$$f'(x_1) =_{Df} (\imath y)f(y, x_1) \tag{4}$$

we can write (3) in the form

$$y_1 = f'(x_1) \tag{5}$$

Putting variables in the places of the constants, we obtain

$$y = f'(x) \tag{6}$$

which is but another form of the functional (2). We may say that we have *solved* the functional '$f(y, x)$' for the argument 'y'.

The function '$f'(\hat{x})$' is called a *descriptional function*, since its special values, resulting from specialization of 'x', are descriptions. This kind of function, which we indicate by the prime mark, is to be distinguished from propositional functions, whose values are propositions. If we want to construct a proposition by means of a descriptional function, we must not only specialize the argument but also add a symbol like '$y_1 = $'. The symbol '$f'(\hat{x})$' expresses a descriptional function in the form of a contracted term, since the bound variable and the iota-operator are not indicated; the explicit form of a descriptional function is indicated by the right-hand side of the definition

$$f'(x) =_{Df} (\imath y)f(y, x) \tag{7}$$

There are also descriptional functions of more than one variable. For instance, the description 'the man who walks between Peter and Paul' can be used for the construction of the descriptional function 'the z walking between $\hat{x}$ and $\hat{y}$', symbolized by '$f'(\hat{x}, \hat{y})$'.

Descriptional functions can be extended to functions of higher types. Thus the description 'the color which is spectrally between red and yellow' determines the color *orange;* and it may be written in the form

$$\varphi'(r, s) =_{Df} (\imath f)\beta(f, r, s).\gamma(f) \tag{8}$$

where 'r' stands for 'red', 's' for 'yellow', 'β' for 'spectrally between', and 'γ' for 'color'. $\varphi'(r, s)$ represents a descriptional function when the arguments r and s are conceived as variables.

The distinction between proper and improper use of descriptions (cf. § 47) applies likewise to descriptional functions. In general a descriptional function will be properly used for some arguments, improperly for others. Thus the descriptional function 'the brother of $\hat{x}$', applied to persons as arguments, will be properly used when x has one and only one brother; in all other cases it will be improperly used. When a descriptional function is properly used for all values of its argument within a certain range, it will be called a *functor* with respect to this range. Thus the descriptional function 'the father of $\hat{x}$' is a functor with respect to the range given by all human beings.

Among the functors, the unique mathematical functions are of particular importance. They are descriptional functions of a higher type, having numbers as arguments and as descripta. Usually they are functors with respect to real numbers as their range. Thus when we write a mathematical equation in the form [1]

$$y = f'(x) \tag{9}$$

the symbol 'f'' represents a functor with respect to real numbers as the range of its arguments. The range of the descriptum y may be more comprehensive; it may include complex numbers, for instance when the functor 'f'' is given by the square root. Functors whose descripta are numbers may be called *numerical functors.*

---

[1] In order to conform to the usual mathematical notation we do not express the type of the variables by Greek letters; we deviate from the mathematical notation only in using the prime mark on 'f' for the indication of the descriptional character.

Equation (9) as a whole represents a propositional functional which can be written

$$f(y, x) \qquad (10)$$

The propositional character is clear from the fact that, for special values of x and y, (10) is either true or false. In mathematics, propositional functionals are always written in the form (9), i.e., in terms of descriptional functions, or functors. The restriction to functors is the reason that symbols of propositional functions are not needed in the usual mathematical notation.

Unique mathematical functions with several variables represent functors of these variables. Thus the product of two numbers x and y constitutes a functor '$f'(\hat{x}, \hat{y})$'. A nonnumerical mathematical functor is the relation *differential quotient*, which coordinates a mathematical function to a mathematical function.

The descriptional functions so far considered determine descripta of the same type as their arguments and may be called *homogeneous*. There is a second kind of descriptional function which determines a predicate in terms of an argument of this predicate; such descriptional functions will be called *heterogeneous*.

These functions are derived from corresponding descriptions of predicates. Consider, for instance, the description 'the color of this house'; it is written

$$(\imath f)f(x_1).\gamma(f) \qquad (11)$$

where 'x_1' stands for 'this house', and 'γ' for the predicate pertaining to color predicates. This description can be abbreviated by

$$f''(x_1) \qquad (12)$$

Putting for 'x_1' the variable 'x', we obtain the heterogeneous descriptional functional

$$f''(x) \qquad (13)$$

meaning in our example 'the color of x'. Formula (13) presents the functional in the form of a contracted symbol; its explicit form is given in the right-hand side of the definition

$$f''(x) =_{Df} (\imath f)f(x).\gamma(f) \qquad (14)$$

A heterogeneous descriptional function is also presented by a form of description in which the definite article is combined with a plural noun. Thus the phrase 'the brothers of x' determines the class of brothers of x. The condition of uniqueness required for descriptions is satisfied for class descriptions *ipso facto* since there is always one and only one class of the described kind, even if it is the null-class. However, we shall speak of a proper use of the description only if the class is not empty, since only on this condition contraries (cf. 13, § 19) like 'the brothers of x are Germans' and 'the brothers of x are not Germans' are not both true. In class descriptions the definite article assumes the meaning of 'all'; thus 'the brothers of x' means the same as 'all the brothers of x'. The meaning of 'all' is found also in simple combinations of the definite article with a class name in the plural form, as in 'the Greeks were polytheists'.

The use of numbers as descripta of descriptional functions is not restricted to mathematics. There are properties of physical things which are expressed by means of numbers. Thus an individual motion-property f can be characterized by a number indicating the speed. This method has great advantages over the use of names for predicates; it presents the different properties in numerical order. Usually the numerical predicate is defined for a range of physical objects, and the descriptional function then is a numerical functor with respect to this range. The range itself then consists in nonnumerical objects, and the functor is of the heterogeneous kind.

As an example of a numerical predicate, let us consider the sentence 'x_1 moves at 50 miles an hour'. It can be written, by analogy with (2, § 53),

$$m_{50}(x) =_{Df} (\exists f)f(x_1).\mu(f).(f = 50) \qquad (15)$$

Here the term '$(f = 50)$' takes the place of the adverbial term '$\sigma(f)$' in (2, § 53). Solving the statement (15) for 'f', i.e., introducing descriptional notation, we obtain

$$(\imath f)f(x_1).\mu(f) = 50 \qquad (16)$$

Introducing the abbreviation (14) we obtain

$$f''(x_1) = 50 \qquad (17)$$

Here the numerical functor '$f''(\hat{x})$' signifies 'the speed of $\hat{x}$'.

Numerical functors of the heterogeneous form are widely used in the natural sciences. Terms like 'velocity', 'temperature', 'pressure', and 'weight' are functors of this sort. The numerical values so introduced are then used for the construction of mathematical relations; i.e., the physical functors enter as arguments into mathematical functors, namely, mathematical functions.

Conversational language employs numerical predicates to a large extent, but frequently with the restriction that only relations of order are asserted, the numerical values themselves being left open. Such use is expressed in the three *comparative forms* which language has developed for functional terms. The first two of these forms may be comprised under the name *comparative relations*. One represents an *equality relation* and is expressed by means of the phrase 'as . . . as'; the second constitutes a *difference relation* and is given by what grammar calls the comparative degree. Both forms represent instruments of constructing from a certain function another function which has one argument more, for instance, of constructing a two-place function from a one-place function. These two forms, therefore, should be coordinated in a grammatical classification. The existence of the first of these forms is overlooked in traditional grammar, which mentions only the comparative degree in its three degrees of adjectives. The third comparative form, which is represented by the superlative degree, is of a different nature; it constitutes a description and may therefore be called a *comparative description*. Incidentally, the comparative forms are not restricted to adjectives but can be applied to verbs and nouns, though then frequently the word 'much' is added; thus, we say, 'Peter resembles you more than John', or 'Peter admires you as much as John'. In these examples, the comparative relations are used to construct a three-place function from a two-place function.

The construction of comparative forms is restricted to predicates that admit of a serial order, and it can be analyzed as follows. In stating the sentence 'Peter is as tall as Paul', we assume that all specific properties of tallness can be ordered in a linear series, and then assert that the specific tallness is the same for both Peter and Paul. In symbols, with 'τ' for 'tall',

$$(\exists f)(\exists g)f(x_1).g(y_1).\tau(f).\tau(g).(f = g) \qquad (18)$$

Similarly, the sentence 'Peter is taller than Paul' can be written in the form

$$(\exists f)(\exists g)f(x_1).g(y_1).\tau(f).\tau(g).(f > g) \qquad (19)$$

The statement (18) can be regarded as a definition of the two-place relation 'as tall as', which we can write in the form

$$ats(x_1, y_1) \tag{20}$$

This means we can define 'as tall as' by using (20) as the definiendum and (18) as the definiens. Likewise statement (19) can be regarded as a definition of the complex function 'taller than', written in the form

$$tlr(x_1, y_1) \tag{21}$$

A similar analysis can be given for the above examples in which the comparative relations constitute three-place functions.

The third comparative form, or comparative description, called the superlative degree by grammarians, reveals its different nature in the following symbolization of the phrase 'the tallest man', with 'm' for 'man':

$$(\imath x)m(x).(y)[m(y).(y \neq x) \supset tlr(x, y)] \tag{22}$$

This form is sometimes used in a different way, namely, as a description, not of an individual argument-object, but of a specific property. Thus we say, 'Jean is most happy', meaning that Jean rejoices in the highest degree of happiness. By such a statement we wish to say that Jean is happier than *or* at least as happy as anyone else; and we therefore determine, by this description, not an individual person, but a specific form of happiness. The statement can therefore be symbolized in the form, with 'δ' for 'happiness',

$$(\exists f)f(x_1).\delta(f).\{f = (\imath g)(\exists x)g(x).\delta(g).(y)(h)[\delta(h).h(y) \supset (h \leqq g)]\} \tag{23}$$

The statement can also be symbolized in a different form, when we use 'ahs' for 'as happy as', 'hr' for 'happier', and 'm' for 'human being':

$$(\exists f)f(x_1).\delta(f).(y)[m(y) \supset hr(x_1, y) \vee ahs(x_1, y)] \tag{24}$$

The two formulations (23) and (24) are equivalent.

For the transition from relations of order to numerical functors language offers specific noun forms, such as the functors 'the height of ...' and 'the weight of ...'. Nonnumerical functors, too, are expressed by nouns, or by forms of verbs èquivalent to nouns. In languages possessing a definite article the functor is indicated by the definite article that precedes the description; but other languages also employ the noun form. Descriptional functions determining arguments can be constructed from every propositional function which can be made univocal with respect to one argument. Functions like 'father', 'king', and 'author' are almost exclusively used as descriptional functions; i.e., statements containing such terms have the form (5). Regarding descriptional functions of the heterogeneous kind we must consider the fact that such functions are of a lower type than the propositional function from which they are derived. Thus in the expression 'the color of this house' the phrase 'the color' is a descriptional function of the first type, whereas the word 'color' is a propositional function of the second type.

For descriptional functions derived from verbs, language provides special verb forms, the verbal nouns, which are constructed by means of participles or infinitives. These descriptional functions may be divided into two groups. In the first group are phrases like 'the dancing of x', which can be regarded as describing either a property of x or the event-argument coordinated to the sentence 'x dances'. In the latter interpretation the phrase 'dancing of x' represents a fact-function, which by the addition of the definite article 'the' is used for the description of the fact-argument. For functions of two variables such descriptional functions can also be constructed, as is seen from examples like 'the defending of x by y'. The *gerund*, as the verbal noun constructed from the present participle is called, is therefore a descriptional function. The German language uses the present active infinitive for such functions, as in 'das Tanzen von x'. Latin, like other ancient languages excelling in exceptions to its rules, expresses a descriptional function of this kind by the present active infinitive, if it is a nominative, and by the gerund, if it is another case of the declension; thus 'scribere' is 'the writing', 'scribendo' is 'by the writing'. The fact that the descriptional functions of this group admit of two interpretations, namely, as describing either a function or an event, is perhaps explicable on the ground that the distinction is irrelevant for many purposes. Thus we saw in § 53 that the adverb can be regarded either as a property of a property or as a property of an event. On the other hand, there are languages that have different forms interpretable as indicating this distinction. We find such forms in the Altaic languages. Turkish, for instance, possesses two such kinds of verbal noun. Thus from the verb 'yürümek', meaning 'to run', the two forms 'yürümesi' and 'yürüdüğü' are derived. The first means 'his running' and refers to the property; the second means 'his having run' and refers to the act of running, i.e., to the event.

In the second group are descriptional functions determining a thing-argument. Here all forms of the participle are used, in English by the addition of the word 'one', indicating the thing-variable. Thus we say 'the beaten one' and 'the beating one'. The German language uses the corresponding participles, without the addition of a term expressing the variable: 'der Schlagende' and 'der Geschlagene'. Latin, which has a future participle, can use this form also as a descriptional function, as in 'morituri', meaning 'those who will die'.

In addition to the verbal nouns in the grammatical sense there are other nouns derived from verbs, used for descriptional functions; logically speaking, they are of the same nature as the verbal nouns and should therefore be incorporated into the same category. Nouns belonging in the first group are forms ending in the suffix 'ion', like 'annihilation'. In the second group we have forms ending in 'er' or 'or'; thus we have the form 'painter' in English, the form 'victor' in English and Latin, the form 'Sieger' in German.

In descriptional functions derived from a one-place function the place of the variable is indicated in many languages by the genitive case of the argument. If two-place functions are used, one argument has the genitive, and the other has the case employed with the passive form of the verb. Thus we say 'the defending *of* the country *by* the soldiers'; 'das Verteidigen des Vaterlands durch die Soldaten'. The genitive has developed this signification from its original meaning of possession, evident in phrases like 'the house of *x*'. Not all languages, however, apply the genitive for this purpose. Latin, for instance, uses, in the argument of a descriptional function, the case employed with the propositional function, as in 'defendere patri*am* militi*bus*'. [Ex.]

§ 55. Logical terms in a syntactical capacity

Of the three categories into which we divide the parts of speech of the object language — argument terms, functional terms, and logical terms — we have so far dealt with the first two. Turning now to the third, we begin our analysis with a definition of this category.

We shall call a term *denotative* when it stands in the place of an argument variable, or a functional variable, or a propositional variable. In the symbolism of logic, this definition means that special values of variables, such as 'x', 'f', 'a', are denotative. The different types of variables are distinguished by the use of different groups of letters for each type. In conversational language we use only one sort of symbol for all kinds of variables, namely, pronouns such as 'any', 'one', and the indefinite article 'a'. The distinction of types, then, either is indicated indirectly by the nature of the function to which the argument belongs — we say, for instance, 'a man' and also 'a color' — or is indicated directly by the addition of a term denoting the class constituted by all special values of the variable, as in phrases like 'a property', 'a thing', 'anything'. In these forms, the indefinite article means the same as the indefinite pronoun 'one' (not the number); namely, 'a man' means 'one among the men', and 'a property' means 'one among the properties'. Similarly, we also say 'a situation', or 'a state of affairs'. This usage makes the nature of the variable sufficiently clear, and we can therefore apply our definition of denotative terms also to conversational language.

In addition to denotative terms, language possesses a great variety of *expressive* terms. We shall say that a term is expressive when it is not used as a denotative term. Let us make clear the nature of these terms by means of an example. In the sentence 'Peter is tall' the words 'Peter' and 'tall' are denotative; the argument term 'Peter' denotes Peter, and the functional term 'tall' denotes the property of tallness. The fact that Peter has the property of tallness constitutes a relation between Peter and tallness; let us call it the function-argument relation. In order to state this fact in a sentence we use the two names 'Peter' and 'tall' and put them into a relation, which, being a relation between signs, is a syntactical relation. We could simply use juxtaposition for this purpose, and say, as children do, 'Peter tall'. The spatial sequence of the two words then constitutes a relation between these signs which portrays the function-argument relation holding between the corresponding objects. To show the sign relation more clearly, the

word 'is' is used in conversational language. This word portrays
the function-argument relation in the same way as juxtaposition.
The function-argument relation holds between the objects of the
words between which the word 'is' stands. The use of a special
sign, like 'is', carries an advantage in that it makes the notation
unambiguous; juxtaposition, in other sentences, does not play the
role of a portrait of the function-argument relation.

Although the word 'is' *portrays* the function-argument relation,
it does not *denote* it. The reason is that, according to the rules of
language, the denotation of relations is accomplished in a different
way. Let us make this clear by the use of the symbolic notation,
in which the distinction is more easily recognizable.

The sentence 'Peter is tall' is symbolized by

$$f(x) \tag{1}$$

Here the parentheses play the part of the copula 'is'; they portray
the function-argument relation. A sentence in which a relation is
denoted is 'Peter loves Mary'; it is symbolized by

$$f(x, y) \tag{2}$$

Here the symbol 'f' denotes the relation love. We see that denotation of a relation is achieved by putting the names of the arguments
into parentheses and placing the name of the relation, as the functional term, before the opening parenthesis. When we wish to
introduce a denotative symbol for the function-argument relation,
we must write the relation in the form

$$\alpha(f, x) \tag{3}$$

Here the sign 'α' denotes the function-argument relation. We used
this way of writing in (6, § 41), for technical reasons. But we see
that in order to introduce a name of the relation we must also use
parentheses, i.e., a sign that merely portrays the function-argument
relation.

We see now why we regard a sign like the parentheses, or the
copula 'is', as an expressive sign. Because it portrays the function-
argument relation, it will *express* it; i.e., it will remind us of this
relation. But it does not *denote* it because denotation, in our language, is accomplished in another way, namely, by putting a sign
for the object in the place of a functional or argument variable.

The above example exhibits three important features. First, it shows what an expressive sign is in contradistinction to a denotative sign. Second, it shows that an expressive sign can be supplanted by a denotative sign. Third, it demonstrates how such supplanting may be accompanied by the reappearance of a corresponding sign in the form of an expressive sign. With reference to the last property we shall speak of an *indispensable* expressive sign.

Not all expressive signs are indispensable. We might, for instance, replace the denotative sign 'red' by an expressive sign; we might write the argument '*x*' in red ink, and introduce the rule that such usage means '*x* is red'. It is clear that this expressive sign is dispensable since a corresponding sign does not reappear when we use the form of writing '*x* is red'.

To regard the copula 'is' as an expressive sign like the parentheses in (1), and not as a denotative sign, requires some justification. The syntactical rules of conversational language are not so clearly given as those of the symbolic notation. In symbolic notation, a functional sign is indicated by its position before the opening parenthesis. In conversational language, the functional signs are indicated by the fact that they are verbs, or combinations of a noun or adjective with a form of the verb 'to be', which for this reason is called an auxiliary verb. We may therefore say that it is the syntactical nature of the verb to indicate a function. On the other hand, since the copula does not constitute a function when used alone, but does so only in combination with adjectives or nouns, it appears justifiable to regard the copula as performing only the syntactical task of indicating the function, while the content of the function is stated in the adjective or noun. We thus consider the forms '*x* sleeps' and '*x* is sleeping' as of a similar logical nature. The copula then appears as the analogue of the suffix 's' used in English for the third person of the singular. We may mention here that the Turkish language adds the copula as a suffix to the adjective in the same way as it adds the personal suffix to a verb root; thus 'güzeldir' means 'he is beautiful', where 'dir' is the copula; and 'gidiyor' means 'he goes', where 'iyor' is the suffix of the third person. The copula, originally an independent verb, has entirely degenerated into a suffix.

Conversational language also possesses the means of supplanting the copula by a denotative sign. Thus we can say: 'Peter has the argument-function relation with respect to tallness'. The phrase 'argument-function relation' is the sign denoting the corresponding relation. That in the given formulation we have the analogue of (3) is seen from the suffix 'ness' in 'tallness', which indicates that 'tallness' is used here, not as an independent function, but as an argument of another function. As in (3), an expressive term for the function-argument relation reappears, namely, in the verb 'has'.

On the other hand, the use of verbs as functional terms shows that we cannot always separate an expressive term from a denotative term. In the sentence 'I see the tower' the verb 'see' has no suffix which we can regard as expressing the functional character. That character is rather expressed by the fact that the word 'see' belongs to a category of words which are used to indicate the tense. We may compare this mode of expressing a structural feature to a notation which uses letters 'f', 'g', for functions, and letters 'x', 'y', for arguments, but does not use parentheses. There are other forms of combining a denotative and an expressive nature in one term. Actually, most denotative terms have a pragmatic component of an expressive nature. Thus the word 'shambles' expresses a derogatory connotation, which, however, does not belong to its denotative meaning. The terms that interest us here are terms that are *merely expressive*.

The merely expressive terms perform a certain linguistic function, which, however, does not consist in denoting. They produce certain sign structures and thus *do* something; but they do not *say* something. They can only *express* features existing outside the realm of signs. Such signs are usually constructed, in symbolic language, not in the form of letters, but in the form of special designs, such as parentheses or the symbols of propositional operations. The sign 'ϵ' is an exception to this rule. In conversational language, suffixes or word order frequently take the place of such signs. In written language, again, special designs are used, like punctuation; in spoken language, gestures and intonation are applied. But we also employ words as expressive signs, such as the word 'is', and the words 'and', 'or', used for propositional operations. The separation of merely expressive terms from denotative terms by special sign patterns appears advisable from the viewpoint of the logician.

Because of their expressive nature, expressive signs are not included in the theory of types. Thus the copula 'is' does not belong to any type level. This is a further reason for regarding the copula as an expressive term. Were we to regard it as denoting the function-argument relation, the sentence 'Peter is tall' would have the form (3) and therefore would belong in the higher calculus of functions. The use of expressive terms enables us to incorporate into a lower level certain terms which, when used in a denotative way, would belong to a higher level.

We may now proceed to the definition of logical terms. We shall call a term *logical* when it is merely expressive and indispensable, or when it is definable by means of such terms alone.

The addition has the important consequence that it includes among logical terms certain denotative terms also. Thus the denotative term 'function-argument relation' is a logical term because it can be defined by indispensable and merely expressive terms, as is shown in (3). We may call it the denotative correlate of the copula. We shall later give further examples of such *derivative* logical terms which are denotative.

In our definition of logical terms we attempt to overcome an ambiguity which so far has been attached to this name.[1] Usually the definition of logical terms is given by the enumeration of such terms, without a statement of a common property, the incorporation of further terms being left to judgments by analogy. We know that our definition, too, does not completely remove this ambiguity. In particular, the word 'indispensable' in our definition may be applied in various ways; what is regarded as indispensable will depend on the purpose for which a language is constructed. We shall presently have some occasion to state more precisely what purposes we have in mind when we speak of indispensable expressive signs.

Let us make this clear by means of another example. The all-operator is an expressive sign, since it does not stand in the place of a functional or argument variable. As for the function-argument relation, we can introduce for the all-operator a corresponding denotative sign; this is done by the use of an α-function of the

---

[1] This was remarked by R. Carnap, *Introduction to Semantics*, Harvard University Press, Cambridge, 1942, § 13.

form (18, § 41). Let us use the abbreviation '*Un*' for the word 'universal'; we then can regard formula (18, § 41) as defining the statement that a property is universal, and we can write this definition in the form

$$Un(f) =_{Df} (x)f(x) \qquad (4)$$

The function '*Un*' is a denotative term; it is a function of the second order.

Now it is possible to supplant the all-operator throughout by the sign '*Un*'; and, since free variables can always be eliminated by the use of an all-operator, we can thus construct a notation in which generality is always expressed by a denotative term. For instance, instead of the relation

$$(x)[f(x) \supset g(x)] \supset [(x)f(x) \supset (x)g(x)] \qquad (5)$$

we would write

$$Un[f(\hat{x}) \supset g(\hat{x})] \supset \{Un[f(\hat{x})] \supset Un[g(\hat{x})]\} \qquad (6)$$

When we do not use an all-operator in our notation, the function '*Un*' must be regarded as a primitive term. But it will not be a logical term in the sense of our definition, since then it is not definable by expressive terms. On the other hand, formula (6) will be true only for the function '*Un*'; it will not remain true when we put a functional variable in the place of '*Un*'. We thus have in (6) a formula that can be verified without empirical observation, but that does not remain true when a denotative term occurring in it is replaced by a corresponding variable. Only if we could regard the function '*Un*' as a logical term would this consequence represent no inconvenience. In order to reach this objective we introduce the following rule for our notation:

Notational principle. When a formula can be verified without empirical observation, it must be possible to write the formula in such a form that, if all the denotative constants are replaced by corresponding variables, the formula remains true.

This rule makes it necessary to introduce an expressive sign for generality, such as the all-operator. Formula (6) will not be regarded as impermissible in our notation; but it is permissible only because we have an equivalent way of writing, given in (5). The term '*Un*' is thus made a logical term since we can use now the definition (4). We see that the notational principle represents a postulate which determines what we regard as indispensable expressive terms. Some additions concerning the term 'indispensable' will be given later.

The use of an expressive sign for generality carries the advantage that it allows us to write a formula like (5) in the simple calculus of functions. Formula (6) is in the higher calculus since the function '*Un*' is of the second type. We see again that the use of an expressive sign frees us from the restrictions established for functional terms in the theory of types. All these considerations hold similarly for the existential operator.

The copula and the all-operator, which we have used so far as illustrations of our discussion, constitute logical terms of different kinds. The copula acts in a *syntactical capacity* since its task consists in establishing a relation between signs. It indicates which sign is the argument and which the function. The all-operator, on the other hand, acts in a *semantical capacity* since it extends the truth range of the statement to all things. Both kinds of terms, however, are expressive terms of a *cognitive* nature. Logical signs of a third kind, which act in a *pragmatic capacity*, will be discussed later; they constitute expressive terms of an instrumental nature.[1] Let us use this division for a classification of logical terms.

The first class to be considered is composed of *logical terms acting in a syntactical capacity*. Let us divide this class into subclasses and thus enumerate the various kinds of such terms. The two other classes will be dealt with in the following sections.

In the first subclass we incorporate terms *referring to arguments*. Among these we first find terms indicating the place of the variable, including inflections of nouns, and prepositions. Prepositions usually not only have this logical function, but also take over a part of the meaning of the statement; thus 'it lies on the table' and 'it lies under the table' differ as to the space indication. Some prepositions are therefore better incorporated into propositional functions of a higher type, along with the adverbs. Second, we have terms

---

[1] Professor R. Carnap has suggested, in a letter to me, that the distinction between syntactical and semantical capacity be dropped and that these two kinds of logical signs be regarded as forming one group. We then would distinguish only between cognitive and instrumental logical signs. I should like to leave this question open for discussion.

representing variables. They are expressive terms, according to our definition, because only special values of variables denote. In conversational language, variables are represented by pronouns. They include the personal pronouns 'he', 'she', 'it' [1] (but not pronouns like 'I' and 'you', which are token-reflexive descriptions, to be classified among arguments), relative pronouns, and possessive pronouns. Possessive pronouns express, jointly, a function, namely, possession; they are therefore contracted symbols. Furthermore, we incorporate here the indefinite pronouns, such as 'somebody', 'any', 'one'. Among these, the word 'any' expresses generality in terms of a free variable, as explained in § 21, and therefore belongs partly in the semantical category. The same holds for the word 'somebody', which may express an existential operator. The examples show that these categories are not mutually exclusive.

The next subclass is composed of terms *referring to functions*. Here we have terms expressing the converse of a function, as the passive voice; furthermore, terms expressing functional modifiers, such as the suffix 'ly' and the term 'in such a way'.

We then find a subclass including terms *referring to argument and function*. Among them is the copula 'is', which we discussed above as a means of expressing the function-argument relation. In combination with the indefinite article 'a', i.e., in the form 'is a', it expresses class membership, as in 'John is an American citizen'; in the combination 'a . . . is . . . a' it expresses class inclusion, as in 'a person born in the United States is an American citizen' (cf. p. 198).[2] If, however, the indefinite article is used without the copula, it belongs among semantical terms, since it indicates an existential operator (cf. 6, § 18). Other terms of this subclass are given by terms indicating event-splitting, such as the conjunction 'that', or the suffixes 'ion', 'ing' (cf. § 48).

Another subclass of syntactical terms is given by some marks of *punctuation*, in particular the comma and the semicolon (the period

---

[1] These pronouns constitute variables with a restricted range, like our variable 'x'' used in (20, § 30), since the gender expresses a restriction to a certain category of objects.

[2] It is perhaps more correct to say that the whole combination 'a . . . is . . . a' expresses class inclusion. The copula 'is' within this combination may be regarded as expressing class membership, namely, when the sentence is construed in the form 'an x which is a member of the first class *is* a member of the second class'.

belongs to pragmatic terms, as will be explained below). They indicate the units of which sentences are composed; but they do so simply by expressing these units, not by speaking about them. Punctuation, therefore, belongs in the object language. In fact, it would be impossible to construct an object language without punctuation. Punctuation is necessary also for symbolic language, where its function may be taken over by parentheses.

Finally, there is a subclass of denotative syntactical terms, i.e., terms which denote and by logical analysis are shown to be reducible to expressive syntactical terms. Here we have the function-argument relation, which we denoted by 'α' and defined in (6, § 41). More terms of this kind will be found in the semantical category.

§ 56. Logical terms in a semantical capacity

We now turn to the second class of logical terms, which consists of terms acting in a *semantical capacity*. The first subclass of this kind is given by the *propositional operations*. We regard these operations as semantical because of their definition by means of the truth tables, which confers on them a truth-functional character. The semantical nature of the propositional operations is also evident from the fact that we can construct true formulas which contain only variables and propositional operations. That, in particular, connective operations are semantical is seen from the fact that according to their definition only true statements can contain connective operations (cf. § 9 and § 61). On the other hand, this interpretation of connective operations shows that these operations should be regarded as improper object terms; their precise definition is given in the metalanguage. Nevertheless we classify such operations here because they are used in the object language.

Let us now turn to the question whether the propositional operations should be used as expressive or as denotative terms. The symbolic way of writing corresponds to an expressive usage. If we were to write a propositional operation, for instance the 'or', as a denotative term, we would have to use the form

$$or(a, b) \qquad\qquad (1)$$

The word 'or' is then regarded as denoting a function which has the situations a and b as arguments. Functionals of the form (1) would have the peculiarity that they could be used as arguments in functions of the same kind; thus we could write

$$or[a, or(b, c)] \qquad (2)$$

This would be the same as

$$a \vee (b \vee c) \qquad (3)$$

The mode of writing used in (2) does not violate the theory of types. Although the function 'or' is of a higher type than its arguments, the expression '$or(a, b)$' is of the same type as 'a' or 'b', since the type of the function is not transferred to the proposition in which it occurs (cf. § 40).

The difficulties connected with the use of propositional variables as arguments of functions were discussed in § 42. But we can adduce even better reasons in favor of an expressive use of propositional operations. It is clear that when we write a tautology in a form that uses the propositional operations as functional constants, as in (1) and (2), it is not permissible to replace these constants by functional variables since the formula would then become false. Our notational principle (cf. § 55), therefore, compels us to introduce the propositional operations as expressive terms. The mode of writing used in (1) and (2) is then also permissible; only we must regard the functional constants of this kind as defined by means of the expressive terms. They are therefore also logical terms. Thus we must write

$$or(a, b) =_{Df} a \vee b \qquad (4)$$

The expressive nature of the propositional operations is shown in symbolic logic by the use of signs of special design put between the letters 'a', 'b'; the omission of parentheses around these letters indicates that we do not use here the function-argument relation. In conversational language the absence of the function-argument relation is indicated by the fact that the conjunctions are not verbs, adjectives, or nouns, i.e., that they are not subject to verbal inflection nor construed with the copula. At least this rule holds for the most important conjunctions, such as 'and', 'or', 'if . . . then', 'unless'.[1] In some established forms, however, we do use verbs;

---

[1] The word 'unless' is usually regarded as meaning 'if not'; it then means the same as the inclusive 'or', since 'a unless b' then is the same as '$\bar{b} \supset a$' or '$a \vee b$' (cf. 6a, § 8). Sometimes, however, 'a unless b' is used with the additional meaning '$b \supset \bar{a}$', or '$\bar{b} \vee \bar{a}$'. Because of (2, § 10), 'unless' then has the meaning of the exclusive 'or'.

thus we formulate implication by the verb 'imply', as in 'that the man was working on his job implies that he is not the murderer'. The verb 'implies', therefore, can be regarded as a denotative form definable in terms of the expressive 'if . . . then', i.e., the logistic 'כ', by analogy with (4).[1] Similarly it would appear logical to say: 'that the man was working on his job excludes that he is the murderer', the verb 'excludes' representing a denotative form of the stroke operation, which means 'not both', or 'not . . . or not . . .' (cf. § 8). That English grammar rules out such usage for the verb 'exclude' and requires the insertion of a noun, as in '. . . excludes the possibility that . . .', indicates an antagonism against a denotative form of propositional operation.

Among the propositional operations we shall therefore have not only the conjunctions of grammar, but also verbs like 'imply' and 'mean', and the adjective 'equivalent'. On the other hand, the grammatical conjunctions include terms which are distinguished from one another, not by their semantical functions, but by additional pragmatic connotations of an instrumental nature. Thus 'furthermore' means 'and' with the indication 'the following statement is similar to the preceding one'; 'but' means 'and' with the indication 'the following statement seems to contradict the preceding one without doing so'; 'although' means the same as 'but', but is added to the other of the two connected sentences, so that 'a but b' means the same as 'although a, b', or 'b although a'. The sentence 'b' carries the emphasis in this usage, both for the 'but' and the 'although'; i.e., 'b' is the unexpected sentence. The hints to the reader expressed by such pragmatic connotations make it easier to understand the text.

There are some other grammatical conjunctions whose additions to the logical conjunctions are not pragmatic but of a different nature. Thus the word 'because' adds to the meaning of 'and' a statement about a causal relation between the two facts to which

---

[1] When the clauses introduced by 'that' in the above example are interpreted as descriptions of events (cf. § 48), the verb 'implies' can be regarded as stating a relation between events. Some logicians prefer an interpretation in which a verb expressing a propositional operation is regarded as performing an operation in the metalanguage. Thus W. V. Quine wishes the word 'implies' to be used only between names of sentences, not between sentences. Cf. his *Mathematical Logic*, Norton, New York, 1940, p. 29.

the sentences refer. Such conjunctions would be better classified as contracted terms combining the expression of a logical conjunction with the denotation of a function. Thus 'because', in addition to expressing 'and', denotes the fact-function of causality, like the verb 'to cause'. A similar meaning is given by the word 'since' in its nontemporal usage.

The word 'except' is not a proper conjunction. The analysis of this word is rather involved. Consider the sentence: 'all soldiers except the wounded were withdrawn', which has the form: 'all F except the G are H', or written in functions, '$(x)\{[f(x)$ except $g(x)] \supset h(x)\}$'. Using a free variable instead of the all-operator and introducing the abbreviation 'ex' for except, we can define this term as follows:

$$[f(x) \ ex \ g(x)] \supset h(x) =_{Df} [f(x).\overline{g(x)} \supset h(x)].[f(x).g(x) \supset \overline{h(x)}]$$

Applying (6d, § 8) to each of the brackets, and using (6e and 6a, § 8) and (2, § 10), we can introduce the exclusive 'or', and thus arrive at the definition

$$[f(x) \ ex \ g(x)] \supset h(x) =_{Df} f(x) \supset [g(x) \wedge h(x)] \tag{5}$$

When we now regard the term 'ex' as a conjunction, we find the following two peculiarities. First, this conjunction is used only between functionals of the same argument, not between other forms of propositions or between propositional variables. Thus we do not write '$(a \ ex \ b) \supset c$'. Second, even if the latter form of writing were used, it would not be possible to define the conjunction 'ex' by means of a truth table. That is, if 'c' is true, the expression '$(a \ ex \ b) \supset c$' will be true whatever the truth value of '$a \ ex \ b$' may be, while the expression '$a \supset (b \wedge c)$', if 'c' is true, will be false if both 'a' and 'b' are true. The term 'except' may therefore be called a *pseudo-conjunction*. It is an abbreviation defined as explained above.

A second subclass of logical terms in a semantical capacity is given by the operators. We classify the operators 'all' and 'there is', or 'some', in this category because they can be conceived as extensions of the operations of conjunction and disjunction, as was shown in (5 and 6, § 19). They therefore belong in the same group as the propositional operations. The iota-operator must also be incorporated in this group because it is defined in terms of the other operators. For this reason the definite article 'the' belongs in this category, along with the indefinite article 'a'. Furthermore, we classify the word 'any' here since it expresses generality in terms of a free variable (cf. § 21). As stated above, this word belongs partly in the syntactical category. The plural suffix, too, belongs in this group; it expresses the condition that there is more than one satisfying argument.

A third subclass is given by *denotative semantical terms*, i.e., terms which denote and by logical analysis are shown to be reducible to expressive semantical terms. In this group we have, first, the denotative terms constructed by means of propositional operations, like the verb 'imply'. It is convenient, however, to classify these terms along with the expressive forms of propositional operations, as was done above, since these terms are used interchangeably. Second, we have the denotative terms constructed from operators. A term of this kind is the word 'universal' introduced in (4, § 55). Another denotative logical term in a semantical capacity is the term *physical object*, which we use to include both things and facts. Since we have shown in (12, § 41) that every physical object has a property f, we can define a physical object as something having a property f. When we abbreviate the phrase 'physical object' by '*phob*' we can therefore define:

$$phob(x) =_{Df} (\exists f)f(x) \tag{6}$$

Since the definiens is always true, the statement 'x is a physical object', or

$$phob(x) \tag{7}$$

is a tautology. Although the function '*phob*' is, of course, a constant and cannot be replaced by arbitrary functions 'f', the form of (7) does not contradict our definition of tautology given in § 41, since '*phob*' is a logical constant, not an empirical one. Furthermore, because of (4, § 55) and formula (12, § 41), we have the tautology

$$Un(phob) \tag{8}$$

which states that being a thing is a universal property.

Formulas (7) and (8) are to be understood as follows. We do not mean to say that we can put for 'x' in (7) any term we like, and then will arrive at a true statement. Thus, although music is loud and so has a property, it would not be permissible to say that music is a physical object. Music is a property of a higher type, namely, a class of tones between which certain relations exist, and therefore the term 'music' must not be substituted for 'x' in (7). If we did so, we would arrive, not at a false statement, but at a meaningless sign combination. Calling (7) a tautology we mean: any admissible substitution for 'x' will make (7) true, and no substitution can make it false.

The term *property* can be defined by the relation

$$Pr(f) =_{Df} (\exists x)[f(x) \vee \overline{f(x)}] \tag{9}$$

If we were to choose here as definiens the expression '$(\exists x)f(x)$', by analogy with (6), the totality of properties would be restricted to nonempty properties. The adoption of definition (9), therefore, extends the totality of properties to the inclusion of empty properties.

Denotative terms meaning existence can be constructed in various ways. Introducing a definition analogous to (4, § 55) in terms of the statement '$(\exists x)f(x)$' we define the *nonemptiness* of a property; synonymously we shall employ the word *occurrence*. We thus have

$$Oc(f) =_{Df} (\exists x)f(x) \tag{10}$$

This states that a property occurs when there is a thing having this property. Thus we say that malaria does not occur in England, meaning there are no cases of malaria in England. Occasionally the word 'existence' is used in this sense; thus the preceding sentence is sometimes phrased in the form 'malaria does not exist in England'. It appears advisable, however, not to use the term 'exist' here because the existential operator combined with the argument sign 'x' in (10) is transformed into a property of the property f, and not of the argument. An existential property holding for things can be defined by means of the relation (19, § 43):

$$Ex(x) =_{Df} (\exists y)(y = x) \tag{11}$$

The function 'Ex' represents the denotative term 'existence'. It is this term that we have in mind when we speak of the existence of things. Such a predicative use of the term 'existence', therefore, is not meaningless; only it is tautological. We proved the tautological character of the definiens of (11) in (19, § 43).[1]

The existence of properties is defined in a similar way by

$$Ex(f) =_{Df} (\exists g)(g = f) \tag{12}$$

This function *existence* is, of course, one type higher than that appearing in (11). While '$Oc(f)$' is a synthetic statement, '$Ex(f)$' is analytic.

We see that denotative terms of an all-embracing generality, like 'existence', 'physical object', 'property', are legitimate and can be defined in symbolic logic. They satisfy our definition of logical terms, and are thus classified in accordance with general usage. In fact, the classification as logical terms appears justifiable because these terms merely make certain features of our language explicit.

Let us add a remark about the predicative use of the term 'existence'. It has sometimes been said that such usage is incorrect and leads to the fallacy committed in the ontological demonstration of the existence of God. This demonstration, so runs the argument, includes the existence of the object in the definition of the term, and then derives the conclusion that the object exists. But this criticism misses its point. There is no danger in including in a definition the property of existence. Thus we can define a sea serpent as a snakelike animal living in the ocean and having the property of existence. From this definition we can derive the conclusion that, if a thing is a sea serpent, it exists — which is a correct statement. In other words, we can derive the conclusion that all sea serpents exist. The fallacy comes in only when we make the inference from 'all' to 'some', and infer that there is a sea serpent.

We saw in our discussion of (13 and 14, § 36) that this inference can be made only when an additional premise is given which asserts the nonemptiness of the implicans of the all-statement. This premise, however, cannot be derived from a definition. To give an illustration, let us write '$ss(x)$' for 'x is a sea serpent', and '$sm(x)$' for 'x is a snakelike marine animal', then we may set up the definition in the form

$$ss(x) =_{Df} sm(x).Ex(x) \tag{13}$$

[1] Cf. also our remarks at the end of § 46.

The existential addition '$Ex(x)$' is quite harmless because it follows tautologically from '$sm(x)$' according to (5, § 18). From (13) we can derive

$$ss(x) \supset Ex(x) \tag{14}$$

$$ss(x) \supset Oc(ss) \tag{15}$$

These implications are tautologous when the meaning of the terms 'Ex' and 'Oc' is substituted according to (11) and (10). There is no way, however, to derive from (14) or (15) the unconditional statement '$(\exists x)ss(x)$', or '$Oc(ss)$', except for the case that the truth of the synthetic statement '$ss(x)$', for an individual instance, is established by observational methods. The ontological fallacy, therefore, does not originate from a predicative use of the term 'existence', but consists in the impermissible manipulation of an inference.

The definition of *occurrence* for properties of higher types is constructed by analogy with (10), with the proviso, however, that the properties which constitute the arguments of the property of a higher type must themselves be nonempty. We thus define

$$Oc(\varphi) =_{Df} (\exists f)\varphi(f).Oc(f) \tag{16}$$

Here the two signs 'Oc' are of different types. This definition can be written for all types of properties. It constitutes a recursive definition of occurrence, since it defines the occurrence of one type in terms of the occurrence of the next lower type. If a property of a higher type occurs, therefore, this occurrence is reducible to the existence of objects on the lowest level. We thus introduce for higher functions a distinction between occurrence and nonemptiness; for nonemptiness the condition '$(\exists f)\varphi(f)$' is sufficient. Thus a class which contains the null-class as its only member is not empty; but it does not have the property of occurrence.

Further examples of denotative semantical terms are given by the numerals, which denote numbers, i.e., classes of classes (cf. § 44). They are abbreviations for complicated logical expressions, as was shown in § 44. Furthermore, we have in this subclass the sign of the identity relation, which is a function of two variables definable in logical terms alone (cf. § 43). In conversational language, this relation is expressed by the word 'is'. We thus find four different meanings of the word 'is': first, it means the relation of class membership; second, it means the relation of class inclusion;[1]

---

[1] Cf., however, footnote 2 on p. 326.

third, it means existence, as in the phrase 'there is'; fourth, it means identity. In the last meaning, it is a denotative term.

Finally, we come to a subclass composed of *sign-denominative symbols*. It includes quotation marks since they transform a sign into a name of that sign, and thus perform a denominative function on signs; however, they do not say what they do. When we wish to say what they do we use such words as 'name', or 'denote', which denote. But the quotes themselves do not denote; it is their product which denotes, namely, the combination of word and quotes. We therefore can write

$$ 'x' = (\imath y)\nu(y, x) \tag{17} $$

where 'ν' means 'name constructed by means of quotes'. Here 'ν' is a denotative term; but the quotes are not. That they are not is evident also from the fact that we cannot regard (17) as a definition of the quotes in terms of the function 'ν'. On the right-hand side of (17) we have a description of the name of x; on the left-hand side, however, we have the name itself. The name is not an abbreviation standing for a description. Thus, although Napoleon is the victor of Austerlitz, the word 'Napoleon' is not an abbreviation standing for 'the victor of Austerlitz'. The quotes create a sign which is the name of a sign; in doing so, they express the relation of denotation, but they do not denote it.

These considerations make it clear that the quotes are an expressive sign. That they are indispensable does not follow from our notational principle (§ 55), but must be justified on other grounds. If we did not use the quotes operation we would have no linguistic device for the introduction of a name of a sign. We would have to learn names of signs by heart, as we learn the names of physical objects by heart. The technical limitations of this method are obvious. When we regard the quotes as indispensable we do so for reasons of linguistic technique. But even if a linguistic system without quotes were constructed, the task of the quotes would be taken over, not by terms of the language, but by the process of learning the names of signs. This consideration shows that the quotes are indispensable expressive signs in the sense that their task cannot be transferred to denotative terms.

A sign-denominative task similar to that performed by the

quotes is performed by the token-reflexive word 'this' (cf. § 50), which we therefore include in the same group. The word 'this' is thus classified as a semantical logical term, although in its usual meaning it is a contracted term, including in addition to the meaning of 'this token' that of certain propositional functions. Thus in the phrase 'this table' the word 'this' stands for the token-reflexive descriptional function 'the table pointed to by a gesture accompanying this token' (cf. p. 284). Other token-reflexive descriptions, which do not explicitly contain the word 'this' (though they do so implicitly), like 'I' and 'you', and the tense inflections of verbs, are incorporated by us in the class of arguments.

§ 57. Logical terms in a pragmatic capacity

The third class of logical terms consists of signs acting in a *pragmatic capacity*.[1] They make the sign combination an instrument of the speaker. Because they *do* this, they do not *say* it; they are, therefore, merely expressive of their instrumental function. That they are also indispensable follows from the fact that we have no other means of making linguistic expressions instruments. It is clear that this task cannot be done by denotative terms, since they are within the cognitive sphere and cannot transform a linguistic expression into something noncognitive. Let us analyze this in the example of the assertion sign.

In written language the assertion sign is supplied by the period at the end of a sentence, meaning: the writer asserts the sentence. In spoken language assertion is frequently expressed by a falling inflection, as in English. In this group belong also the two words 'yes' and 'no', which assert, or deny, another person's utterance; we shall explain this below. Russell has used the sign ' $\vdash$ ' in his symbolic language; he writes

$$\vdash p \tag{1}$$

in order to indicate that 'p' is asserted. That this sign is expressive, and not denotative, can be seen as follows. The sentence 'I assert 'p'' may be written in the form

$$as(I, \text{'}p\text{'}) \tag{2}$$

Now, if the assertion sign were denotative, (1) would mean the

[1] Cf. footnote 1 on p. 325.

same as (2), and could be defined by (2). But this is not possible
because (1) cannot be negated, whereas (2) can. We can only
write

$$\vdash \bar{p} \tag{3}$$

but not

$$\overline{\vdash p} \tag{4}$$

Combination (4) must be regarded as a meaningless expression.
We could, of course, introduce a rule according to which (4) would
mean the same as the negation of (2); then (1) would be the same
as (2). But such a rule would not correspond to the usage of the
assertion sign in conversational language. Furthermore, if we
should introduce such a usage, we would be compelled to introduce
another assertion sign, expressing that (4) is asserted. Language
cannot dispense with a merely expressive assertion sign (cf. also
§ 4). In symbolic logic, Russell's sign ' $\vdash$ ' is now usually abandoned;
instead, we follow the rule that a formula written on a separate
line is asserted. We can express assertion by this simpler device
because we usually do not write down false formulas. If, excep-
tionally, we write a false formula for the purpose of studying it, the
procedure is explained in the text, i.e., by an asserted statement of
the metalanguage which cancels the assertion sign for the other
statement.

This analysis also makes it clear that expressions including a
pragmatic sign are not propositions. They are not true or false, as
is shown by the fact that they cannot be negated. They are instruments constructed by the help of propositions, and they therefore
belong in language; this is what distinguishes them from other
instruments devised to reach a certain aim. They must therefore
be dealt with in a theory of language. But they cannot be substituted for the symbols '*a*', '*b*', etc., used in logical formulas. We
shall say that a sentence transformed by a pragmatic sign represents an expression in a *pragmatic mood*. Thus (1) is an expression
in an *assertive mood*, or an *assertive* expression.

Since assertive expressions are not propositions, they cannot be
combined by propositional operations. Therefore, two assertive
expressions in sequence do not represent a conjunction of these
expressions. Rather we must say that juxtaposition of two assertive expressions amounts to the same as the assertion of the con-

junction of the two respective statements. Thus the sentences in a book, each asserted by the period sign, follow one another; this arrangement amounts to the same as asserting the conjunction of all these sentences, each taken without the assertion sign. Juxtaposition, therefore, constitutes the pragmatic analogue of conjunction. The other binary operations do not have such analogues.

The negation, however, possesses a pragmatic analogue; it is the word 'no', employed in a corrective sense after an assertion is made, either by the speaker or another person. The word 'no' cancels the preceding assertion and replaces it by the assertion of the negation of the respective sentence; it thus expresses the transition from (1) to (3).

The *assertive* terms constitute the first subclass of pragmatic logical terms. While the period sign and the words 'yes' and 'no' constitute the first group within this subclass, the second group is given by the moods of verbs: the indicative, the subjunctive, and the conditional. Of these, the indicative expresses assertion, whereas the subjunctive and conditional express either absence of assertion or the assertion that the clause is false, i.e., the assertion of the negation of the clause. We have the first meaning of the subjunctive in a statement like 'if he be your friend, he will help you'. There is a tendency to abandon the subjunctive mood in such statements, at least in colloquial English, and to use the indicative. The absence of assertion is then regarded as sufficiently expressed by the 'if'. An analogous rule holds in the French language, which requires that the connective 'si' be construed with the indicative. Modern languages thus reduce the richness of expressions which they took over from the ancient languages. When the clause is denied, however, English still keeps to the subjunctive; that mood is then used in the past tense, as in 'if he were your friend, he would have helped you'. The second clause, which is in the conditional, is also denied.

Other languages are richer in moods than English. Turkish, for instance, possesses, besides subjunctive and conditional, a special mood expressing probability, i.e., a mood indicating that the truth of the sentence is none too well established; this mood is expressed by the suffix 'miş' (pronounce: mish). Thus 'gitmiş' means 'he probably went away', whereas 'gitti' means 'he went away'.

Of the historical origin of the moods expressing absence of assertion we have little knowledge. We know, however, that a past tense in an indicative mood, in combination with an 'if', can assume a modal function, as in 'if he had come'. Although the modal use of the past tense is frequently preceded, historically, by the use of the subjunctive (the French 's'il avait' has replaced the older form 's'il eût'), it may perhaps indicate the way in which a mood can originate. It is not unlikely that the modal meaning derives from the fact that the word 'if' is usually not applied to past events which actually have happened; since such events are known to us, we do not say '$a \supset b$', but '$a.b$'. Thus the combination of 'if' with the past tense is left to cases where the implicans is false.[1] Once this conversion of a past tense into a mood was adopted, the modal meaning was transferred to cases where the truth of the implicans is left open, as in 'if he came now . . .'. This conjecture would also explain the shift of the time value of such moods; 'if he came now . . .' expresses a present tense, and 'if he had come . . .' expresses a simple past.

Another source of moods expressing absence of assertion is indicated by terms that originally expressed obligation or volition, such as 'should' and 'would', now used for the conditional mood. It seems that the uncertainty of fulfillment connected with these terms is utilized to express the absence of assertion. Such usage appears appropriate, in particular, when the obligation refers to the past, since we ordinarily speak of a past obligation only when it has not been met; 'I should have gone' means 'I was obliged to go, but did not do so'. It seems possible that the subjunctive mood, which in addition to its modal capacity is used as an imperative form (to be explained presently), originated as an imperative and assumed its modal capacity by a similar shift of meaning. This supposition would explain the use of the present tense of the subjunctive, while the meaning of the past tense is perhaps reducible to a tense shift as described above.

In the second group of pragmatic logical terms we include *interrogative* terms, such as the interrogative pronouns and adverbs, and the question mark. In addition to special terms, language

---

[1] A somewhat different explanation is given by J. O. H. Jespersen, *A Modern English Grammar*, C. Winter, Heidelberg, 1931, Part IV, p. 114.

possesses another device to express a question: the inversion of the order of words. The English language restricts the application of inversion to the verb 'to do' and therefore expresses questions by a combination of the verb carrying the meaning with the verb 'to do'. The French language prefers to apply inversion to pronoun phrases instead of noun phrases, as in 'ton père viendra-t-il?' In some languages the function of the question mark is taken over by a special word, such as the Latin 'ne', and the Turkish 'mi'. Thus 'gittimi?' means 'did he go?', while 'gitti' means 'he went'.

That questions are in a pragmatic mood is clear from the fact that they express a desire of the speaker, namely, to get an answer from the listener. Their cognitive correlates have the form 'I wish to know the answer to this question'. The question itself, however, merely expresses this desire but does not say it. As before, we show the merely expressive character by referring to the fact that a question cannot be negated. A question is not a statement, but an instrument used for a determinate aim, namely, to acquire a certain knowledge. We therefore say that a question is an expression in an *interrogative mood*.

The knowledge desired is to be formulated in a statement; the question, therefore, must indicate which statement is intended. This indication is achieved by making the statement in an incomplete form; the listener is then invited to make it complete. Usually, the missing term is unknown to the speaker; only in examinations or other tests is it known, though not said. Since the missing term will be in one of the three main categories of the object language, we have three kinds of questions, according as we ask for an argument, a function, or a logical term.

In a question asking for an argument we state a propositional function, expecting the listener to supply the satisfying argument. The propositional function is chosen in such a way that there is only one satisfactory argument. We can symbolize this kind of question by

$$(?x)f(x) \tag{5}$$

The question-mark operator means 'which is the x such that'; it resembles the iota-operator in so far as it can be applied only to functions with precisely one argument.

In conversational language, the question-mark operator is expressed by an interrogative pronoun or interrogative adverb. The pronoun or adverb indicates the kind of argument inquired; thus 'when' stands for a time argument, 'where' for a space argument, 'who' for a person, 'which' for other things, and 'what' for events or event sequences (permanent states). Thus we ask 'who painted the Mona Lisa?', 'which is your car?', 'what do you want?', the last question asking for an action, i.e., an event, which is to be performed. Sometimes the functional '$f(x)$' in (1) is given in the descriptional form '$x = (\imath y)f(y)$', as in 'who is the painter of the Mona Lisa?' The pronoun 'which' has the additional connotation that it indicates selection from a restricted number. The word 'why' also asks for an event; however, it not only stands for the variable but also includes part of the description defining the inquired event, since 'why' means 'what is the cause of'. For the case that the function '$f(\hat{x})$' is satisfied by more than one argument we have no question form, but use other forms of expression, for instance the imperative, as in 'name a drama of Shakespeare'.

The second kind of question is one asking for a function. Thus we ask 'what is the color of your house?' We can express this question in symbolic form as

$$(?f)[f(x_1).\gamma(f)] \qquad (6)$$

Here the argument 'x_1', standing for 'your house', and the second-type function 'γ', standing for 'color', are given in the question. The interrogative words employed for this kind of question are 'what' and 'how'. The word 'how' is used, in particular, if a modifier is asked, as in 'how do you feel?' The same word is used if the numerical value of a predicate is asked, as in 'how many miles is the distance from x to y?'

The third kind is represented by questions asking for a logical term. The most important form of this group is the question asking for assertion. The statement 'p' is uttered without the assertion sign but, instead, with an invitation to the listener to utter the assertion sign 'yes', or the sign of denial 'no'. The sign 'no', as explained above, expresses the assertion of the negation of the statement. We have here a case where the utterance of the statement and that of the assertion sign are made by different persons.

This kind of question, the true-false question, can be symbolized as

$$?p \tag{7}$$

A question of this kind, such as 'did Peter come home?', is expressed by a sentence in which the words are inverted, or, in some languages, by means of a special question word, such as mentioned above.

The third subclass of pragmatic logical terms is composed of *imperative* terms. We have here the imperative mood of the verbs; in addition, we incorporate in this group such terms as 'shall', 'should', and the subjunctive in its imperative application ('be it resolved that . . .', 'qu'il s'en aille'). Let us use the sign '!' for the symbolic formulation of command. An expression like 'go out', then, can be written in the form

$$!p \tag{8}$$

We shall speak here of an expression in an *imperative mood*. Its cognitive correlate is given by the sentence 'I wish that you would go out'; we can write this statement in the form

$$w(I, p) \tag{9}$$

when we use the situation p as an argument. It is also possible, of course, to use the name 'p' in its place; cf. § 49. That (8) is not the same as (9) is shown as before by the fact that the negation of (8) is meaningless, whereas (9) can be denied. We cannot put the word 'not' before an imperative; it can only be added to what is commanded. Thus we cannot say: 'not — go out!'; we can only say 'do not go out'. The latter expression has the form

$$!\bar{p} \tag{10}$$

whereas the former would have the form

$$\overline{!p} \tag{11}$$

It is clear that (11) is a meaningless expression. Similarly, the expression 'you should not do that' has the form (10), but it cannot be interpreted as having the form (11). The term 'should' cannot be denied. Instead of (11) we use a negation of (9), for instance in the sentence 'you need not do that'. Terms like 'need' and 'must' do not act in a pragmatic capacity but are denotative terms.

As for assertive expressions, we have also for imperative expressions analogues of conjunction and negation, expressed likewise by juxtaposition and the word 'no'. Juxtaposition of two commands amounts to the same as commanding the conjunction of the respective propositions. The word 'no', uttered in a corrective sense after a command has been given, expresses the transition to a contrary command, i.e., the transition from (8) to (10). The other propositional operations have no analogues.

The exclamation mark '!' can be used to symbolize the term 'right', discussed in § 4. 'p is right' is symbolized by (8). The term 'wrong' is not given by the negation of 'right' but by the expression (10). We sometimes say 'this is not right' and thus apply the negation directly to the term 'right'; but such usage seems to result from a false analogy with indicative sentences. The meaning of 'this is not right' is 'do not do it', and is therefore given by (10).

A further group included in imperative terms is composed of the terms expressing permission. The word 'permission' has two meanings. First, it may mean the absence of a command to the contrary. This meaning can be stated in an indicative sentence of the form

$$\overline{(\exists x)c(x, p)} \tag{12}$$

where '$c(x, p)$' means 'x commands p'; (12) then means 'p is permitted'. The second meaning of 'permission' is an invitation to follow one's own decision; it then has the form 'do as you like' and is therefore an imperative expression. This meaning is expressed in terms like 'may'. The expression 'you may smoke' means 'make your own decision as to smoking'. Therefore, when we write '$d(x, p)$' for 'x decides about p', we can symbolize the expression 'you may smoke' in the form

$$!d(x, p) \tag{13}$$

For terms of this kind, grammar has constructed a special category, which is called the *optative mood*. This category can be dispensed with; (13) shows that permissive terms can be included in the imperative terms.

The fourth group of pragmatic logical terms comprises *exclamatory* terms. Here we have interjections like 'ah', 'oh', 'ouch' used as an emotional outlet of the speaker. They are expressive terms because they perform this outlet but do not say it. We can, of course, coordinate to the exclamation 'ouch' the sentence 'I have pain'; but the 'ouch' is not equivalent in meaning to this sentence, as is seen from the fact that the 'ouch' cannot be negated. The exclamation 'ouch', rather, is an indexical sign for pain; therefore, when somebody says 'ouch', we know that he has pain. On the other hand, this indexical sign is of a linguistic nature; we are conditioned to this expression by our environment, and people within a different linguistic environment use other expressions. Exclamatory terms are sometimes used with a communicative intention. Thus an admiring 'ah' may be uttered with the intention of having others participate in the admiration.

We conclude our exposition of logical terms at this point. We have given outlines only, knowing that much work is still to be done in this field, in particular, for the analysis and formalization of pragmatic terms. Among such terms, commands require specific elaboration, since they represent the kind of expressions that, in traditional logic, are called *value judgments*.[1]

§ 58. Extraneous terms

The preceding three categories, argument terms, functional terms, and logical terms, contain all terms of the object language proper, that is, of a language system which in a logical sense can be called one language. Conversational language, however, is not one language system in this sense. Its main body represents the logical system classified by us in the above categories; let us call this system the object language in the narrower sense. In addition to this system, conversational language includes terms of the metalanguage; but there are further terms which, though in the object language, cannot be incorporated in the narrower system. Let us classify all such terms in a fourth category which we call *extraneous* terms of the language.

The metalinguistic terms may be subdivided with respect to the

---

[1] A modern treatment of the logic of commands, including suggestions for a symbolization, is presented in the following papers: W. Dubislav, 'Zur Unbegründbarkeit der Forderungssätze', *Theoria*, 3, 1937; J. Jørgensen, 'Imperatives and Logic', *Erkenntnis*, 7, 1938; K. Menger, 'A Logic of the Doubtful: On Optative and Imperative Logic', *Reports of a Math. Colloquium*, 2nd ser., no. 1, pp. 53–64; R. Rand, 'Logik der Forderungssätze', *Zeitschr. f. Theorie d. Rechtes*, 1939; A. Hofstadter and J. C. C. McKinsey, 'On the Logic of Imperatives', *Phil. of Sci.*, 6, pp. 446–457, 1939. Further contributions to pragmatics and grammar are presented in the dissertations of A. Kaplan and W. Holther, University of California at Los Angeles.

three parts of the metalanguage. Let us begin with a collection of metalinguistic terms, constructed from this viewpoint.

In the syntactical subclass we have such words as 'sentence', 'clause', 'word', 'speech', 'letter', 'term'; we have likewise the names of the grammatical categories, such as 'noun', 'propositional operation', and 'propositional function'. In this subclass belong also terms referring to the technique of the logical calculus, such as 'derivable', 'equipollent'.

The *semantical* subclass contains the terms of the *denotation group*, such as the verbs 'denote', 'express', and the phrase 'is called'. Furthermore, we have here the terms of the *truth group*, such as 'true', 'false', 'probable'. In traditional grammar, some of these terms are classified as adverbs. Such classification leads to non-sensical constructions; for instance, the sentence 'Peter will proba-bly come' would then be construed as containing a function 'to come probably'. Actually, the word 'probably' in the above sen-tence does not modify the verb, but refers to the sentence as a whole and states something about the truth-value of the sentence.[1]

Further terms of the semantical subclass are 'tautology', 'analytic', 'synthetic', 'contradiction'. That these terms are semantical, and not syntactical, is evident from our definition of the terms given in § 8, since this definition refers to the truth tables. The generalized definition of tautology given in § 41 also uses the word 'true'.

Finally, we may incorporate into the semantical terms of the metalanguage the modalities, i.e., the terms *necessary, possible, impossible*. These terms qualify sentences; in fact, their precise meanings can be defined only in the metalanguage, as will be shown in chapter VIII. We shall therefore construe, in this interpreta-tion, the sentence 'Peter will possibly come' as meaning that the sentence 'Peter will come' is possible. There is another interpreta-tion of such sentences, given in the object language, with which we shall deal presently.

[1] There is another way of interpreting the word 'probable', namely, as a term of the object language stating a certain frequency of occurrence. In this interpretation, the word 'probable' will be classified as a denotative syntactical term within the class of logical terms, since it expresses numerical relations. Cf. the author's paper 'Über die semantische und die Objekt-Auffassung von Wahrscheinlichkeitsausdrücken', in the *Journal of Unified Science (Erkenntnis)*, 8, 1939, p. 50.

The third subclass consists of pragmatic terms. Here we have, first of all, the verb 'assert' and the noun 'assertion'. Incidentally, 'assertion' is used in three different meanings: it denotes, first, the act of asserting; second, the result of this act, i.e., an expression of the form ' $\vdash p$ '; third, a statement which is asserted, i.e., a statement ' p ' occurring within an expression ' $\vdash p$ '. It should be noticed that it is not possible to define the verb 'assert' in terms of the assertion sign. One might suppose that such a definition could be constructed by regarding the sentence " p ' is asserted' as having the same meaning as the expression ' $\vdash p$ '. But the coordination is not possible because ' $\vdash p$ ' is not a sentence (cf. § 57).

Another pragmatic term of an assertive nature is the term 'presumably'. It has a meaning similar to that of the term 'possibly', but it refers to the speaker and is therefore pragmatic. The sentence 'Peter will presumably come' means 'I presume that Peter will come'. The term can also be interpreted as an improper object term, as will be explained later in this section. Further pragmatic terms of this kind are the terms 'of course' and 'doubtful'.

Furthermore, we have here the term 'meaning' when defined by reference to verifiability, as in the verifiability theory of meaning (cf. § 2). The term 'verifiable' is a pragmatic term because it means 'possibility of verification by an observer'. It is the reference to the observer which makes this term pragmatic, not the reference to possibility, since 'possible' can be defined as a semantical term.[1] In this conception, therefore, 'meaning' is a pragmatic term because it states a property which signs have with respect to the sign user. It is also possible to introduce a semantical definition of meaning, according to which a sentence is meaningful if it is true or false. This definition, however, does not appear advisable because we are interested only in the use of signs which have a meaning for the user.

Further pragmatic terms of the metalanguage are terms like 'command', 'permission', 'exclamation', and 'question'. It should be noticed that all these terms, including the term 'assert', have no instrumental function, but are cognitive terms.

---

[1] Only 'technical possibility' is a pragmatic term since it is defined by reference to the abilities of human beings; cf. the author's *Experience and Prediction*, Chicago University Press, Chicago, 1938, § 6. But technical possibility has scarcely ever been used in the definition of verifiability.

Let us now turn to the consideration of extraneous terms which are not in the metalanguage. We may link up this discussion with our analysis of fictitious existence in § 49. We introduced there qualified existential operators. In doing so we abandoned the logical categories of the object language in the narrower sense. Let us study this procedure by an examination of the form of the definitions used for the qualified existential operators.

In § 22 we presented a set of rules limiting the form of definitions of contracted terms. When, from the viewpoint of these rules, we consider the definition of immediate existence given by the equivalence of (1, § 49) and (3, § 49), we see that that definition violates rules 4–6. The definiendum, presented by (1, § 49), contains a bound variable; furthermore, it contains a function occurring in the definiens (3, § 49) and a propositional operation. Now it is clear that a definition of this kind cannot be set up without qualifications. In (25, § 22) we gave an example in which we derived contradictions from such a definition. To regard the definition of (1, § 49) by (3, § 49) as permissible is possible only because the existential operator in (1, § 49) is qualified by a subscript. We have thus abandoned the kind of existence used in ordinary language, and introduced a new kind of existence.

In doing so we have introduced a language with a different logical category of existence, and therefore a language with a different logic. The two languages are connected by the equivalence of (1, § 49) and (3, § 49). This equivalence shows, furthermore, that what is a logical term in the second language is translatable into an empirical term in the first; immediate existence of things is translatable into the physical existence of a state of the human body. But it seems necessary to separate the two systems as two different languages. It is true that, when we speak, we often switch over from one language to the other without hesitation; but we should not forget that the two languages must not be confused.

These results apply likewise to the further kinds of fictitious existence introduced in § 49. Each such realm of existence must be conceived as belonging to a separate language. Similar considerations hold for the use of the term 'existence' in mathematics. When we say that there exists a solution of an equation, or that two parallel lines intersect in the infinitely distant point, we use languages differing from our language with respect to logical structure. These languages, however, are connected with our ordinary language by certain relations permitting a translation of the fictitious statements into statements using physical existence.[1]

This analysis should be kept in mind when we deal with problems of fictitious existence. We can refer to a plurality of existence only with respect to different languages. Therefore, qualifying existence by the word 'real' is necessary only when we use other languages in the same context and wish to distinguish the ordinary language from the others. Within the ordinary language, the use of the word 'real' is redundant since it means the same as the tautological term 'Ex' defined in (11, § 56). When we say: 'the bear you saw last night was real', this must be interpreted as meaning: 'the statement 'there is a bear which you saw last night' holds within the ordinary language'.

Returning to the problem of a classification of terms, we shall therefore introduce a category of *extraneous terms belonging to different object languages*. Among such terms we have 'immediate existence', 'logical existence', 'unreal', and so on. Furthermore, we incorporate into this class terms which refer to fictitious objects. Of this kind is the verb 'perceive', but also the term 'possible' since when used in the object language it is applied to fictitious objects. Referring to the earlier example we may say that Peter's coming is possible. The event denoted by the phrase 'Peter's coming' is then regarded as belonging in a realm of fictitious existence; the objects of this realm are divided into real (or necessary), possible, and impossible objects, and the event under consideration is incorporated into the second group. It is clear that this construction represents a shifting of the level of language; the term 'possible' in this interpretation is therefore an improper term of the object language. But this mode of speech is actually followed in conver-

---

[1] There are cases in which we can use physical existence with respect to a bound variable in the definiendum. In such cases, it must be proved that this definition does not lead to contradictions. It seems that definitions of this kind are used in physics when the existence of atoms is defined in terms of observational data. In a more precise analysis, such definitions do not represent equivalences, but probability connections. This is the viewpoint taken in the author's *Experience and Prediction*, University of Chicago Press, Chicago, 1938, chapters I–II. An analysis of this kind of language connections has been given in the philosophical dissertation of Dr. Norman Dalkey, University of California at Los Angeles.

sational language. The same consideration holds for the other modalities, i.e., for the terms 'necessary' and 'impossible'. Of a similar nature is the pragmatic term 'presumably' when used as an improper object term.

When we call these terms improper object terms we do not wish to exclude the possibility that a calculus may be constructed which correctly formulates the use of such terms in the object language. In such a calculus, however, the meanings of these terms must be assumed as primitive. Only when a reduction of these primitive meanings to other meanings of a simpler nature is intended will it be necessary to use the metalanguage.

§ 59. Classification of the parts of speech

A classification of linguistic terms usually proceeds in three steps. First, compound sentences are divided into *elementary sentences.* Second, elementary sentences are divided into units called *sentential parts;* they include, for instance, descriptions and similar composite expressions. Third, the sentential parts are divided into ultimate units, the *parts of speech.* These ultimate units are mostly individual words; but sometimes they may consist in parts of words, such as the suffix 'ed' and the plural suffix 's'. For a classification of the parts of speech it is irrelevant whether they are written separately or combined with other units into one word.

These divisions are not always unambiguous. We may define an elementary sentence as one that contains either no propositional operations or that contains different functionals combined by propositional operations and united by the same operator. According to this definition, the sentence, 'it excites me that I have to do this job alone', would be one elementary sentence. It would not appear unreasonable, however, to regard it as consisting of two sentences. That the first sentence, 'it excites me', does not possess a definite meaning until the reference of the pronoun 'it' is added is a feature which this sentence shares with many others. For logical purposes, however, it may be convenient to regard such *intensional* combinations of sentences as one elementary sentence, and to use the above definition.

The second division is more difficult to define. What we call parts of a sentence will depend largely on the form in which the sentence is uttered. Consider the sentence, 'a friend of a soldier came'. We would regard the phrase 'a friend of a soldier' as a unit, and the function 'came' as the other unit, which might be called the *major function*. When we symbolize this sentence, however, we find the following form, with '*s*' for 'soldier', '*f*' for 'friend', and '*c*' for 'came':

$$(\exists x)(\exists y)s(y).f(x, y).c(x) \tag{1}$$

Here the first unit has disappeared and we find three coordinated functions, among which no major function is distinguishable. In fact, the sentence (1) could also be given the wording: 'there was a soldier who was a friend of somebody who came'. Here the two functions 'soldier' and 'friend' might be regarded as major functions.

If we want to construct a symbolization that corresponds more closely to the *pragmatic structure*, i.e., the structure as far as emphasis and other psychological intentions are concerned, we may use descriptions. Thus we can symbolize the sentence by means of indefinite descriptions (§ 47) in the form:

$$c\{(\eta x)f[x, (\eta y)s(y)]\} \tag{2}$$

Whereas the meaning of (2) is the same as that of (1), the form (2) shows the units 'a soldier' and 'a friend of a soldier'. We see that the division into sentential parts refers, not to the meaning of the sentence, but to the *specific version* in which the sentence is formulated.

Similarly, definite descriptions supply us with units such as required for the expression of the pragmatic structure. In the sentence 'the king of England was crowned at Westminster Abbey' we have the two argument units 'the king of England' and 'Westminster Abbey', while the major function is given by 'was crowned'. When we symbolize the sentence in the form (6, § 47) these parts are, in fact, clearly distinguished. In the form (7, § 47), however, they have disappeared. Once more it is only the specific version of the sentence that is expressed by means of descriptions.

The pragmatic structure of a sentence is usually made visible by the time indication expressed in the tense of the verb. The *tense function*, as we may call the verb carrying the tense, is in general the major function in the sense of emphasis. We did not express in (1) the fact that the function 'came', and only it, carries a time argument; it is, however, the time argument that makes the function 'came' stand out as the major function, thus determining the structure expressed in (2). There are, however, exceptions. Thus the phrase 'the king of England' must be regarded as meaning 'the present king of England' and therefore carries a time indication, too, although this function is not emphasized.

Let us now turn to the parts of speech. The viewpoint of the classification here is essentially identical with that of the grouping of sentential parts; namely, it is the position within the sentence which determines each category. We find, however, that, when we break down the sentential parts into ultimate units, the ultimate units are always used in the same category. This result allows us to construct an absolute classification.

For instance, the phrase 'the present king of England' is used as an argument in the above sentence; but this argument has an inner structure which reveals that the word 'king' is a function, while the word 'England' is a proper name and thus ultimately an argument term. Proceeding in this way, we have constructed the following table of the parts of speech. The table does not contain categories like description, tense function, and functor, because these are sentential parts possessing an inner structure; only the categories referring to ultimate parts are given.

Some ambiguities result from the fact that the same term may belong in different categories. The prepositions, for instance, combine the nature of the function with that of a logical term; thus the word 'to' may be regarded as a function expressing direction and at the same time as a logical term indicating the place of a variable. The possessive pronouns express the function of possession and, simultaneously, an argument variable, sometimes even an iota-operator; 'his coat' means 'the coat belonging to x'. Such terms may be called *semi-logical* terms; for the sake of convenience we classify them among the logical terms.

Among the functions we incorporate only independent predicates, thus excluding incomplete symbols which acquire a meaning only in combination with a functional term, such as 'last' (cf. p. 309), or the prefix 'un' used in functions like 'undo' and 'unfold'.

For the subdivision of the functions we do not use the grammatical division into nouns, adjectives, and verbs, for the reasons explained in § 45. Instead, we use the classification given in § 52. A remark about this classification may be in order.

The divisions 1-4 which we mention under category II do not represent subclasses but state only the viewpoints from which the classification can be made. The same holds for the categories mentioned under II, 1, to which we therefore do not assign letters. The categories a-b, occurring under II, 2-4, however, represent subclasses. Let us make this clear by an example. The function 'taller', classified with respect to number of places, is a two-place function; classified with respect to univocality, it is a many-many function; classified with respect to field properties, it is uniform and interconnective; classified with respect to symmetry, it is asymmetrical; classified with respect to transitivity, it is transitive; classified with respect to reflexivity, it is irreflexive; classified with respect to inner structure, it is a simple function; classified by the nature of the argument, it is a thing-function; classified with respect to type, it is of the first type.

The logical terms show a great variety. Here more than in other categories the incorporation of a given term may appear dubious.

In the brackets we indicate, as far as possible, the corresponding categories of traditional grammar, together with some examples. [Ex.]

CLASSIFICATION OF THE PARTS OF SPEECH

   I. Arguments [subjects, objects]
      1. proper names ['John', 'England']
      2. token-reflexive descriptions
         a. of persons ['I', 'you', 'we']
         b. of space-time-points [space-time adverbs: 'now', 'yesterday', 'there'; tenses of verbs, suffix 'ed']
  II. Functions
      1. division by place properties:
         number of places, univocality, field properties, symmetry, transitivity, reflexivity
      2. division with respect to inner structure
         a. simple functions ['$\hat{x}$ sees $\hat{y}$']
         b. complex functions ['$\hat{x}$ is successful']

   3. division by the nature of the argument
     a. thing-functions ['to sleep', 'to crown']
     b. fact-functions ['coronation', 'earthquake']
   4. division by types
     a. functions of the first type
       [nouns, adjectives, verbs: 'house', 'taller', 'give']
     b. functions of higher types
       [nouns, adverbs, modifiers, numerical qualifiers: 'color', 'temperature', 'slowly', 'beautifully', 'frequently']

III. Logical terms
   1. terms in a syntactical capacity
     a. terms referring to arguments
       α. indicating the place of the variable
         [inflections of nouns, prepositions: suffix ''s', 'to', 'of']
       β. indicating the variable
         [pronouns: 'he', 'which', 'any']
     b. terms referring to functions
       [passive voice, suffix 'ly', gerund and infinitive suffixes]
     c. terms referring to argument and function
       [copula 'is', articles 'the' and 'a', suffixes 'ion', 'ing', conjunction 'that']
     d. syntactical parts of punctuation
       [comma, semicolon]
     e. denotative terms
       [function-argument relation]
   2. terms in a semantical capacity
     a. conjunctions
       [propositional operations: 'or', 'and', 'but', 'if ... then']
     b. operators and terms indicating that a variable is free
       ['all', 'there is', 'some', 'any']
     c. denotative terms
       [general terms like 'physical object', 'existence'; numerals; 'is' in the meaning of identity]
     d. sign-denominative symbols
       [quotation marks; token quotes: 'this']
   3. terms in a pragmatic capacity
     a. assertive terms
       [period sign, moods of verbs, 'yes', 'no']
     b. interrogative terms
       [interrogative pronouns and adverbs, question mark]
     c. imperative terms
       [imperative and optative mood, 'shall' and 'should' in imperative meanings, 'may', exclamation mark]

      d. exclamatory terms
        ['oh', 'ouch']

IV. Extraneous terms
  1. metalinguistic terms
     a. syntactical terms
       ['sentence', 'word', 'propositional function', 'derivable']
     b. semantical terms
       ['denote'; 'true', 'false', 'probable'; 'tautology', 'synthetic';
       'necessary', 'possible']
     c. pragmatic terms
       ['assertion', 'meaning', 'command', 'question', 'presumable']
  2. terms belonging to other language structures
      ['immediate existence', 'perceive'; terms like 'possible' used
      as improper object terms]

VIII. Connective Operations and Modalities

§ 60. Practical reasons for the introduction of connective operations

We now turn to the consideration of connective operations. In § 9 we discussed a special kind of connective operation — the analytic connective operations; we showed there that these operations can be interpreted as tautological adjunctive operations. In the present chapter we shall consider a more general kind of connective operation, which includes synthetic connective operations.

To show the use of nontautological connective operations let us consider an example involving implication. Tautological implication is used in mathematics; it is a relation between structural forms and can be established without reference to the particular empirical meanings of the terms involved. In physics, however, we find another type of implication which is not determined by structural form, which holds only between sentences of particular empirical meanings, and whose establishment in any particular case goes back to experience. This synthetic connective implication is what we usually call a *natural law*, i.e., a physical law, or a biological law, or a sociological law. Thus it is a physical law that heated metals expand. This implication does not follow from the meanings of the terms 'heating', 'metal', and 'expansion', and is therefore not tautological; it is only a matter of fact that this implication holds. Similarly we use other connective operations of the factual kind. Although their validity is not due to tautological character, they are universally true.

It would be mistaken to believe that we could reduce the necessity of natural laws to logical necessity by introducing a tautological implication in place of the synthetic connective implication. It is true that, choosing a suitable definition of 'heat', as for instance 'irregular movement of molecules', we could tautologically deduce from this definition that a heated metal expands. We would never be sure, however, whether a body to which the predicate 'being

355

heated' applies in the ordinary sense deserves this predicate in the new sense. In other words, we would then have to make use of a synthetic connective implication stating that a body showing the usual symptoms of heat is also heated in the sense defined as irregular movement of molecules. Introducing tautological relations at one point in the system of physics results only in shifting the synthetic connective relations to other points.

Although we emphasize the indispensability of synthetic connective operations, we should like to make it clear that we reject any kind of mysticism in the interpretation of these relations. We do not wish to say that physical necessity is due to invisible forces tying things together, or that it is apriori, or that it is a law of reason projected into nature, or whatever else has been subtly devised by certain metaphysicians. We agree with Hume that physical necessity is translatable into statements about repeated concurrences, including the prediction that the same combination will occur in the future, without exception. 'Physically necessary' is expressible in terms of 'always'. However, we have to find out how this definition can be given without misinterpretation of the language of physics.

Russell has suggested that natural laws be expressed as general implications. Thus, if we understand by '$f(x)$': 'x is a metal being heated', and by '$g(x)$': 'x expands', the above law would be expressed by

$$(x)[f(x) \supset g(x)] \tag{1}$$

The merely factual character of the implication is expressed here by the fact that the formula is synthetic, whereas its universal validity is expressed by the all-operator.

Now it is true that, to a certain extent, the use of a general implication eliminates the paradoxes of the adjunctive implication. For instance, considering the adjunctive implication 'sugar is black implies sugar is sour', which is true owing to the fact that the implicans is false, we may argue that the implication is unreasonable because the corresponding general implication is not true. Namely, the statement 'for all x, x is black implies x is sour' is not true, not even in the adjunctive sense. We might therefore attempt to define a reasonable implication as an adjunctive implication which holds, not only in the individual case considered, but also in all similar cases. Thus we would say that heating a particular piece of metal implies its expansion in a reasonable sense of the word 'implies', because the corresponding adjunctive relation holds for all pieces of metal.

Although this conception certainly represents a step toward a definition of a reasonable implication, it cannot be regarded as a solution of the problem because of some difficulties remaining. A closer consideration shows that sometimes a general implication like (1) is true although we would not consider it a reasonable implication. First, as Russell has pointed out, the paradoxes of the individual adjunctive implication are repeated for the general adjunctive implication if the implicans is false for all x or if the implicate is true for all x. Thus the implication 'x is a man living on the moon today implies x has a red necktie' is true for all x. We would not consider it a reasonable implication because we know that its validity derives only from the fact that there are no men on the moon; and we would add that if *there were* a man on the moon he need not wear a red necktie. The last statement, however, cannot be expressed by the adjunctive implication since this implication does not permit us to differentiate between what would be true if the implicans were true, and what would be false under those circumstances. The implication 'for all x, x is a man living on the moon today implies x does not wear a red necktie' is also true, as well as the first.

It must be added here that not all implications of this kind are unreasonable; thus there are also reasonable implications with an always-false function in the implicans. The implication, 'for all x, x is a man living on the moon today implies x uses an oxygen respirator', is a reasonable implication because we know that if *there were* a man on the moon he could not live otherwise, since there is no atmosphere on the moon. The distinction between these two kinds of implication, and thus the formulation of a *conditional contrary to fact*, is a problem from which logicians like to turn away, considering it irrelevant for scientific language. This attitude may originate from a preoccupation with problems of mathematical deduction, behind which the needs of empirical sciences disappear. We cannot pretend to have a logistic theory of physics unless we are able to define a synthetic connective implication. Moreover, logicians should not forget that they themselves use connective implication in their scientific writings as soon as they use the metalanguage. A statement like 'if a sign is printed in Gothic type it is a sentence name variable' is meant in the sense of a connective implication since it is considered reasonable even with respect to a text which does not contain Gothic letters.

The second difficulty with the interpretation of physical laws as general implications is this: even if we consider mixed cases, i.e., cases in which the implicans is not always false and the implicate not always true, we may obtain unreasonable implications. Assume that Peter's two dogs have white spots on their foreheads, and that Peter has only two dogs; then the implication 'x is Peter's dog implies x has a white spot on its forehead' would be true for all x. We would not call this implication a physical or biological law, but a fact due to chance. We do so because we would say that if Peter had a third dog it would presumably have no white spot on its forehead. This addition, however, cannot be expressed by means of an adjunctive implication, for the reasons explained above. The adjunctive statement 'for all x, x is Peter's third dog implies x has no white spot on its forehead' is as true as the opposite statement 'for all x, x is Peter's third dog implies x has a white spot on its forehead'; both these implications are true because the implicans is always false. A contrary instance would be given by the implication 'for all x, x is Peter's dog implies x has a heart'. This is a reasonable implication because we know that it would be true even if Peter had a third dog. We see once more that we must distinguish between two kinds of general implications, only one of which corresponds to what we call a natural law. The definition of a connective implication, therefore, although it will make use of a reference to general implication, will require a specification of the kind of general implication to be used here.

Our example, on the other hand, should make it clear that we cannot dispense with the use of adjunctive implication for the symbolization of conversational language. The first statement about Peter's dogs appears as a reasonable statement when we formulate it in the usual form: 'all Peter's dogs have white spots on their foreheads'. Here the implication is not expressly stated; but it is contained in the structure of the all-statement since the symbolization of class inclusion requires the use of an implication, as was shown in (20–22, § 35). When an adjunctive implication is expressed in the form of class inclusion it is regarded as a legitimate instrument of language (cf. also our remarks on p. 199). Only when we use the word 'implies' explicitly do we wish to restrict it to a connective meaning.

We shall deal with the problem of a connective meaning, however, not as a problem of implication alone. We shall try to solve it in its full generality, i.e., with respect to all kinds of propositional operations. In § 7 we introduced the distinction between adjunctive and connective operations, which applies to all operations. It therefore is the general problem of defining connective operations that constitutes the subject of our inquiry.

Let us add a remark about the method to be used. We said that for connective operations the T-cases of the truth tables mean only cases of confirmation, not of verification, whereas an F-case would mean a proof of falsehood. It follows that, whenever a connective general operation is true, a corresponding adjunctive general operation of the same kind also holds; but the reverse cannot be maintained. The question, therefore, arises: how do we know whether a connective operation is true? If the connective operation is of the analytic kind, the answer is easily given since we have simple rules defining a tautology. If the connective operation is synthetic, however, the answer cannot be given within a theory of deductive logic since this determination requires the whole apparatus of induction. Statements saying that all bodies expand on heating, or that all dogs have hearts, are demonstrable only with the use of inductive inference. We therefore leave to inductive logic the task of answering the question of verification of connective operations of the synthetic kind.

In fact, the problem of connective operations, in particular, of a reasonable implication, so far as it overreaches the scope of merely analytic operations, constitutes a problem of induction, and not of deduction. What we mean by a reasonable synthetic implication is an implication associated with *inductive confidence*, i.e., an implication which we are ready to assert also for individual instances not yet experienced. Since the present book is not concerned with problems of induction we cannot discuss methods of verifying statements of this kind of generality. What we can do, however, is to introduce conditions which rule out all noninductive forms of verification. This is the method to be followed by our investigation. It permits us to define connective operations by establishing certain structural conditions so devised that the assertion of such statements is necessarily left to the methods of induction. On the other hand, it will be clear that such procedure, ultimately, transfers the establishment of synthetic connective operations from deductive to inductive logic.

§ 61. Formal characterization of connective operations

The connective operations which we want to define include both analytic and synthetic operations. We shall therefore speak of connective operations without special reference to this distinction. Since, however, the difficulties of our problem originate from synthetic connective operations, we shall use such operations in the examples to which we refer; the proof that analytic operations, defined as tautologies, satisfy the same conditions will be given later. The notions to be defined will apply both to definite and indefinite synthetic statements; i.e., they will be applicable also when the synthetic statements contain free variables. Such statements, however, are seldom used in the empirical sciences; and if they are, they usually contain only free argument variables, while free functional or propositional variables scarcely ever occur. It will therefore be convenient to restrict our examples to definite synthetic statements.

Let us begin with some remarks clarifying our terminology. As before (§ 9), we shall introduce connective operations as operations substituted for the major adjunctive operations in formulas of a certain kind; while for analytic connective operations these formulas are tautologies, the appropriate definition of such formulas for the general case constitutes the very problem of connective operations. We shall call these formulas *nomological formulas*. The term 'nomological', derived from the Greek word 'nomos', meaning 'law', is chosen to express the idea that the formulas are either laws of nature or logical laws. Analytic nomological formulas are tautological formulas, or logical laws; synthetic nomological formulas are laws of nature. The term 'nomological' is therefore a generalization of the term 'tautological'.

In order to state the formal properties of nomological statements we shall divide these statements into two classes. In the first class we include all those statements which can be immediately recognized as having 'reasonable' major operations; in the second class we place those statements that are derivable from the first and that therefore owe the 'reasonableness' of their major operation to this derivation. We call the statements of the first class *original nomological statements;* those of the second class, *derivative nomological statements*. Correspondingly we speak of *original connective operations* and *derivative connective operations*.

As an instrument to be used in the definition of original nomological statements we shall introduce the concept *exhaustive*. This term will be so defined that it applies only to true statements and delimits among them a subclass of statements whose structure satisfies certain requirements. For the characterization of these structural properties we shall use the method of expanding a formula in T-cases, introduced in (21, § 11).

This method was used in § 11 only within the calculus of propositions; in order to apply it to formulas containing functions we must add some considerations concerning the occurrence of operators, If a formula contains no major operators, its expansion in major T-cases (cf. 24, § 11) can be constructed as in the calculus of propositions. It is different when the formula contains one or more major operators. We set up the rule that in this case we shall expand the major operand in its major T-cases, and put the set of major operators before this expansion. The formula so constructed will be called the expansion of the formula in its major T-cases. A similar extension of the definition for elementary T-cases will be given later.

Let us consider, for instance, the formula

$$(x)(\exists y)[f(x, y) \supset g(x, y)] \qquad (1)$$

in which the terms '$f(x, y)$' and '$g(x, y)$' may be abbreviations of complicated expressions, including further operators. The expansion of this formula in major T-cases is given by

$$(x)(\exists y)[f(x, y).g(x, y) \lor \overline{f(x, y)}.g(x, y) \lor \overline{f(x, y)}.\overline{g(x, y)}] \qquad (2)$$

We now introduce another term referring to structure. Let us regard a statement which results from (2) when we cancel one, or several, of the terms of the disjunction, for instance the statement

$$(x)(\exists y)[\overline{f(x, y)}.g(x, y) \vee \overline{f(x, y)}.\overline{g(x, y)}] \tag{3}$$

We call such a statement a *residual statement in major terms* of the considered statement; the considered statement may be called the *main statement*. Let us assume that the main statement (1) is true; the question arises: is one of its residual statements, for instance (3), also true? The answer will depend on the statements considered. We now define:

A true statement is *exhaustive in its major terms* if none of its residual statements in major terms is true.

If all the major operators of the main statement are all-operators this condition is equivalent to saying that none of the terms of the disjunction in which the operand is expanded is an always-false function. Thus the statement

$$(x)[f(x) \supset g(x)] \tag{4}$$

which expanded in its major terms has the form

$$(x)[f(x).g(x) \vee \overline{f(x)}.g(x) \vee \overline{f(x)}.\overline{g(x)}] \tag{5}$$

is exhaustive in its major terms if the three statements

$$
\begin{aligned}
&(\exists x)f(x).g(x) \\
&(\exists x)\overline{f(x)}.g(x) \\
&(\exists x)\overline{f(x)}.\overline{g(x)}
\end{aligned} \tag{6}
$$

are true, since then none of the terms of the disjunction in (5) can be canceled. However, if existential operators are among the major operators, the condition of exhaustiveness cannot be stated in this way, but must be stated as given above.

The significance of the condition of exhaustiveness for the problem of connective operations may be made clear by examples. An implication of the form (4) with an always-false implicans is not exhaustive, because the first of statements (6) does not hold; neither is an implication with an always-true implicate exhaustive, since the last of statements (6) then is not true. Thus the condition of exhaustiveness rules out unreasonable implications of this kind.

We obtain an example of an unreasonable implication of the form
(1) when we interpret '$f(x, y)$' as meaning 'x is the father of y',
and '$g(x, y)$' as meaning 'x is taller than y'. Since in this interpretation the statement

$$(x)(\exists y)\overline{f(x, y)} \tag{7}$$

holds, the implication (1) holds also; but it seems to be an unreasonable implication since it owes its truth only to statement (7). [That
is, if a father is not taller than his child, we choose as y a person who
is not the child of x; then the implicans in (1) is false, and thus the
implication holds.] Now the validity of (7) has the consequence
that one of the residual statements of (1) is true, namely, statement
(3). Thus the condition of exhaustiveness rules out unreasonable
implications of this kind.

If an all-statement is not exhaustive in its major terms, its major
operation is used out of place, so to speak; the operation then
suggests connections which do not exist. Unreasonable statements
of the considered kind are therefore made reasonable if they are
stated as exclusive disjunctions of T-cases in which superfluous
terms are omitted. Thus instead of the implication 'for all x, x is a
man on the moon today implies x wears a red necktie' we can say
'for all x, either x is not a man on the moon today and x wears a red
necktie, or x is not a man on the moon today and x does not wear
a red necktie'. In this form, in which the first term of the disjunction in (5) is omitted and the exclusive 'or' is used, the statement
does not appear unreasonable. For an exhaustive implication like
'for all x, x is a piece of metal being heated implies x expands' we
must use the complete form (5) and cannot omit one of the terms.
These considerations indicate that exhaustiveness is one of the
features to be used for the definition of reasonable operations; we
shall see that some other features must be added.

If the major operand contains no operation at all, or if the major
operation is the 'and' or the negation, as in (7), the statement is
always exhaustive in its major terms. This consequence holds under
the same conditions for a formula which contains no major operators,
i.e., if this formula contains either no propositional operations or
has the 'and' or the negation as its major operation.

When a statement contains no variables at all, but only constants, it cannot be exhaustive if it contains binary operations other than the 'and'. Thus, when the symbols '*a*', '*b*', '*c*' represent constants, formula (22, § 11) is not exhaustive in its major terms because in (24, § 11) only one of the two terms can be true. The other term then can be canceled, and the true term will represent a true residual statement.

We shall now extend our considerations about exhaustiveness to apply to an expansion in elementary *T*-cases (cf. 23, § 11). When we wish to construct this kind of expansion for a formula containing operators, we must first transform it into a one-scope formula. We showed in § 20 that this transformation is always possible. We then expand the operand in its elementary *T*-cases and put the set of operators before the formula; the resulting formula will be called the expansion in elementary *T*-cases. Thus the statement (28, § 20), which we proved to be equivalent to (30, § 20), would be written in the form

$$(x)(\exists y)(z)[g(y).h(x, y).f(x).k(y, z) \vee \ldots \vee \overline{g(y).h(x, y).f(x).k(y, z)}] \quad (8)$$

It is easy to put into the brackets all those combinations of elementary terms for which the statement is true. In accordance with our previous definition, we understand by a *residual statement in elementary terms* any statement obtained from the main statement by canceling one of the terms of the disjunction presenting the expansion of the statement in elementary *T*-cases, such as (8). We now define:

A true statement is *exhaustive in its elementary terms* if none of its residual statements in elementary terms is true.

For a statement which in its one-scope form contains only all-operators we can use, as before, the interpretation of exhaustiveness by existential statements such as (6). Thus our statement 'for all *x*, *x* is a piece of metal and *x* is heated implies *x* expands', which possesses three elementary terms, is exhaustive in these terms, as there are things which are pieces of metal and are heated and expand, and things which are pieces of metal and are not heated and expand, etc.

A formula may be exhaustive in its elementary terms although it is not exhaustive in its major terms. Assume, for instance, that formula (4) is exhaustive in '$f(x)$' and '$g(x)$'. It then is exhaustive both in its elementary and in its major terms, which in this instance coincide. Formula (4) is tautologically equivalent to

$$(x)[\overline{f(x) \supset g(x)} \supset \overline{g(x) \supset f(x)}] \tag{9}$$

Because of the equivalence, (9) is also exhaustive in its elementary functions '$f(x)$' and '$g(x)$'. However, the case that both the implicans and the implicate of (9) are true is a contradiction; therefore (9) is not exhaustive in its major terms. Formula (9) may be called an *inflated form* of (4). By 'inflated form' we shall understand a one-scope statement that is not exhaustive in its major terms but is exhaustive in its elementary terms.

It can be proved that, if a formula is exhaustive in its elementary terms, it can always be written in a form in which it is also exhaustive in its major terms. We prove this as follows. We write the operand of the formula as an expansion in the elementary T-cases $e_1 \ldots e_m$:

$$e_1 \vee e_2 \vee \ldots \vee e_m \tag{10}$$

Since the $e_1 \ldots e_m$ are mutually exclusive, (10) can be written with the sign of the exclusive 'or':

$$(e_1 \vee e_2 \vee \ldots \vee e_{m-1}) \wedge e_m \tag{11}$$

or if we prefer to eliminate the sign of the exclusive 'or' (cf. 2, § 10):

$$(e_1 \vee e_2 \vee \ldots \vee e_{m-1}) \equiv \bar{e}_m \tag{12}$$

With both the last forms of the operand (which are tautologically equivalent to the original operand as soon as the $e_1 \ldots e_m$ are written in the elementary terms) the formula is exhaustive also in its major terms. Of course there may be other forms of the formula having the same property.

Furthermore, it can be shown that in an inflated form of a statement each dispensable T-case of the major terms is a contradiction. We shall first prove this theorem on the assumption that the original formula is a one-scope formula, and then extend it to all other formulas. We write the operand of the formula as an expansion in its major T-cases $a_1 \ldots a_k$ (k is at most $= 4$):

$$a_1 \vee \ldots \vee a_k \tag{13}$$

Each of the $a_1 \ldots a_k$ which is not a contradiction [1] can be written as a disjunction of elementary T-cases $e_1 \ldots e_n$; taken altogether,

---

[1] A contradiction cannot be expanded in T-cases because it is not true for any combination of the elementary terms.

they must be the same as occur in the expansion of the whole formula in elementary T-cases. If, however, 'e_m' is one of the terms of the disjunction that constitutes 'a_n', 'e_m' cannot be contained in one of the disjunctions representing any other 'a_i'; otherwise the various 'a_i' would not be mutually exclusive. Now let 'a_i' be a dispensable T-case. If 'a_i' is a contradiction it satisfies the theorem to be proved. On the other hand, assuming that 'a_i' is no contradiction, we can derive the consequence that it must be a contradiction; in other words, we prove the theorem by *reductio ad absurdum*. This proof is constructed as follows. If 'a_i' is non-contradictory we can expand it in its own elementary T-cases and write:

$$a_i \equiv e_r \vee \ldots \vee e_s \qquad (14)$$

Now it is clear that the $e_r \ldots e_s$ must belong to the F-cases of the operand. If, for instance, 'e_r' were one of the T-cases of the operand, e_r would be contained in (10); but then 'e_r' could be canceled in (10), because 'a_i' can be canceled in (13), and then the formula would not be exhaustive in its elementary terms. We therefore have

$$a_1 \vee \ldots \vee a_k \supset \overline{e_r \vee \ldots \vee e_s} \qquad (15)$$
$$a_1 \vee \ldots \vee a_k \supset \overline{a_i} \qquad 1 \leqq i \leqq k \qquad (16)$$

Now (15) and (16) must be tautologies in the elementary terms since all T-cases or F-cases are mutually exclusive in the sense that they contradict each other; every statement will tautologically imply that its F-cases are false. But (16), which contains 'a_i' not only on the right-hand side but also in the implicans, can be transformed by means of (6a, 5b, and 4e, § 8) into the statement '$\overline{a_i}$'; therefore '$\overline{a_i}$' must be a tautology, or, what is the same, 'a_i' must be a contradiction.

In order to extend the proof to other formulas than one-scope formulas, we proceed as follows. We need not regard formulas whose major operation is the negation or the conjunction, since such formulas are exhaustive in the major terms. Let 'A' and 'B' be the major terms; each of these may include operators of a limited scope, and there may or may not be, besides, major operators before the whole formula. Expanding the formula in major T-cases we obtain a major operand consisting of several, or all, of the terms

'$A.B$', '$A.\bar{B}$', '$\bar{A}.B$', '$\bar{A}.\bar{B}$'. We now transform both 'A' and 'B', separately, into one-scope terms; they may be called, respectively, 'A'' and 'B'', and we have the tautological equivalences '$A' \equiv A$' and '$B' \equiv B$' (assuming that all variables are bound). The original expansion will now be replaced by a second expansion in which the terms 'A'' and 'B'' occupy the places, respectively, of the terms 'A' and 'B'. After introducing different names for variables bound by all-operators, we can merge the operands, putting all the operators in front of the whole formula and using (11b and 12b, § 25). The operators will remain unchanged thereby; and the terms of the disjunction will also be the same as before, except for the elimination of the operators. These terms may be called, respectively, '$a.b$', '$a.\bar{b}$', '$\bar{a}.b$', '$\bar{a}.\bar{b}$'. Now assume that, for instance, '$A.B$' can be canceled in the original expansion in major T-cases; then '$A'.B'$' can be canceled also in the second expansion. But then it must be possible also to cancel the corresponding term '$a.b$' in the one-scope expansion, since when we divide the operand after the canceling of '$a.b$' we obtain the same residual formula as in the second expansion. When, however, the formula is exhaustive in the elementary T-cases, it follows from the above proof that '$a.b$' must be a contradiction. But then '$A'.B'$' must be a contradiction also since it differs from '$a.b$' only by the addition of operators. Furthermore, the term '$A.B$' of the original expansion in major T-cases must also be a contradiction because it is tautologically equivalent to '$A'.B'$'. This concludes the proof.

We can also have the opposite case, i.e., a formula which is exhaustive in its major terms, but not in its elementary terms. Assume, for instance, that we have two functions for which the condition

$$(x)\overline{f(x).g(x)} \tag{17}$$

holds. Then the formula

$$(x)[f(x) \vee g(x) \equiv h(x)] \tag{18}$$

is not exhaustive in its elementary terms. However, this formula will be exhaustive in its major terms if it is true and '$h(\hat{x})$' is a mixed function. Of course it is not possible to transform such a formula into a tautologically equivalent formula which is exhaustive in its elementary terms.

If a statement is exhaustive both in its elementary and its major terms, we shall call it *fully exhaustive*.

The requirement that a statement be fully exhaustive eliminates some difficulties of the following kind. We saw that a statement having the 'and' or the negation as major operation is always exhaustive in its major terms. Now it is possible to combine two implications with an empty implicans by the sign 'and'; the resulting conjunction is exhaustive in its major terms. But it will not be exhaustive in its elementary terms; statements of this kind are therefore excluded by the requirement of full exhaustiveness.

We now turn to the definition of original nomological statements. The first requirement for these statements is that they be *demonstrable as true*, since nomological statements are to represent either laws of nature or laws of logic. By the phrase 'demonstrable as true' we mean that there is a method, of the deductive or the inductive kind,[1] proving the statement to be true. The word 'demonstrable', therefore, does not presuppose the definition of possibility to be given below; such procedure would be circular. We do not require, however, that the proof be actually given. If it is given, we shall know that the statement is nomological; but there may be nomological statements that are not known to be such. In requiring that the statement be not only true but also demonstrable as true, we exclude statements that are true but that cannot be proved to be true. Thus the statement 'all gold cubes are smaller than one cubic mile' may be true; but since we cannot prove it as true it is no nomological statement. We thus exclude statements that are true by chance. Such exclusion is in accordance with our maxim of reducing the definition of connective operations to the methods of inductive evidence (cf. the remarks at the end of § 60).

The second requirement is that these statements be *all-statements*, i.e., statements beginning with an all-operator whose scope is the whole formula. The statements may have further operators, including existential operators, either as major operators or in other places; however, a beginning with a major all-operator will be sufficient to effectuate that kind of generality which we demand for natural laws. A third requirement consists in the demand that

---

[1] We include here a verification as *practically true*, since inductive methods never lead to absolute truth.

the statements be fully exhaustive, for the reasons mentioned above. We must now explain a fourth requirement, which will make our definition complete.

Looking at the examples of unreasonable operations given above, we see that there are all-statements which, although exhaustive, do not contain the kind of operations which appear to us reasonable. These formulas, however, can be established as all-statements only because of some limitations imposed upon the propositional functions. Thus in the example 'for all x, x is a dog of Peter's implies x has a white spot on its forehead' the first function is limited because it is stated in terms of the individual person Peter. There are other forms of limitation, given by restrictions as to space or time. If during a certain time interval $t_2 - t_1$ there are no persons in a certain room, we can establish the exhaustive adjunctive implication 'for all x, x is a person implies x is not in that room during the time $t_2 - t_1$'. This implication is unreasonable because it carries no necessity with it; and we see that the absence of necessity is due to the fact that the implication is stated with reference to the individual time points t_1 and t_2, and the individual location 'this room'.

At the end of § 47 we introduced the term 'universal statement' for statements containing no individual-signs. We shall therefore formulate our fourth requirement as the condition that original nomological statements be universal. With this condition we exclude all-statements that, though exhaustive, possess only a limited number of satisfying arguments. This exclusion makes it impossible to verify the considered all-statements by complete enumeration, and thus eliminates the use of the second direction of reading the truth tables (§ 7) for the verification of the all-statements; this is the reason that, if such an all-statement is true, its major operation can be interpreted as a connective operation.

Combining our four requirements, we define original nomological statements as follows:

An original nomological statement is an all-statement that is demonstrably true, fully exhaustive, and universal.

Let us now consider statements derivable from original nomological statements. By 'derivable' we mean here deducible, i.e., derivable by means of deductive operations, excluding inductive inferences. As to the result of the derivation we must distinguish

several cases. We may obtain either an original nomological statement, or a statement that does not satisfy one, or several, of the
requirements of these formulas. Only the property of being demonstrably true, of course, will always be preserved in the derivation.
The case that the derived statement is not fully exhaustive, in
particular, may be subdivided into the two cases of reasonable and
unreasonable operations. Thus from the original nomological
statement

> 'for all x, x is a body lighter than water and x is
> put into water implies x floats' (19)

we can deduce the statement

> 'for all x, x is a piece of iron lighter than water
> and put into water implies x floats' (20)

which may be called a reasonable implication although the implicans is always false. Adding to (20) the original nomological
statement

> 'for all x, x is a piece of iron implies x is heavier
> than water' (21)

we can deduce that the implicans of (20) is always false, and thus
also the unreasonable implication

> 'for all x, x is a piece of iron lighter than water
> implies x is green' (22)

The difference between the two cases can be explained as follows.
In (20) we derive a statement with an always-false implicans, but
we do not make use of this fact in the derivation, whereas in the
derivation of (22) we do make use of it. In order to define derivative
nomological statements that can be called 'reasonable', we shall
therefore have to add a qualification stating that, if a resulting
formula is not exhaustive in some sense, this fact is not to be used in
the derivation of the formula. This seemingly psychological idea
can nevertheless be given a purely logical formulation. For this
purpose we need only demand that the statement be derivable from
a set of nomological formulae $a_1 \ldots a_n$ from which the emptiness
of the implicans cannot be inferred; we then are sure that in a
derivation based on $a_1 \ldots a_n$ this emptiness has not been used.
Thus the reasonable implication (20) is derivable from (19) alone,

whereas for the derivation of the unreasonable implication (22) we need (21), that statement being necessary to derive the emptiness of the implicans in both (20) and (22). In order to express this idea for all kinds of operations and all kinds of statements we shall formulate our condition as requiring that no residual statement of the considered statement can be derived from $a_1 \ldots a_n$. This more general condition includes the case of an empty implicans considered in our example.

Now it is a matter of definition whether we include statements of the kind (22) in the class of nomological formulas. These statements are not quite so unreasonable as other statements containing adjunctive implications with an always-false implicans, like 'x is a man living on the moon today implies x has a red necktie', because in statements of the kind (22) the implicans is not only always false, but — in a terminology to be explained later (§ 65) — even impossible. It turns out that scientific language is ambiguous as to the acceptance of statements like (22); although in most cases only statements like (20) are considered nomological formulas, there are other cases in which statements like (22) are also dealt with as nomological formulas. We shall distinguish these two meanings of 'nomological' by the respective additions 'in the narrower sense' and 'in the wider sense'. Therefore we now define:

A statement 'p' is a *nomological statement in the wider sense* if it is derivable from a set '$a_1 \ldots a_n$' of original nomological statements.

A statement 'p' is a *nomological statement in the narrower sense* if there is a set '$a_1 \ldots a_n$' of original nomological statements from which the statement 'p' is derivable, and from which no residual statement of 'p', either in major or in elementary terms, is derivable.

Applications of both kinds of nomological statements will be given later. We shall see that nomological statements in the wider sense will be used to define modalities, whereas nomological statements in the narrower sense will be used to make precise what ordinary language considers reasonable operations. Our definition shows that statements of the latter kind are characterized by means of a qualification added to the condition of derivability from original nomological statements; we call it the *qualification of derivation*.

It is obvious that the qualification of derivation does not demand that nomological statements in the narrower sense be exhaustive.

A statement that is nomological in the narrower sense but not exhaustive is shown in (20). However, the qualification of derivation does exclude inflated forms of statements from nomological statements in the narrower sense. Since the dispensability of a T-case of the major terms in an inflated statement is due to a contradiction, this exclusion is always demonstrable without the use of synthetic nomological statements; we may therefore say that, if an inflated statement is derivable from a set of original nomological statements, the' dispensability of one of its T-cases is derivable from the same set. (This amounts to saying that in a derivation it is always permissible to add tautologies.) We see that, if a statement is nomological in the narrower sense, a tautologically equivalent form of it may not be nomological in the narrower sense. This consequence of our definition, however, appears reasonable since the word 'nomological' includes reference to the syntactical form of a statement. On the other hand, the definition of 'nomological in the wider sense' is so given that, if a statement is nomological in any sense, all tautologically equivalent forms of this statement are at least nomological in the wider sense, including inflated forms.

By the process of derivation it is possible to introduce nomological statements containing no all-operators, but free variables or only existential operators; and it is also possible to introduce nomological statements containing individual-signs, even of such a kind that they contain no operators at all. Since the property of exhaustiveness has been defined for all such statements, the distinction of nomological character in the wider and in the narrower sense can be applied to all derivative nomological statements. It follows from our considerations concerning exhaustiveness that a fully specialized nomological statement, i.e., a nomological statement containing only constants, will be fully exhaustive only if it contains nothing but conjunctions and negations, or no operations at all; in all other cases it must always be not fully exhaustive. Such a statement, however, can be nomological in the narrower sense; it is so if it is derivable from a set of original nomological formulas from which we cannot infer which of the T-cases of the statement holds.

An example of a specialized nomological statement in the narrower sense is the statement 'for all x, x is a dog of Peter's implies x

has a heart'; or the statement 'for all x, if x was a learned man of the Middle Ages, and if at that time all scientific knowledge was transmitted in Latin, then x knew Latin'. In the latter example the terms 'Middle Ages' and 'Latin' are individual-signs; the statement may be considered as derived from a general nomological statement 'for all x, if x was a learned man living at the time t and if at that time all scientific knowledge was transmitted in a certain language, then x knew that language'. In a fully specialized nomological statement containing an implication we frequently indicate by the use of pragmatic means of expression whether or not the implicans is true. Following this usage we derive from our second example the fully specialized nomological statements in the narrower sense, 'since William of Occam was a learned man of the Middle Ages and since at that time all scientific knowledge was transmitted in Latin, he knew Latin', and 'if Plato had been a learned man of the Middle Ages, at which time all scientific knowledge was transmitted in Latin, he would have known Latin'. Using this mode of speech, however, we have added to our statements some knowledge which is not derived from nomological statements, namely, the statements that the Middle Ages were a period of the kind described, that William of Occam was a learned man of the Middle Ages, and that Plato was not. The correct form of the considered statements would therefore use the 'if' instead of the 'since', and the 'was' instead of the 'had been'. Only in such a form are these statements nomological. This form will also make it clear that the source of knowledge from which we derive these statements does not inform us about the truth or falsehood of the implicans, and that therefore the qualification of derivation is satisfied; the statements are therefore nomological in the narrower sense.

An example of a specialized nomological statement in the wider sense is the statement 'if the Eiffel tower is made of iron and is lighter than water it is 1000 miles high'. Not all unreasonable implications with an empty implicans fall under this category, however. Thus the statement 'for all x, x is a sea serpent living today implies x is 30 feet long', though true, is not a nomological statement even in the wider sense, since it is not derivable from original nomological statements. It would be so only if the statement 'there is no sea serpent living today', which is specialized by

the individual sign 'today', were derivable from an original nomological statement. But the corresponding universal statement,
which contains no restriction as to time, cannot be maintained as
true and is therefore not nomological.

Our terminology is explained by the following diagram:

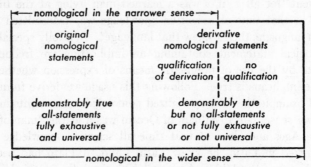

FIG. 8.   Classification of nomological statements.

After the definition of nomological statements we proceed to the
introduction of connective operations by the following rule:

*The major operation of a nomological statement can be replaced
by the corresponding connective operation.*

As before (§ 9), we shall use an accent to denote connective operations. Thus if the statement

$$(x)[f(x) \supset g(x)] \tag{23}$$

is a nomological formula, the statement

$$(x)[f(x) \,\bar{\supset}\, f(x)] \tag{24}$$

will be true. The following considerations will make our terminology
clear. A nomological formula, such as (23), contains only adjunctive operations, as is seen from the implication sign occurring in (23).
The corresponding formula containing a connective operation, such
as (24), is not a nomological formula; we shall call it a connective
formula. The major operation of a nomological formula, such as the
implication in (23), will be called a nomological operation. Nomological operations are therefore adjunctive operations of a special
kind, namely, adjunctive operations which are major operations of
nomological formulas. It would be false to say that a nomological
operation *is* a connective operation. The correct formulation is:

a nomological operation *can be replaced by* a connective operation, or *can be interpreted* as a connective operation. This terminology corresponds to our terminology developed for analytic connective operations (§ 9), except that the word 'tautological' is replaced by 'nomological'.

The difference between the two kinds of nomological statements is transferred to nomological and hence to connective operations. Thus we distinguish between connective operations in the wider and in the narrower senses. The operation in (24) will be a connective operation in the wider or narrower sense according as (23) is a nomological statement in the wider or narrower sense. Connective operations in the narrower sense are those operations that are considered as reasonable operations in conversational language. Similarly, the distinction between original and derivative nomological formulas applies also to nomological and connective operations. This distinction, however, has only a formal significance and is of no import with respect to conversational language.

It is easy to show now that tautological operations satisfy our conditions for nomological operations and can therefore be interpreted as analytic connective operations. They are partly original, partly derivative nomological operations. Among the conditions determining this classification, the condition of exhaustiveness, in particular, must be taken into consideration. If a tautology is not specialized, i.e., if it contains only variables, it will always be exhaustive in its elementary terms. This is clear because, when the formula is written as an expansion in elementary terms, we cannot cancel one of these terms without destroying the tautological character. Thus when we cancel a term of the tautologous expansion

$$a.b \vee a.\bar{b} \vee \bar{a}.b \vee \bar{a}.\bar{b} \tag{25}$$

the formula is no longer true for all values of 'a' and 'b'. It is different with the expansion in major terms; in this form some tautologies are exhaustive, but others are not. For instance, the formula

$$(x)[f(x) \vee g(x) \equiv \overline{\overline{f(x).g(x)}}] \tag{26}$$

is exhaustive in its major terms when 'f' and 'g' are free variables, since in the expansion

$$(x)\{[f(x) \vee g(x)].\overline{\overline{f(x).g(x)}} \vee \overline{f(x) \vee g(x)}.\overline{f(x).g(x)}\} \tag{27}$$

no term can be canceled. On the other hand, the tautology

$$(x)[f(x).\overline{f(x)} \supset g(x)] \tag{28}$$

is not exhaustive in its major terms since it has an always-false
implicans. When we write it in a form corresponding to (5), the
first term can be canceled. But it is exhaustive in its elementary
terms since, when the operand is expanded in elementary terms, it
assumes the form (25) with '$f(x)$' standing for 'a' and '$g(x)$' standing for 'b'; (28) is therefore an inflated form. That it is possible to
transform (28) into a fully exhaustive formula follows from the
proof given above for inflated forms. For tautologies, however,
such transformation is trivial since all tautologies are equivalent to
one another; for this very reason, on the other hand, it is not clear
what we mean by saying 'the same tautology in another form'.
However, our rules make it clear that (28), like all other inflated
forms of tautologies, is nomological only in the wider sense.

The tertium non datur

$$(x)[f(x) \vee \overline{f(x)}] \tag{29}$$

is not exhaustive in its major terms and is therefore an inflated
formula. This fact may perhaps explain why some people consider
it unreasonable to write (29) with the inclusive 'or', and want to
have it written with the sign of the exclusive 'or':

$$(x)[f(x) \wedge \overline{f(x)}] \tag{30}$$

in which form it would be fully exhaustive and therefore nomological
in the narrower sense. We see that not all tautological operations
can be interpreted as connective operations in the narrower sense.
This is the reason that not all tautological operations are considered
reasonable operations.

Let us add a remark about specialized tautologies. They may be
nonexhaustive even when they result by specialization from a
tautology that is fully exhaustive. But then their nonexhaustive-
ness is due to empirical facts and thus not derivable from tautologies;
the formulas, therefore, satisfy the qualification of derivation and
are nomological in the narrower sense. Thus, when the function
'$f(\hat{x})$' occurring in (26) means '$\hat{x}$ is a sea serpent', and '$g(\hat{x})$' means
'$\hat{x}$ is an inhabitant of the moon', the formula is no longer exhaustive
in its major terms since the first term in (27) can be canceled.
But the inexhaustiveness is demonstrable only by reference to

empirical fact, and the formula, therefore, remains nomological in the narrower sense.

The inclusion of tautological operations in nomological operations, however, is of secondary importance since tautological operations can be defined independently, and in a simpler way. The practical significance of the given definition of nomological operations is that it furnishes an instrument for interpreting the synthetic connective operations occurring in scientific and conversational language, in particular the concept of natural law. Thus the proof that a certain general implication is a natural law comes down to a demonstration, first, that the implication is true in the adjunctive sense, and second, that it satisfies the conditions for nomological operations in the narrower sense.

§ 62. The logical nature of connective operations

We introduced connective operations by the following procedure. Employing adjunctive operations alone we defined a certain group of formulas, which we called the group of nomological formulas. We then defined nomological operations as the major operations of these formulas. We finally established the rule that nomological operations can be replaced by the corresponding connective operations.

When we translate this procedure into a *definition* of connective operations, we come to the result that a connective formula means the same as the corresponding nomological formula with the addition that this formula *is* a nomological formula. Thus we can write:

$$(x)[f(x) \ni g(x)] =_{Df} \begin{cases} (x)[f(x) \supset g(x)] \\ \text{`}(x)[f(x) \supset g(x)]\text{' is a nomological formula} \end{cases} \quad (1)$$

Since the addition is a statement of the metalanguage, we see that connective operations are improper object terms; they are introduced by a *shifting of the level of language*. This result holds for both synthetic and analytic connective operations. With this conception we follow ideas first developed by Carnap [1] with respect to tautological operations.

The shifting of the level of language is only the first peculiarity of connective operations. We must explain now a second peculiarity, which concerns a *transfer of meaning from the general to the*

---

[1] Cf. footnote on p. 42.

particular case. Nomological character is originally definable only
with respect to general statements and is then transferred to
specialized statements. Now the facts to which the specialized
statement refers cannot verify it in the sense of a connective statement; what these facts can verify is only the adjunctive statement.
Let us consider as an illustration the implication 'if Peter takes
poison he will die', which derives from the general nomological
implication 'for all x, if x takes poison he will die'. We have no
means of verifying the individual nomological implication for the
case of Peter; even if he takes the poison and dies, it will be only a
case of confirmation, not a verification of the general implication.
We see that, as far as the individual case is concerned, the connective statement does not say any more than the corresponding
adjunctive statement. The surplus meaning of the particular connective statement concerns facts other than those spoken of in the
statement. Saying that Peter's taking poison connectively implies,
or causes, his death means two things: first, that there is an individual adjunctive implication between the two sentences, and second,
that there is a corresponding general implication which is an original
nomological statement. Statements of the kind mentioned are
therefore improper particular statements; their complete meaning
is given only with the addition of a general statement. This is what
we call transfer of meaning from the general to the particular case.
It is the same thing that is meant by saying that necessity is expressible in terms of 'always'.

The definition of a connective implication in a specialized statement would therefore be as follows:

$$f(x_1) \mathbin{3} g(x_1) =_{Df} \begin{cases} f(x_1) \supset g(x_1) \\ (x)[f(x) \supset g(x)] \\ \text{'}(x)[f(x) \supset g(x)]\text{' is a nomological formula} \end{cases} \quad (2)$$

Since the first line of the right-hand side is deducible from the
second, it follows that the meanings of the specialized connective
implication (2) and the general connective implication (1) are
identical. The independent use of the specialized form with an
apparent meaning of its own is what we call transfer of meaning
from the general to the particular case. The accent implication on
the left-hand side of (2) is therefore an *improper individual operation.*

The considerations presented apply also to tautologies, as tautologies are a subclass of nomological formulas. If we interpret a tautological operation as connective we make use of a shifting of the level of language; and if we do so for a specialized tautology we add to this a transfer of meaning from the general to the particular case. Thus, in the statement 'neither Caesar nor Napoleon reached an age of 60 years implies Napoleon did not reach an age of 60 years', the word 'implies', if it is meant in the sense of a connective operation, is an improper object term and an improper individual-operation.

It may seem strange that well-known terms of physical language are regarded by us as being used out of place, so to speak; that they require an interpretation, in which, first, a certain general statement is to be added, and second, even a statement of the metalanguage appears. The question arises whether this interpretation is inevitable. Even if connective operations, in ultimate analysis, must be given this peculiar interpretation, it may still appear possible to construct a calculus which deals with the accent operations like operations of the object language and sets up the rules for the manipulation of such signs. Although this calculus would be constructed in accordance with the principle that all its formulas and rules must be translatable into general formulas of the object language and statements of the metalanguage, such translatability need not be stated in the calculus, which rather would represent an independent logical system. In fact, the calculus of *strict implication* developed by C. I. Lewis [1] constitutes a system of this kind, although when this calculus was constructed it was not known that a metalinguistic interpretation of a reasonable implication can be given. The reason we regard such a calculus as dispensable can be made clear as follows.

If we symbolize connective implications of conversational language by adjunctive implications of the usual calculus, we shall never arrive at unreasonable consequences when these implications are used as the major operations of nomological formulas. We

---

[1] C. I. Lewis and C. H. Langford, *Symbolic Logic*, Century, New York, 1932. The strict implication, however, differs from our accent operations in many respects; thus it does not include the synthetic connective operations, and furthermore, as Lewis has pointed out (pp. 174–175), it leads into paradoxes analogous to those of adjunctive implication.

therefore can apply any of the formulas (6a–6h, § 8) for a transformation of such an implication; the transformed formula will then also be a nomological formula, at least in the wider sense. Difficulties arise only when we use a negation of an adjunctive implication. Thus the implication in '$\overline{a \supset b}$' is not a connective implication, even if the formula is a nomological formula, because it does not constitute the major operation of this formula (which is given by the negation). Consequently, transformations of this formula will not express the meaning of a denied connective implication. In particular, the formula '$a.\bar{b}$', which according to (6b, § 8) is equivalent to '$\overline{a \supset b}$', cannot be asserted when the existence of a connective implication is denied. Thus, when a chemist says: 'if this salt is soluble in water this fact does not imply that it is a sodium compound', he does not wish to say: 'it is soluble in water and it is not a sodium compound'. This transformation is false because he does not say '$\overline{a \supset b}$', but asserts: '$a \overline{\supset} b$', which amounts to the same as "$a \supset b$' is not a nomological formula'.[1] It is therefore only when a connective implication is denied that our calculus does not offer the means of a symbolization of the statement.

This limitation of the calculus, however, is not serious. When we look more closely at the language of science, we find that denied connective implications are never used as premises of deductive operations. What a scientist wishes to say when he denies a connective implication is expressed rather by the statement: 'do not use this implication as a premise'. That is, his statement is actually in the metalanguage and expresses a rule for his derivations. Thus in the above example the chemist wishes to say: 'should this salt be soluble in water, do not make the inference that it is a sodium compound'. This means: 'do not use the implication '$a \supset b$' as a premise'. It does not mean that the adjunctive implication is false; it merely means that we cannot derive its truth from a law of nature, and that we can therefore assert the implication only after the truth of 'a' and 'b' has been ascertained separately — in which case it is no longer necessary to use the implication for an inference. The impossibility of expressing denied connective implications in

---

[1] The negation concerns here the conjunction of the two statements '$a \supset b$' and "'$a \supset b$' is a nomological formula'. Now, if '$a \supset b$' is false, it is not a nomological formula. Therefore the negation results in the above statement. We have '$\overline{p.q} \equiv \bar{p} \lor \bar{q}$'; if '$\bar{p} \supset \bar{q}$', '$\overline{p.q}$' is the same as '$\bar{q}$'.

the calculus, therefore, does not restrict us as far as the symbolization of deductive operations of science is concerned; it rather compels us, in a wholesome way, to separate the deductive part of scientific thought from metalinguistic additions by which this deductive procedure is analyzed and controlled.

When the negated implication is expressed in class notation, on the other hand, there is no difficulty in interpreting it as an adjunctive implication and applying the tautological transformations pertaining to implication, since class notation employs adjunctive implications (cf. p. 199). We say 'not all Cretans are honest', and are willing to regard this statement as equivalent to 'some Cretans are not honest'. Were we to use a form containing the word 'implies' we might hesitate to accept the equivalence. The statement 'that somebody is a Cretan does not imply that he is honest' would be regarded as true even if, by chance, Crete were a happy island inhabited by honest men only. The negation then would deny the existence of a connective implication between being a Cretan and being honest, and thus would require the metalanguage for its formulation like the preceding example of an individual implication.

The metalinguistic elements of conversational and scientific language should not be regarded as irrelevant; on the contrary, they may contribute their share to the system of knowledge and may formulate results which cannot be stated in another way. In § 55 we introduced the distinction between denotative and expressive terms. We may add here that an expressive nature is not restricted to specific terms; a statement as a whole may express something in addition to what it denotes. The content of a statement is not always exhausted by what the statement *says;* in addition there may be a content which the statement *shows* through its structural form.[1] A tautology like

$$\overline{a \vee b \supset \bar{a}} \tag{3}$$

says 'not *a*-or-*b* implies not *a*', but it does not say that this is a tautology. That it is tautologous is *shown* by the formula but is

---

[1] The distinction between what a statement *says* and what it *shows* originates from Wittgenstein, *Tractatus Logico-Philosophicus*, Harcourt, Brace, New York, 1922, p. 79. Wittgenstein believed, however, that what a statement shows cannot be said. He did not see that it can be said in the metalanguage. This possibility was pointed out by Russell in his introduction to Wittgenstein's book; *op. cit.*, p. 23.

not *said*. If we want to say it, we must use the metalanguage. This *dual function of saying and showing* is the very reason that the metalanguage cannot be dispensed with even in conversational language; it formulates contents which cannot be said *in* the object language, although they may be given *with* the object language.

These considerations hold also for synthetic and therefore empirical statements. Although empirical science tries to formulate empirical content, as fully as possible, in the object language, there remain empirical contents which can be said only in the metalanguage. They are those contents which enter into language, not through the meaning, but through the structural form of statements of the object language. If the formula

$$(x)[f(x) \supset g(x)] \tag{4}$$

is an original nomological formula, this fact may be seen from the statements

$$(\exists x)f(x).g(x)$$
$$(\exists x)\overline{f(x)}.g(x) \tag{5}$$
$$(\exists x)\overline{f(x)}.\overline{g(x)}$$

which for this case express the exhaustiveness, and from the fact that the functions 'f' and 'g', which we assume to be given in their wording, do not contain individual signs. Thus if statements (4) and (5) are given, we can prove that (4) is an original nomological formula; but we cannot derive this from those statements by means of operations remaining in the object language. Statements (4) and (5) together *show* the original nomological character of (4), but they do not *say* it. They show it through their syntactical structure, but they do not say it as part of their meaning.

Now statements (4) and (5) are verified by experience; the relation between the functions 'f' and 'g' is therefore empirical. We see that this empirical relation, though completely *presented* through statements (4) and (5), is not completely *said* by these statements; in order to say it fully we must add a statement in the metalanguage saying that (4) is an original nomological formula. This addition in the metalanguage can be made without further inquiry into experience, and in this sense all empirical content of (4) and (5) is given in these statements; but, if we want to express this content completely, we must use the metalanguage. It is this kind of empirical content which can be said only in the metalanguage.

To see the relevance of such empirical statements in the metalanguage for science, let us consider our example of Peter's dogs. Both the general implications, 'for all x, x is Peter's dog implies x has a heart' and 'for all x, x is Peter's dog implies x has a white spot on its forehead', are true, and are not original nomological implications since they contain a proper name. Now the relation between Peter's dogs and hearts differs relevantly from the relation between Peter's dogs and white spots: the first embodies a natural law; the second is a matter of chance. We formulate this difference by saying that the first statement is derivable from an original nomological statement, the second, not. The difference is therefore expressed by the fact that, when we eliminate the reference to Peter, the relation between dogs and hearts satisfies formulas (4) and (5), whereas these formulas do not hold for the relation between dogs and white spots. This empirical difference, although *given* in formulas of the object language, is not *said* in such formulas; it can be said only by statements in the metalanguage referring to the syntactical structure and the derivative relations of the statements established in the object language. Such a statement is expressed by the use of a connective implication.

The foregoing analysis makes it clear that it is the dual function of statements — the joint function of saying and showing — that makes it impossible to define all concepts of empirical science in the object language. In particular, it is the formulation of a conditional contrary to fact that requires metalinguistic terms. Fortunately, however, the use of concepts defined in the metalanguage can be confined to statements in which we speak about the method and the results of scientific inquiries, whereas these concepts can be dispensed with during the process of derivative operations leading to those results. Thus it is sufficient for the purpose of logical deductions if the statements used as premises are nomological statements; within the process of deduction, however, it need not be said that they are so. This is the reason that the adjunctive calculus of logic and mathematics can be used for all deductions in empirical sciences, and that a connective calculus can be dispensed with. In other words: empirical science can be constructed throughout by means of an adjunctive calculus, in which nomological operations occupy the place of connective operations.

The interpretation of connective operations by means of nomological statements and statements in the metalanguage has the advantage that the connective operations so defined are truth-functional (cf. § 7). They are so because the statements on the right-hand side in (1) and (2) contain only adjunctive operations. However, if this translation of connective into adjunctive operations is to be more than a merely fictitious coordination, it must be shown that the truth-functional character of adjunctive operations can be retained also for general universal statements. Since such statements cannot be verified by complete enumeration of individual cases, but only by means of inductive inferences, this problem leads into questions of the theory of probability.[1] We cannot enter here into these investigations but must restrict ourselves to the statement that a theory of probability can be developed in which, by means of the frequency interpretation, probability statements are reduced to a class of elementary statements which are true or false, or, more precisely, which are approximately true or false. Thus the ultimate answer to the question of truth-functional character of connective operations, given in the theory of probability and induction, is that these operations are truth-functional to the same degree of approximation as can be secured for the application of two-valued logic.

§ 63. Relative nomological statements

We now must add some remarks about a usage of nomological operations that is somewhat different from the one so far considered. Let us illustrate this usage by an example. With respect to the button of an electric bell, we can say: if you press this button the bell will ring. Using '$f(x)$' for 'x is pressed', and '$g(y)$' for 'y rings', we can write this implication with respect to the individual button x_1 and the individual bell y_1 in the form

$$f(x_1) \supset g(y_1) \tag{1}$$

This appears a reasonable implication; but it is not nomological since it includes not only the presupposition of the laws of elec-

---

[1] Cf. the author's *Experience and Prediction*, University of Chicago Press, Chicago, 1938, chapter V.

tricity but also the assumptions that the button is wired to the bell, that the network is supplied with current, and so on. Assumptions of this kind concern matters of fact that are not statable in nomological statements. Let us summarize all these assumptions in the statement 'p'; we then obtain a nomological statement only if we include 'p' in the implicans, i.e., if we set up the implication

$$p.f(x_1) \supset g(y_1) \qquad (2)$$

Statement (1), therefore, appears as an elliptic form of speech since the condition 'p' is omitted. It is obvious that many forms of what we regard as reasonable implications fall in this category. Of this kind is, for instance, the implication cited above according to which a man on the moon would use an oxygen respirator, since it is only a matter of fact, and not a nomological statement, that there is no atmosphere on the moon.

A somewhat different interpretation may be constructed as follows. (1) is derivable only if, in addition to nomological statements, the nonnomological statement 'p' is used. We may therefore say that (1) is a nomological statement *relative to* 'p'. We thus introduce a category of *relative nomological statements*, compared with which the statements so far defined may be called *absolute nomological statements*. We call a statement 'q' *nomological relative* to 'p' when there exists a set 'n' of nomological statements so that 'q' is derivable from the conjunction '$p.n$'.

This condition is the same as saying that '$p \supset q$' is a nomological statement. For if 'q' is derivable from '$p.n$', the implication '$p.n \supset q$' is a tautology (we assume the formulas to be closed); but this is the same as '$n \supset (p \supset q)$', from which form we see that '$p \supset q$' is derivable from the nomological statement 'n' and thus is a nomological statement.

The introduction of the distinction between nomological statements in the narrower and wider sense involves some complications. In order that 'q' be nomological in the narrower sense relative to 'p' we require first that the statement '$p \supset q$' be nomological in the narrower sense. Second, however, we require that it must not be possible to derive from 'p' a residual statement of 'q' in major or elementary terms, in other words, to derive from 'p' the result that 'q' is not fully exhaustive. This addition is necessary because otherwise the truth of 'p' would enable us to construct an un-reasonable statement 'q', while '$p \supset q$' is reasonable. The statement '$a \supset a \vee b$', for instance, is nomological in the narrower sense; how-ever, the statement '$a \vee b$' is nomológical in the wider sense rela-tive to 'a' because from 'a' we can derive the residual statement '$a.b \vee a.\bar{b}$' of '$a \vee b$'. This terminology corresponds to the usual conception: if 'a' is known to be true we regard '$a \vee b$' as an un-reasonable statement.[1]

Let us illustrate the use of relative nomological statements by an extension of a former example. We said that the statement 'for all x, if x is a dog of Peter's it has a white spot on its forehead' is not a nomological statement and represents an unreasonable implica-tion. Now assume that somebody tells me: 'there is a dog in your garden, I think it is Peter's dog'. I may then answer: 'If it is Peter's dog, it will have a white spot on its forehead'. This implica-tion appears now quite reasonable. It may be used for the purpose of identification; for instance, finding that the dog has no white spot on its forehead we may conclude that it is not Peter's dog.

To analyze this case we must use time arguments. At the time t_1 all Peter's dogs had white spots on their foreheads; so we may write, with 'f' for 'is a dog of Peter's' and 'g' for 'has a white spot on its forehead',

$$(x)[f(x, t_1) \supset g(x, t_1)] \tag{3}$$

Furthermore, we know that Peter did not acquire another dog, or lose one of his; and so, when t_2 is the present time, we have

$$(x)[f(x, t_1) \equiv f(x, t_2)] \tag{4}$$

Using the nomological statement

$$(x)[g(x, t_1) \equiv g(x, t_2)] \tag{5}$$

which says that dogs do not change the pattern of their skin, we can derive

$$(x)[f(x, t_2) \supset g(x, t_2)] \tag{6}$$

and thus, when x_1 is the specific dog considered,

$$f(x_1, t_2) \supset g(x_1, t_2) \tag{7}$$

---

[1] Since '$p \supset p$' is a tautology, it is a consequence of our definition that every state-ment 'p' is nomological relative to itself. But it is so only in the wider sense since '$p \supset p$' is an inflated form, the case '$\bar{p}.p$' being a contradiction.

This implication can be regarded as a nomological statement relative to (3) and (4), since (7) is derivable from these statements by the addition of nomological statements alone. Regarding (7) as a reasonable implication appears to be based on the fact that we refer this implication to some antecedent knowledge, i.e., that it represents a nomological statement relative to our knowledge of matters of fact.

Now assume that we see the dog in the garden has no white spot on its forehead. Will it be reasonable to say, with respect to this dog, 'if it were Peter's dog it would have a white spot on its forehead'? Although this implication is true in the adjunctive sense, the implicans being false, we would not agree with such a statement if it is regarded as meaning: 'if this dog which now has no white spot belonged to Peter, it would have a white spot on its forehead'. Thus we do not wish to assert that if Peter had meanwhile bought this dog it would have changed the pattern of its skin. This shows that with the acquisition of further knowledge about the dog the implication has changed its meaning.

This objection can be analyzed as follows. Since the dog now has no white spot we infer that at the time t_1 it did not have a white spot either. Introducing the latter statement as a new assumption, i.e., adding the statement

$$\overline{g(x_1, t_1)} \tag{8}$$

we can prove, using (5), that the implicans of (7) is false; i.e., we can derive a residual statement of (7). Therefore, the qualification of derivation is not satisfied. While (7) is nomological in the narrower sense relative to (3) and (4), it is only nomological in the wider sense relative to (3), (4), and (8). When our knowledge includes the statement (8), therefore, we no longer regard the implication (7) as reasonable. The same result obtains when we regard our implication as constructed from (7) by the addition of (8) in the implicans.

This analysis shows that what we regard as reasonable operations may vary with the extent of our knowledge. The reason is that we often use relative connective operations, applying them in the form of an elliptic mode of speech in which we do not mention the matters of fact to which the statements are referred.

The notion of relative nomological statement permits us to

clarify a problem that has been discussed occasionally. When we have a nomological statement in the narrower sense, of the form,

$$a \supset (b \supset c) \tag{9}$$

the first implication will be a connective implication, whereas the rules for nomological statements do not confer this property on the second implication. But when we are confronted by formulas of the form (9) we regard also the second implication as reasonable. This cannot be explained by assuming that also '$b \supset c$' is a nomological statement; for if it were, the implication (9) would be unnecessary since then we would know that the implicate is true. The explanation is rather given by the fact that then '$b \supset c$' is a relative nomological statement, namely, relative to 'a'. The notion of relative nomological implication, therefore, allows us to account for cases where two implications are arranged in series.

As an example, consider the statement 'if a number is even, then if it is divisible by 3, it is also divisible by 6', which as a mathematical statement is a tautology. Now when we know of a certain unknown number that it is even, the statement 'if this number is divisible by 3 it is also divisible by 6' appears as a reasonable implication, although it is neither a tautology nor a synthetic nomological statement. But it is a relative nomological statement with respect to our knowledge; this explains why we regard it as a reasonable implication.

On the other hand, in the statement

$$\bar{a} \supset (a \supset b) \tag{10}$$

only the first implication is a reasonable implication, the statement (10) being nomological in the narrower sense. The second implication is not reasonable because '$a \supset b$' is nomological in the wider sense relative to '$\bar{a}$'; this follows because from '$\bar{a}$' we can derive the residual statement '$\bar{a}.b \vee \bar{a}.\bar{b}$' of '$a \supset b$'. We see that the fact that a false proposition implies every proposition, expressed in the second implication of (10), can be stated only in terms of an implication which in our terminology is not a reasonable implication.

§ 64. The semi-adjunctive implication

While nomological statements, of the absolute or relative form, suffice to define the connective operations used in conversational language, there remain linguistic forms of a different nature which require a somewhat different kind of operation for their interpretation.

Consider an instance where a gambler says: 'if I cast the die this time, face 'six' will turn up'. Assume that you ask him to cast the die, and that, in fact, he throws a 'six'. He then will regard his statement as verified. In doing so, however, he has used the first row of the truth tables of implication in the direction from left to right (cf. § 7), i.e., he regards his implication as adjunctive. He may insist that he is entitled to do so because he referred only to one individual case, and did not claim generality for his statement. In fact, he certainly could not; we know that if he were to repeat throwing the die he would arrive at different results in most of the cases, the result 'six' being restricted to one sixth of his throws.

On the other hand, if he had not thrown the die at all, he would not regard his statement as verified, although the adjunctive implication furnishes the value 'true' also in this case. The implication used here, therefore, is not fully adjunctive. We shall call it a *semi-adjunctive* implication. An implication '$a \supset b$' is semi-adjunctive if it is regarded as verified when 'a' is true and 'b' is true, as falsified when 'a' is true and 'b' is false, while its truth-value is left open when 'a' is false, whether or not 'b' is true.[1]

This definition does not mean that the implication must forever remain unverified if the implicans is false. We have other ways of

---

[1] An implication of this kind was analyzed in the author's paper: 'Über die semantische und die Objekt-Auffassung von Wahrscheinlichkeitsausdrücken', *Journal of Unified Science (Erkenntnis)*, Vol. 8, 1939, pp. 61–62. It was called there *operation of selection (Auswahl-Operation)*, for reasons pertaining to the theory of probability, and defined by means of a truth table corresponding to that of adjunctive implication, with the difference, however, that the symbols 'T' of the third and fourth row were replaced by a question mark. This table was originally given in the author's *Wahrscheinlichkeitslehre*, A. W. Sijthoff, Leiden, 1935, p. 381, table IIc. The question mark means that the truth-values of the respective cases can be 'T' or 'F'. When we read this table from left to right, therefore, we cannot say that the implication is verified in these cases. On the other hand, reading the table from right to left, we must take into account that the question mark may mean a 'T'; therefore, when we know by other means that '$a \supset b$' is true, we cannot say which of the three cases 'T–T', 'F–T', 'F–F', will hold, as in the case of the fully adjunctive implication. The question mark, therefore, is a device which excludes the adjunctive interpretation of the respective cases. A fully connective operation could be defined by putting a question mark in place of every 'T' of its truth table. An implication similar to the semi-adjunctive one, which, however, is three-valued, is used in the author's *Philosophic Foundations of Quantum Mechanics*, University of California Press, Berkeley, 1944, § 34.

verification, the same as used for connective implications. In the example given, such verification may be technically impossible, although in principle it should be possible to foretell the results of a throw of a die from the initial conditions, given in the position of the die, the physiological status of the person considered, and other factors. Let us say that Laplace's superman could do it. There are instances in which even we could. Imagine that a chemist is given a salt of unknown nature and wishes to know whether this salt is soluble in water; i.e., he wishes to know the truth-value of the implication, 'if this salt is put into water, it will dissolve'. If he throws the salt into water and it dissolves, he will regard the implication as true. If he did not put the salt into water, he could employ other means of verification; for instance, he could inquire from which bottle the salt was taken and apply his general knowledge of chemical laws in order to answer the question. The implication that he considers, therefore, is semi-adjunctive, since the first row of the truth tables is read from left to right.

Implications of this kind are frequently used — in fact, wherever individual events are considered and no claim is made that the same result will obtain on repetition. These conditions are realized for the example of the die because there we know that repetition in general would not furnish the same result. In the second example, the implication must be carefully worded; namely, the implicans must be so expressed that it does not describe conditions for which a connective implication can be established. When we say, 'if sodium chloride is put into water, it will dissolve', the implication is connective and known to be true. A single instance where sodium chloride was dissolved in water cannot be regarded as a strict verification; only in combination with other inductive laws, for instance, such as saying that one sample of salt reacts like every other one, could such inference be made. This is a case belonging in the theory of inductive evidence; in fact, there are cases known in this theory where one instance is regarded as sufficient for an induction. In our example, we deal with a different kind of implication; we wish to know whether this particular sample of salt of an unknown nature will dissolve if put into water. The question will be definitively answered in the affirmative when implicans and implicate are both observed to be true. Implications of this kind, therefore, are usually not stated in the form of predictions but in the form of questions.

It appears that the confusion of connective and semi-adjunctive implications is one of the factors obscuring the discussion of reasonable implication. It is clear that in statements of physical laws connective implications are meant even when they are applied to single cases; the meaning of implications used in prophecies and other mystical predictions, however, is not clearly defined. Imagine that a soothsayer predicts 'if you meet a white cat on New Year's Eve, you will become rich during that year'. Assume that you do meet the white cat and you inherit a fortune during the year; would this prove that the implication stated by the soothsayer was true? You may find out that the man was a rather bad prophet, that he made similar but unfulfilled predictions to many other persons. You then would argue that the man's success in your case was due to chance and that the appearance of the white cat did not *imply* your inheritance; i.e., you interpret the implication as connective. The soothsayer, I think, would defend himself by saying that at least in your case the implication stated by him was verified as true; in other words, he interprets it as a semi-adjunctive implication. There is no way, therefore, of proving that the man's prediction in this case remains unverified, since he did not explicitly state beforehand what sort of implication he meant, and therefore can easily retreat behind the ambiguities of conversational language. It is a different question whether in the face of his other failures you will ever consult the man again. The truth of the individual prediction will depend on the interpretation of the implication and must be distinguished from the prophet's general reliability.

§ 65. Modalities

The use of connective operations is closely connected with that of modalities. We, too, have used the words 'necessary', 'possible', 'impossible', in the interpretation of connective operations; the definition of these operations, however, has been given without any reference to these concepts (cf. § 61). We shall now show that both derive from the same root, and that not only connective operations but also modalities can be defined in terms of nomological operations.

The duality of analytic and synthetic nomological operations is repeated in the definition of modalities. Using analytic nomological operations we obtain *logical modalities;* by means of synthetic nomological operations we define *physical modalities.*

Before entering into the discussion of these definitions let us make some introductory remarks. Modalities are used with reference either to facts or to situations; in both uses they are terms of the object language, not of the metalanguage. Thus we say 'it is possible that European culture may vanish from this world', or 'the vanishing of European culture is possible'. This sentence can be regarded as having either the form '$(w)p^*(v)$ is possible' (cf. § 48) or 'p is possible'; but it does not have the form "'p' is possible". In correspondence with our treatment of connective operations we shall interpret this mode of speech, therefore, as a shifting of the level of language. We shall thus consider modalities as improper object terms; i.e., we shall use them in the object language, but define them by reference to the metalanguage. From the meaning of the term 'impossible' it is clear that when we apply this term to facts, or situations, we use a fictitious existence, since real facts, or situations, are not impossible (cf. § 49). In order to be consistent, we shall therefore regard all three modalities as referring to fictitious objects, and thus as extraneous terms (cf. § 58).

We must add a remark concerning the word 'possible'. This term is usually so defined that possibility includes necessity; i.e., what is necessary is also possible. We distinguished above (p. 128) between a wider and a narrower meaning of possibility and said that we prefer to use a definition of modalities such that the meanings do not overlap. We shall use the term 'possible', therefore, in the narrower sense, and shall say that what is necessary is not possible. In cases of ambiguity the phrase 'merely possible' may appear convenient for the narrower meaning; the wider meaning may be covered by the term 'not impossible'.

A further remark concerns a distinction between *absolute* and *relative* modalities similar to the corresponding distinction introduced for nomological statements in § 63. We say 'it is necessary that two masses attract each other', 'it is possible that there will be a storm tomorrow'. As no reference to other facts is involved in these statements, we use here absolute modalities. An analysis of the meteorological conditions of today may lead us to the result that there will necessarily be a storm tomorrow; in saying so we use a relative modality since we mean 'relative to the meteorological conditions of today a storm is necessary tomorrow'. The necessity of this example is different from the necessity of the first; that two masses attract each other is necessary in a sense statable without reference to special conditions. Both kinds of modalities are used, and it would be a mistake to believe that one of them is eliminable. The fact that a storm tomorrow is necessary relative to the meteorological conditions of today does not make it false to say that in another sense a storm tomorrow is only possible. Logicians should not forbid the use of such concepts which may be expedient for many purposes; they should restrict their task to clarifying the meanings of the concepts used.

We turn now to the definition of absolute modalities. Let 'p' be a formula composed of elementary sentences 'a', 'b', ... For the sake of simplicity we shall conceive the modalities as referring to the situation p, thus disregarding the second interpretation according to which they refer to the fact $(w)p^*(v)$. We define absolute logical modalities by the definitions: [1]

p is logically necessary $=_{Df}$ 'p' is a tautology
p is logically impossible $=_{Df}$ 'p' is a contradiction
p is logically possible $=_{Df}$ 'p' is synthetic

Using these terms we say, for instance: it is necessary that the sum of the angles in a Euclidean triangle be equal to two right angles; it is possible that the energy of a closed system increases; it is impossible that a triangle has four angles.

The definition of absolute physical modalities is as follows:

p is physically necessary $=_{Df}$ 'p' is a nomological statement
p is physically impossible $=_{Df}$ '$\bar{p}$' is a nomological statement
p is physically possible $=_{Df}$ neither 'p' nor '$\bar{p}$' is a nomological statement

By 'nomological' we mean here 'nomological in the wider sense'. This interpretation seems to be required by the usage of language.

---

[1] The use of the sign '$=_{Df}$' between two propositions of different languages requires some qualifications. We have here only equipollent expressions, and we cannot always substitute one expression for the other unless certain precautions are taken.

Otherwise a statement which follows tautologically from a necessary statement might not be necessary. Thus the statement 'for all x, x is a piece of iron lighter than water implies x is green' follows tautologically from the physically necessary statement 'no piece of iron is lighter than water', and is therefore considered physically necessary although it represents an unreasonable implication. If we had required nomological character in the narrower sense in the definition of necessity the implication would be only possible. Furthermore, the negation of the implication would not be impossible either, and we would come to the consequence that 'there is a piece of iron lighter than water and not green' is possible, whereas 'there is a piece of iron lighter than water' is impossible.

We see that nomological character in the narrower sense is appropriate only for the definition of reasonable operations, not for the definition of reasonable modalities.

The analogy between logical and physical modalities is seen from the fact that the definition of the first can be obtained from the definition of the second if we replace, in that definition, the word 'nomological' by 'analytic nomological', i.e., if we exclude synthetic nomological statements in the definiens. The difference between the two kinds of modalities is expressed in their extensions. Physical possibility differs from logical possibility in that its domain is more restricted; i.e., not all logically possible situations are also physically possible. Thus it is not physically possible that the energy of a closed system increases. Consequently both the domains of necessity and impossibility are wider for physical than for logical modalities. What is logically necessary, or impossible, is also physically necessary, or impossible, because the word 'nomological' in the definiens includes both kinds of nomological character; however, the converse does not hold. Thus it is physically, but not logically, necessary that two masses attract each other; it is physically, but not logically, impossible that a signal moves faster than light. On the other hand, what is physically possible is also logically possible. These relations of extensions are made clear in Fig. 9.

Synthetic existential statements may be special cases of nomological formulas, and then they are also nomological. Thus the statement 'there is a thing which, if it is metallic and heated, will expand' is nomological and necessary. Usually we do not make such statements, as they are trivial in comparison with the important all-statements from which they are derived. Of practical importance are existential statements that are not derivable from nomological formulas and therefore are not nomological; these statements have the absolute modality of possibility. Thus it is not necessary in the

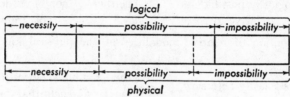

FIG. 9.   Modalities.

absolute sense that there are metals lighter than water. It would be correct to say that this is possible; however, in conversational language we use the word 'possible' only as long as the truth of a sentence is not yet established. Negative synthetic existential statements are equivalent to all-statements, and therefore are absolutely necessary if they are nomological formulas. Thus it is physically necessary in the absolute sense that there are no signals faster than light.

Sometimes we speak, not of physical possibility, but of technical possibility. This is a still narrower concept; there are situations which are physically possible but not technically possible. Thus it is physically possible, but not technically possible, to go to the moon. We shall not consider technical possibility here because it depends on human abilities and varies with the development of civilization; moreover, it applies only to actions performed by men, not to happenings in general.

We turn now to relative modalities. If we say, for instance, that with respect to the weather conditions of today it is physically necessary that there be a storm tomorrow, we state this necessity with reference to some facts which, on their part, are not necessary in the absolute sense. What we want to say is that, given these conditions, it is derivable by means of physical laws that there will be a storm. Similarly we shall have relative logical necessity if the derivation can be made by means of tautological formulas alone.

We therefore now define relative modalities, first of the logical kind, as follows:

$$q \text{ is logically necessary relative to } p \quad \Big\} =_{Df} \text{ '}q\text{' is derivable from '}p\text{'}$$

$$q \text{ is logically impossible relative to } p \quad \Big\} =_{Df} \text{ '}\bar{q}\text{' is derivable from '}p\text{'}$$

$$q \text{ is logically possible relative to } p \quad \Big\} =_{Df} \Big\{ \begin{array}{l} \text{neither '}q\text{' nor '}\bar{q}\text{' is deriv-} \\ \text{able from '}p\text{'} \end{array}$$

The definition of relative modalities of the physical kind is as follows:

$$q \text{ is physically necessary relative to } p \quad \Big\} =_{Df} \left\{ \begin{array}{l} \text{there is a nomological state-} \\ \text{ment '}n\text{' such that '}q\text{' is} \\ \text{derivable from '}p.n\text{'} \end{array} \right.$$

$$q \text{ is physically impossible relative to } p \quad \Big\} =_{Df} \left\{ \begin{array}{l} \text{there is a nomological state-} \\ \text{ment '}n\text{' such that '}\bar{q}\text{' is} \\ \text{derivable from '}p.n\text{'} \end{array} \right.$$

$$q \text{ is physically possible relative to } p \quad \Big\} =_{Df} \left\{ \begin{array}{l} \text{there is no nomological state-} \\ \text{ment '}n\text{' such that '}q\text{' or '}\bar{q}\text{'} \\ \text{is derivable from '}p.n\text{'} \end{array} \right.$$

The letters 'p', 'q', and 'n' are used here as abbreviations of complicated formulas; frequently 'p' and 'n' will be conjunctions of formulas. The formulas 'p' and 'q' are not nomological formulas, whereas 'n' is nomological. As before, the word 'nomological' here is meant to include both analytic and synthetic nomological statements, so that physical necessity and impossibility include the corresponding logical modalities.

We give some examples. That a triangle has an angular sum of 180° is logically necessary relative to the axioms of Euclid. That a triangle has an angular sum of less than 180° is logically impossible relative to the axioms of Euclid. That a triangle has three equal angles is logically possible relative to the axioms of Euclid. Turning to physical relative modalities we see that 'p' formulates what is usually called *initial conditions* of closed physical processes. This term denotes the physical conditions existing at the beginning of the closed process, i.e., of a process that in its further development is practically closed to influences from its environment. Given the initial conditions and the physical laws concerned, we can calculate the course of the process. Thus the movement of a projectile is determined by the laws of gravity, the laws of motion in a resisting medium like air, and the initial conditions of shooting, such as the inclination of the gun's barrel, the powder charge, and so on. The last conditions would be formulated as 'p' in our definition, whereas the two physical laws mentioned would be included in 'n'; by 'q' we might abbreviate a statement concerning the point of arrival of the projectile. For instance, meaning by 'q' the statement that the projectile will come down within a certain area, we may say that q is necessary relative to p; meaning by 'q' the statement that the projectile will hit a definite point within the considered area, we may say that q is possible relative to p; meaning by 'q' the statement that the projectile will come down 'outside the considered area, we may say that q is impossible relative to p. Physicists speak of initial conditions in a somewhat wider meaning when they refer to a cross-section of physical conditions existing while the process is going on. Thus, considering the present position and velocity of the earth, with reference to the sun, as initial conditions p, we shall say that an elliptic orbit of the earth ($= q$) is necessary relative to p, whereas relative to other initial conditions a hyperbolic shape of the orbit is necessary. The nomological formula 'n' in this example is Newton's law.

From our definitions we see immediately that relative physical modalities can be transformed into relative logical modalities if we consider the modality of q relative to $p.n$. However, this transformation is not expedient since it is the relation between the nonnomological statements 'q' and 'p' which we want to characterize. These statements describe individual events which may be chosen accessible to observation, whereas the nomological formula 'n' expresses relations between all things.

The difference between logical and physical modalities may be illustrated by Fig. 9. For relative modalities this diagram may be interpreted as giving the modalities for a chosen set of initial conditions p. If we enlarge this set by adding more true statements about initial conditions, the domain of physical possibility shrinks more and more. The same happens if the set of nomological formulas 'n' is enlarged. If it is true that all laws of nature are ultimately deducible from the principles of relativity and atomic physics, these principles would constitute the widest possible set 'n'. It is the idea of determinism that if we choose as 'n' this widest set, and as 'p' the totality of true statements about initial conditions at one moment concerning the whole universe, the domain of possibility is reduced to zero, and only the two domains of necessity and impossibility are left. Upon this dream of physicists, modern quantum mechanics has put a check in Heisenberg's principle of indeterminacy, which states that in the described case the domain of relative physical possibility would not be zero. Apart from this restriction it is questionable, however, whether there is an ultimate set 'n' of physical laws from which all other natural laws can be deduced. It may be that in this field also, as in the domain of initial conditions, we can speak only of a continuous process of approximation leading to more comprehensive but never to ultimate principles of nature.

In specialized nomological statements we often omit some parts which are true for the individual considered, and then consider the statement as necessary in the absolute sense. Thus we say that it is necessary that the sun loses its energy by radiation. We refer this assertion to the corresponding universal nomological statement: 'if a body emitting an amount of radiation does not receive from its environment an equal amount of radiation, it will lose its energy'. The sun satisfies the implicans; therefore, we omit this condition. Since this omission means an inference in which we use the non-nomological statement 'p', stating that the implicans holds for the sun, we obtain here only a necessity relative to p; this reference, however, is usually dropped. If we call specialized statements of this kind necessary, we must interpret this term as meaning 'necessary relative to our knowledge'. The general reference 'relative to our knowledge' is very often to be added to modal statements if they are to be meaningful.

It will appear pertinent to discuss the relation of absolute modalities to the division of propositional functions into always true, always false, and mixed functions. We showed in § 23 that this division can be used for the definition of *extensional modalities*. Let us call the modalities defined in the present section *nomological modalities*. The two kinds of modalities coincide only for universal functions, i.e., functions containing no individual-signs. Otherwise it may happen that an always-false function is not impossible, but possible in the nomological sense, or that an always-true function is not necessary, but merely possible in the nomological sense. Examples are given by the functions '$\hat{x}$ is a man on top of Mount Everest in the year 1940', which is always false but not impossible, and '$\hat{x}$ is a planet of the solar system at present known implies $\hat{x}$ revolves in the same direction as the earth', which is always true but not necessary. A mixed function, however, is always a possible function.[1]

These relations are sometimes expressed by the statement: existence implies possibility, but possibility does not imply existence. In this formulation the term 'possibility' is used in the sense of 'not-impossibility', i.e., as including necessity. This interpretation, at least, is required for the first part of the statement.

The superiority of the nomological modalities is visible in the fact that for the extensional modalities the distinction between physical and logical modalities cannot be defined since it is based on the distinction between synthetic and analytic nomological statements, a distinction that finds no expression in the extension of functions. In spite of the difference between extensional and nomological modalities, the tables IIIb, § 23, apply also to nomological modalities; that is, these tables remain valid when the symbols 'A', 'M', 'E' are interpreted as nomological modalities. From the nonadjunctive character of the tables IIIb it follows, then, that neither the extensional nor the nomological modalities are adjunctive.

We must now add a general remark concerning both nomological statements and modalities. We have included in the definition of original nomological statements the requirement that these statements be true; derivative nomological statements are con-

---

[1] It should be kept in mind that extensional modalities apply only to propositional functions, and cannot be unambiguously applied to individual-statements, since the same individual-statement may be derived from different propositional functions and then may have different extensional modalities. Thus the statement 'Hannibal was a man on top of Mont Blanc in the year 218 B.C.' may be derived from the function '$\hat{x}$ was a man on top of Mont Blanc in the year 218 B.C.', or from the function '$\hat{x}$ was a man on top of Mont Blanc in the year $\hat{i}$'. Accordingly different modalities would result for the specialized statement, since the first function is always false, whereas the second is a mixed function. For absolute nomological modalities we have no such difficulty, since the question whether the specialized statement is derivable from original nomological formulas can be answered unambiguously.

sequently also true. Now there are many universal statements whose truth or falsehood is unknown, and others which though at present considered true may later turn out to be false. We therefore do not know all nomological statements, and we may even be mistaken about those statements that today we consider nomological. For all practical applications of the preceding considerations we shall therefore have to replace the word 'nomological' by the phrase 'known as nomological'.

This indeterminacy is transferred to modalities, since modalities are defined by means of nomological statements. We call modalities defined in terms of statements known to be nomological, *epistemic modalities;* [1] in contradistinction to them the other modalities are called *objective modalities.* For all practical purposes we are concerned with epistemic modalities.

We do not know to what extent the domains of epistemic modalities differ from those of objective modalities, as we do not know whether the universal statements considered true are actually true. However, assuming that they are actually true, and considering the fact that we certainly do not know all true universal statements, we may say that the domain of epistemic possibility is wider than that of objective possibility, whereas the domains of both epistemic necessity and epistemic impossibility are narrower than the corresponding domains of objective necessity and objective impossibility, respectively. For the phrases 'epistemically necessary' and 'epistemically impossible' we may use, respectively, the synonymous phrases 'known as necessary' and 'known as impossible'. For 'epistemically possible', however, we have only the synonymous phrase 'not known as necessary or impossible'.

§ 66. Modal interpretation of connective operations

We have mentioned on several occasions that the meaning of connective operations is often stated in terms of modalities. In our construction we preferred to follow the reversed procedure: we gave an independent definition of nomological formulas and then defined both connective operations and modalities in terms of these formulas. After giving these definitions, however, we shall now use

---

[1] The word 'epistemic' is derived from the Greek 'ἐπιστήμη', which means 'knowledge'.

the defined modalities for an interpretation of connective operations, not with the intention of giving another definition of these operations, but to make their meaning clearer. This interpretation concerns general connective operations, i.e., connective operations occurring as major operations of an all-statement.

An all-statement can be written in the form

$$(x)f(x) \tag{1}$$

where '$f(x)$' is an abbreviation including the operand and, possibly, some further major operators among which there may be existential operators. (1) is equivalent to

$$\overline{(\exists x)\overline{f(x)}} \tag{2}$$

Now if (1) is a nomological formula, the term under the long negation line in (2) expresses an impossible situation (in the absolute sense). This means: it is impossible to find an x which does not have the property f. We can express this idea by saying: the statement (1) must hold for all possible x. We formulate this result as follows:

In a general nomological statement the range of the all-operator is given by all possible argument-objects and is not restricted to all real argument-objects.

If the statement has more than one major all-operator, but no major existential operator, the given rule can be applied to all these operators. If the statement contains a major existential operator, the given rule applies only to those major all-operators which precede the major existential operator. In order to make our form of speaking complete we then may be obliged to interpret an existential assertion as meaning the existence of a possible object. Thus if we specialize the statement

$$(x)(\exists y)f(x, y) \tag{3}$$

for a possible object x, the coordinated object y whose existence is stated may be also only a possible object. Let, for instance, (3) be used as an abbreviation for the statement 'for all x there is a y such that, if x is a human being, y is the father of x'. We then shall say that in the domain of possible things there is a y coordinated to every x such that for these two possible things the functional '$f(x, y)$'

holds. This is but another way of expressing the statement: 'if *there were* an x, *there would be* a y such that '$f(x, y)$' holds'.

If all the major operators of a nomological statement are all-operators, and if the major operand contains propositional operations, we can also use the following form of speaking. Instead of expanding the operand in the form of a disjunction of T-cases of the major operation we can also expand it in the form of a conjunction of the denied F-cases. For instance, the statement

$$(x)[f(x) \equiv g(x)] \tag{4}$$

can be expanded either in the disjunction

$$(x)[f(x).g(x) \vee \overline{f(x)}.\overline{g(x)}] \tag{5}$$

or in the conjunction

$$(x)[\overline{f(x).\overline{g(x)}}.\overline{\overline{f(x)}.g(x)}] \tag{6}$$

Because of (9a, § 25) this is the same as

$$(x)\overline{f(x).\overline{g(x)}}.(x)\overline{\overline{f(x)}.g(x)} \tag{7}$$

Now, if (4) is a nomological statement, (7) is also nomological. Then (7) can be interpreted as meaning that the F-cases of (4) are impossible.

A similar interpretation can be given to nomological statements which contain no major operators at all. Since the major operation of a nomological statement can be replaced by a connective operation, we can formulate our result as follows:

If a connective operation occurs in a statement whose major operators are all-operators, or which has no major operators at all, the F-cases of the operation are impossible.

It would be wrong to infer from this result that all the T-cases of a connective operation were possible, since a nonexhaustive statement may have impossible T-cases, as for instance the statement (20, § 61).[1] The relation between connective operations and possible cases must be derived from the preceding rule, and can be stated as follows:

---

[1] It might be suggested that reasonable operations be defined in such a way that, if the statement is nonexhaustive, each excluded T-case should at least be possible. Such a definition, however, leads to difficulties, as for instance neither the inference from 'a implies b' to '$a.c$ implies b', nor the inference from 'a implies $\bar{a}$' to '$\bar{a}$', could then be applied to an implication so defined.

A general connective operation holds for all possible objects, and is not restricted to real objects.

This is a very plausible interpretation of connective operations. It expresses very clearly what we mean by a reasonable general operation, in particular by a reasonable implication.

To illustrate our rule let us take our previous example, '*x* is a man living on the moon today implies *x* uses an oxygen respirator'. It is possible that there is a man on the moon today, as there is no nomological formula excluding this situation. We know that our connective implication would be true if there were a man on the moon; therefore it holds for all possible *x*. We cannot say the same for an adjunctive implication such as the one used above, '*x* is a man living on the moon today implies *x* wears a red necktie'; this implication does not hold for all possible *x*.

We have not specified, in this section, whether we are speaking of logical or physical modalities. This distinction applies as follows: if the connective operation is an analytic operation, the range of the all-operator is given by all logically possible arguments *x;* if it is a synthetic operation, the range of the all-operator is given by all physically possible arguments *x*. Thus the range of the operator in the implication 'for all *x*, *x* is a Euclidean triangle implies *x* has an angular sum of 180°' is given by all logically possible triangles; the range of the operator in the implication 'for all *x*, *x* is a physical body not retained by other bodies implies *x* moves according to the laws of gravity' is given by all physically possible bodies.

We may use the modal interpretation of nomological statements if we want to explain why a given statement is not nomological. Thus we may say that the statement 'for all *x*, *x* is not a man living on the moon today' is not nomological because it does not hold for all possible arguments. We must not forget, however, that with our definition of terms it is more correct to say: it is possible that there is a man on the moon today because the above statement is not nomological. It is not nomological because it cannot be proved as true, in the adjunctive sense, if we state it without limitations as to time. For similar reasons, the statement 'there is no chemical element with an atomic weight greater than 239' is not nomological; it holds only with the qualification 'among the elements at present known'. We therefore say it is possible that there are heavier elements. By considerations concerning the stability of the atom, physicists may at some time be induced to say that it is impossible that there are heavier elements; the statement would then be maintained as true without qualification, and thus be made nomological.

In addition to implication, synthetic connective equivalence plays an important role in physics. We use this equivalence when we say that two statements have the same meaning, physically speaking. Thus when Einstein says that the two statements, 'a body is in a gravitational field' and 'a body is in accelerated motion', have the same meaning, he uses a synthetic connective equivalence. His contention means: it is physically impossible that one of the statements is true, and the other false, or vice versa. He does not speak of a tautological equivalence here; he says explicitly that this equivalence is a physical law based on experience, and that we can imagine a world in which this equivalence does not hold.[1] On the other hand, if a synthetic equivalence, although adjunctively true, is not nomological, we do not speak of equisignificance. Thus the equivalence between 'x is a human being' and 'x is a featherless biped' is not a physical equisignificance, as this equivalence is not a nomological formula. It is not, because it is proved to be true only for the present time, and is therefore to be written as a formula containing an individual-sign, namely, 'the present time', which is not derivable from an original nomological formula. In other words: it is possible that one day there will be featherless bipeds which are not human beings.

[1] As to this, cf. the distinction of physical meaning and logical meaning in the author's *Experience and Prediction*, University of Chicago Press, Chicago, 1938, § 6.

Exercises

I. INTRODUCTION

§ 3. Punctuate the following sentences, and insert quotation marks where necessary:

3-1. Red designates a color but red indicates danger, even though the name of danger is colored black.

3-2. The Roman numeral IIII means the same as IV and both designate the number 4.

3-3. A word is a phrase consisting of two words, although every English phrase consists of three words.

3-4. John where Jack had had had had had had had had had had had a better effect on the teacher.

3-5. A word that by the addition of some letters becomes shorter is short.

II. THE CALCULUS OF PROPOSITIONS

Translate the following propositions into symbols, using one letter for each elementary proposition. Use the adjunctive implication for expressions like 'if . . . then'. Use the period sign not only for the 'and' but also for words like 'but' and 'although'. (As to the latter words, cf. the remarks on p. 329.) Always use the inclusive 'or'. Express phrases like 'it is false that . . .' by a long negation line.

§ 7. A. Formulate in symbols:

7-A-1. The fire was produced by arson or by an explosion.

7-A-2. There will be a better future or there will not.

7-A-3. Either you did not send me the book or it has been lost in the mail.

7-A-4. We cannot both dance and not pay the piper.

7-A-5. If Peter comes I shall not go.

7-A-6. Whether you wish to enlist in the army or in the navy, you must be physically fit.

7-A-7. If you wash and iron that dress, you will ruin it.

7-A-8. Neither Napoleon nor Hitler could invade Britain.

7-A-9. In case neither John nor Mary goes, Peter will go.

7-A-10. You are not wood, you are not stones, but men.

7–A–11. We have not asked for help; neither do we desire it.

7–A–12. The board will not take action unless it sees its way to do something effective towards reestablishing order or preventing further disruption.

B. Formulate in symbols and state whether the operation is adjunctive or connective. The classification as adjunctive or connective is achieved by answering the question: would you regard the compound statement as verified if you knew the truth-value of one or both of the elementary statements and if you found the value *truth* for the compound statement by reading the truth tables from left to right?

7–B–1. The speech of the Prime Minister will be transmitted by Station KNX or by Station KFI.

7–B–2. Neither Plato nor Aristotle was an empiricist.

7–B–3. We do not recognize a change of opinion but frankly admit a change of emphasis.

7–B–4. The compass needle points to the north or there is some iron in the environment.

7–B–5. If the volume of a gas is decreased, the pressure increases.

7–B–6. If the good people have not made the world better, they have surely made it duller.

7–B–7. When a plant is shut off from the light, it loses its green color.

7–B–8. Only if we first consider a large variety of instances can we form a trustworthy opinion.

7–B–9. As thy soul liveth, I will not leave thee.

7–B–10. If, and only if, a number is positive, its square root is a real number.

§ 8. Determine by case analysis whether the following formulas are analytic, synthetic, or contradictory.

8–1. $a \supset \bar{a}$

8–2. $(a \supset b) \vee (b \supset a)$

8–3. $\bar{a}.\bar{b} \equiv \bar{a} \supset b$

8–4. $\overline{(\bar{a} \vee b).\bar{b}.\bar{a}} \supset (a \equiv \bar{b})$

8–5. $(a \equiv b) \supset a.\bar{b}$

8–6. $a . \overline{a \supset b} \vee b$

8–7. $a \vee b \supset (b \vee c \supset a \vee c)$

8–8. $(a \supset b) \supset (a \vee c \supset b \vee c)$

8–9. $\bar{a} \vee b.c \supset (a \vee b).\bar{c}$

§ 11. The remark 'Tr: . . .' means: 'transform by means of . . .'. When this remark is added to a problem, use the respective formula of the list on p. 38 for the transformation. Notice that an equivalence can be used in both directions for a transformation. When several numbers follow the sign 'Tr:', as in 'Tr: 6a, 6f, 4c', the meaning is: apply these formulas successively, i.e., apply each formula to the result of the preceding transformation. When the sign 'Tr:' is repeated, the meaning is that the new transformation applies to the original formula. Thus 'Tr: 6a. Tr: 6b' means that the transformation by means of 6b is to be applied to the original formula. The result of each transformation, including intermediary steps, should always be reformulated in words. We omit this reformulation in the solutions in order to save space.

A. Symbolize and transform:

11-A-1. The lilies of the field toil not, nor do they spin. Tr: 5b.

11-A-2. You cannot both have your cake and eat it. Tr: 5a.

11-A-3. It is not true that logic is dry or not exciting. Tr: 5b, 1d.

11-A-4. If you step on the brake, the car slows down. Tr: 6a. Tr: 6b. Tr: 6c.

11-A-5. If a man is not prejudiced he will recognize the value of a liberal education. Tr: 6a. Tr: 6b. Tr: 6c.

11-A-6. Unless you come with me I shall not go. Tr: 6c.

11-A-7. If you neither have the cash nor can obtain a sufficient loan you cannot buy a car. Tr: 6a. Tr: 6c.

11-A-8. Mr. A will run for the presidency, and Mr. A will be the candidate or Mr. B will be the Republican candidate. Tr: 4e.

11-A-9. Throwing with two dice, a man makes the following bet: 'I shall get face 'six' on one die or the other'. Translate the sentence between quotes into symbols. Tr: 5e.

11-A-10. Women are fickle, but either women are not fickle or they are slow of understanding. Tr: 4a, 5d.

11-A-11. Either there is no new thing under the sun or both Copernicus and Einstein are innovators. Tr: 4b, 6a twice, 6e.

11-A-12. If silence is not golden then the radio is both a friend to the lonely and not a friend to the lonely. Tr: 6a, 1d, 5d.

11-A-13. It is not the case that Cleopatra was alive in 1938 and was not married to Hitler and not to Mussolini. Tr: 5a twice, 1d, 6a so as to make the first sentence the implicans.

11-A-14. If Peter or Paul comes, we shall play chess. Tr: 6f.

11-A-15. If Peter and Paul come, we shall play the Schubert trio tonight. Tr: 6d. Tr: 6h.

11-A-16. If John must know Greek in order to understand Plato, he will not understand Plato. Tr: 6a for the first implication, 6f, 6a for the tautological part, 5c.
Notice that the conclusion 'John does not know Greek' cannot be drawn. When we believe that this conclusion follows we have tacitly assumed the further premise: 'if John knows Greek he will understand Plato'. Show that, with this addition, the conclusion can be drawn.

11-A-17. When triangles are congruent, they have equal angles and equal areas; and when triangles have equal angles and equal areas, they are congruent. Tr: 7a, 7b, 5a, 4a.

B. Transform into simplest form, using the formulas of the list on p. 38:

11–B–1. $(a \vee \bar{c}).\bar{b}$

11–B–2. $a \supset (b \supset c)$

11–B–3. $a.c \supset b$

11–B–4. $\overline{a.\bar{b} \vee c}$

11–B–5. $\overline{a \vee b \supset c}$

11–B–6. $(a \vee b).(\bar{a} \vee c)$

11–B–7. $\overline{a.b.(c \supset d)}$

11–B–8. $\overline{(a \vee b).\bar{a}.\bar{b}}$

11–B–9. $\overline{a \supset b} \supset c$

11–B–10. $\overline{a.(b \vee c)}$

11–B–11. $a \supset \overline{b.c}$

11–B–12. $a \equiv a \supset b$

11–B–13. $\overline{a.(b \supset c)} \equiv d$

11–B–14. $\overline{c \equiv d}$

11–B–15. $\overline{a \supset b.\bar{c}}$

11–B–16. $\overline{(a \supset \bar{b}).(c \supset a)}$

11–B–17. $a \equiv \overline{c \vee d}$

11–B–18. $\overline{(\bar{a} \supset b \vee c) \supset a}$

C. Expand in major and elementary T-cases.

11–C–1. $\overline{a.b}$ 11–C–2. $a \supset b.c$ 11–C–3. $a \vee b.\bar{c}$

D. In the following problems the result of the transformation is given, and the individual steps of the transformation must be found, by the use of the list on p. 38.

	Transform	into
11–D–1.	$\overline{a \supset b}$	$a.\bar{b}$
11–D–2.	$a \vee (b \supset \bar{c})$	$\bar{a} \supset \overline{b.c}$
11–D–3.	$a \supset (b \supset c)$	$a.b \supset c$ without using 6d
11–D–4.	$b \vee \bar{b} \supset c$	c
11–D–5.	$a \vee \bar{c} \supset b$	$(\bar{a} \vee b).(c \vee b)$
11–D–6.	$a \vee (\bar{a} \supset c)$	$\bar{a} \supset c$
11–D–7.	$a \supset b.c$	$(a \supset b).(a \supset c)$ without using 6e
11–D–8.	$\overline{a \supset b.c}$	$(a \supset b) \supset a.\bar{c}$
11–D–9.	$(a \equiv b)$	$(a \supset b).(b \supset a)$ without using 7a
11–D–10.	$\overline{a \equiv b}$	$a \equiv \bar{b}$ without using 7c
11–D–11.	$a \equiv \bar{b}$	$(a \vee b).(\bar{a} \vee \bar{b})$
11–D–12.	$(a \equiv b) \supset c$	$\bar{c} \supset (a \vee b).\overline{a.b}$
11–D–13.	$a.(a \equiv b)$	$a.b$
11–D–14.	$\bar{a}.b \vee (a \equiv b)$	$a \supset b$
11–D–15.	$a \vee (b \equiv c.a)$	$b \supset a$

§ 15. In the following problems a conclusion is to be derived. The conclusion will say less than the premises. Use the list of formulas on p. 38.

A. Derive a conclusion by means of a given formula.

15–A–1. If a boy is forbidden to smoke, he will smoke, and if he is forbidden to drive a car, he will drive a car. Use formulas 8j and 8k.

15-A-2. If the earth moves into the shadow of the moon, there will be an eclipse, and if we have an eclipse, the stars become visible. Use formula 8 l.

15-A-3. If an American citizen is put into jail, he is entitled to a hearing within twenty-four hours. Apply 8g, using for 'c' the sentence 'he has no taxable income'.

B. Derive a given conclusion:

15-B-1. If a man is 21 years of age or older, a citizen, and a negro, he cannot vote.

If a man can vote, he is 21 years of age, or older, and a citizen.

Derive: If a man is a negro, he cannot vote.

Notice that, since the conclusion is made illegal by the 15th amendment of the American Constitution, the combination of the two premises is also made illegal by this amendment. The first premise, in fact, is not true, whereas the second is. If, on the other hand, the second premise were false, the first premise need not contradict the 15th amendment. For instance, if the right to vote were reserved to persons under 21 years of age, the second premise would be false while the first premise would be true. It then would not represent an exemption for negroes, however, since the conclusion would not be derivable.

15-B-2. Let us call a certain rat 'Bill'. Now, if Bill is put into maze A and Bill is hungry, then either Bill has learned maze A or he will be able to find food.

If Bill has learned maze A then he will be able to find food

But he was put into maze A, and did not find food.

Derive: Bill was not hungry.

15-B-3. Given: $a \supset b \lor c$

$\bar{b}.\bar{c}$

Derive: $\bar{a}$

15-B-4. Given: $a \lor (b \supset c)$

$b.\bar{c} \lor \bar{d}$

d

Derive: a

C. Find and derive a relevant conclusion.

15-C-1. There had to be heavy taxes in this country or an American army could not be sent to Europe.

An American army was sent to Europe.

Conclusion?

15-C-2. If Napoleon had not been sick during the battle of Leipzig, he would have won the battle.

If Napoleon had won the battle of Leipzig, Central Europe would have remained under his control.

If Central Europe had remained under Napoleon's control, the Continental System would have ruined England economically.

If England had been ruined economically, it finally would have been conquered by Napoleon's armies.

If England had been conquered by Napoleon's armies, it would have been put under a French government.

If England had been under a French government, the decimal system would have been introduced in England.

If the decimal system once had been introduced in England, it would be in use there today.

Conclusion?

15–C–3. If this is a galvanometer and current passes through the wire, then the needle will be deflected.

The needle is not deflected.

But this is a galvanometer.

Conclusion?

15–C–4. The car does not start.

If the car does not start, the failure is caused either by unclean points of the distributor or by an obstruction of the carburetor.

I clean the points and the car does not start.

Conclusion?

15–C–5. When wages and prices are raised, there will be inflation.

When there is inflation, the government will have to abdicate.

The government is able and will therefore manage not to abdicate.

The wages have been raised.

Conclusion?

15–C–6. If this salt is put into water and heated, then, if it dissolves, it is salt X or salt Y.

If it is salt X, it will make a red flame.

If it is salt Y, it will make a yellow flame.

The salt is put into water, heated, and it does dissolve.

The salt makes a yellow flame.

Conclusion?

15–C–7. When that switch is turned on and the vacuum in the glass valve is perfect, there will be a tension of 10,000 volts at the valve and the glass will show a green fluorescent light; and vice versa.

I shall turn that switch and there will be a tension of 10,000 volts at the valve.

Conclusion?

III. THE SIMPLE CALCULUS OF FUNCTIONS

Translate the following propositions into symbols, using the symbolism of the calculus of functions. Regard phrases consisting of verb and adverb, as in 'he feels well', as one function. Adjectives preceding a noun, however, should be regarded as separate functions. Do not express in the symbolism the time indication given in the tenses of verbs or the modal indication given in the moods of verbs. Regard the words 'I' and 'you' as signs of individuals.

§ 17.

17–1. George III was insane.

17–2. George Washington was president of the United States.

17–3. The Bosphorus separates Europe from Asia.

17–4. The Eiffel tower is not as tall as the Empire State Building.

17–5. I did not see Donald or William.

17–6. Oregon is situated between California and Washington.

§ 18.

18–A–1. All crows are black.

18–A–2. All brothers resemble each other.

18–A–3. All is well that ends well.

18–A–4. Barking dogs do not bite.

18–A–5. All mothers love their sons.

18–A–6. All followers of Oscar Wilde imitated his manners.

18–A–7. All men or women working at the Douglas factory are citizens of the United States.

18–A–8. All men and women working as welders receive equal wages. ['working as welders' = one function.]

18–B–1. There are black swans.

18–B–2. There is a tower which is not vertical.

18–B–3. John gave Mary a book.

18–B–4. A boy eats an apple.

18–B–5. A man came and brought Dr. A a briefcase.

18–B–6. Something is rotten in the state of Denmark.

18–B–7. There are logicians who do not admire Aristotle.

§ 19. Symbolize each statement, both with a longest and a shortest negation line.

19–1. Not all American soldiers are American citizens. ['American citizen' = 'citizen of America'; same interpretation for 'American soldiers'.]

19–2. No planet is self-luminous.

19–3. No mother hates her child.

19–4. Not all fathers have sons only.

19–5. There are no men on the moon.

19–6. No cold day is as cold as this one. ['this one' = individual-sign.]

19–7. A healthy man does not take drugs.

19–8. No clever girl marries a man whom she does not love.

§ 20. A. Symbolize:

20–A–1. All mammals breathe through lungs.

20–A–2. Every soldier wears a uniform.

20–A–3. The soldiers followed the order of their general.

20–A–4. All men or women working at the Douglas factory wear a badge.

20–A–5. There are metals lighter than water.

20–A–6. Every man other than John would have hated Margot.

20–A–7. There is a man who is taller than all other men.

20–A–8. Every flea has a smaller flea.

20–A–9. All living beings descend from one primitive species.

B. The problems of this group should first be written with divided operands and then be transformed into one-scope formulas.

20–B–1. Rolling stones gather no moss.

20–B–2. If some mammals live in the ocean, there are water animals which emerge when breathing. ['ocean' = individual-sign; 'water animals' = one function.]

20–B–3. If there are human beings on other planets than the Earth, modern scientists would not be surprised. ['earth' = individual-sign; 'human being' = one function.]

20–B–4. If all things consist of atoms, Democritus was a great philosopher. ['consist of atoms' = one function; 'great philosopher' = one function.]

20–B–5. If all men were heroes and all women were beautiful, the movies would appeal to nobody.

20–B–6. If malaria germs are transferred only by a species of flies, they will disappear when all swamps and stagnant pools are dried up. ['malaria germs' = one function; 'species of flies' = one function.]

§ 21. First formulate the following sentences by means of free variables; then apply the rule for free variables; and finally divide the operand where this is possible.

21–1. If anybody is just, Aristides is just.

21–2. If there were any oil deposits in Germany, there would be men who developed them. ['oil deposit' = one function.]

21–3. It is not true that any oil deposits are situated in Germany. ['oil deposit' = one function.] Transform the sentence into a negated existential statement.

21–4. Any human race is as good as any other one. ['human race' = one function.]

21–5. If anybody knows all the rules of his native language, I shall be surprised. ['native language' = one function.]

§ 22. Classify the following functions with respect to field properties, univocality, symmetry, transitivity, reflexivity.

22–1. to marry.

22–2. to see.

22–3. to include, restricted to closed curves in a plane.

22–4. to include, restricted to a set of concentric circles.

22–5. cousin.

22–6. president.

22–7. as tall as.

22–8. taller.

22–9. taller or as tall as.

22–10. larger by one, restricted to positive integers.

VII. ANALYSIS OF CONVERSATIONAL LANGUAGE

Symbolize by the use of functions.

§ 47. Formulate the definite descriptions by means of the iota-operator; do not use indefinite descriptions. If the resulting expressions are too long, use letters with subscripts as abbreviations for the descriptions and add the definition of these symbols; for instance, write:

$$x_1 = (\imath x)g(x)$$
$$y_1 = (\imath y)h(y)$$
$$f(x_1, y_1)$$

47–1. Robert Browning's wife was ill.

47–2. The father of John Stuart Mill is the author of the *Analysis of the Phenomena of the Human Mind*.

47–3. Bill's twin brother lent him his jacket.

47–4. Thomas Hobbes was born in the year of the defeat of the Armada.

47–5. The president of the United States at the time of the Civil War wrote a letter to Mrs. Bixby.

47–6. The serpent that stung thy father now wears his crown.

47–7. The miserable has no medicine but hope.

47–8. It is not the hen who cackles the most that lays the largest eggs.

§ 48. In order to express event-splitting, use abbreviations such as ' $g(x) =_{Df}$ $(\exists y)f(x, y)$ ', or ' $p =_{Df} (x)(\exists y)f(x, y)$ ', and then indicate the fact-function in the forms ' $[g(x)]^*(v)$ ' or ' $p^*(v)$ '.

48–1. Torricelli discovered that air exerts a pressure.

48-2. The destruction of Carthage made Rome the ruler of the Mediterranean.

48-3. It is not permitted to smoke in a bus.

48-4. It is more blessed to give than to receive.

48-5. It snows.

48-6. The production of synthetic rubber made the United States independent of the importation of natural rubber from the East Indies.

48-7. The discovery of gold brought a large number of immigrants to California [number = class].

§ 49. Indicate fictitious existence by subscripts of the existential operators or superscripts of constants. Do not give the translation of the statements into physical existence.

49-1. Columbus believed that he had discovered India.

49-2. Romeo did not know that Juliet was not dead.

49-3. Socrates refused to escape.

49-4. Pharaoh dreamed that there came up out of the river seven kine.

49-5. The teacher who is attempting to teach without inspiring the pupil with a desire to learn is like someone hammering on cold iron.

§ 50. Use the symbol 'θ' for the expression of all token-reflexive terms.

50-1. My father was a merchant.

50-2. This road goes to Yosemite.

50-3. Thou shalt see me at Philippi.

50-4. Here I shall leave you.

50-5. Beneath this stone, in hopes of Zion,
is laid the landlord of the Lion.

50-6. Beneath those rugged elms the rude forefathers of the hamlet sleep.

§ 51. A. Diagram the tenses for each clause separately, using forms like (1, § 51). Do not symbolize the sentences.

51-A-1. Thou swearest thou hast supped like a king.

51-A-2. If yet I have not all thy love,
Dear, I shall never have it all.

51-A-3. She that, O, broke her faith will soon break thee.

51-A-4. The north-bound train will have left the station when you arrive.

51-A-5. As you have lied once I cannot believe you.

51-A-6. After you lied yesterday I cannot believe you today.

51-A-7. Have you seen but a bright lily grow
Before rude hands have touch'd it?

51-A-8. Think not, leech, or newt, or toad
Will bite thy foot when thou hast trod.

51-A-9. If she love me, this believe
I will die ere she shall grieve.

51-A-10. Young Romeo will be older when you have found him than he was
when you sought him.

B. Symbolize and indicate the time argument expressed in the tenses. Use
't_0' for the point of speech, and write '$t < t_0$' or '$t > t_0$' for the indication
of time order. The general definition '$t_0 = (\imath t)rf(t, \Theta)$' may be regarded
as understood. Use constants, e.g., 'x_1', for token-reflexive terms like 'I'.

51-B-1. When Peter came, Dr. Jones received him.

51-B-2. When Peter came, he showed me the letter which John had written
to him.

51-B-3. When Peter comes, I shall have seen John.

51-B-4. If Peter comes he will give you a letter.

51-B-5. Aristotle was a disciple of Plato, who had been a disciple of Socrates.

51-B-6. The book you looked for will be reprinted.

51-B-7. Gold is heavier than iron.

51-B-8. You can fool all the people some of the time, and some of the people
all of the time; but you cannot fool all the people all of the time.
['can fool' = one function.]

51-B-9. He was a bold man that first ate an oyster.

51-B-10. Rarely, rarely, comest thou, Spirit of Delight.

§ 53. Use higher functions. Do not indicate the tenses of the verbs.

53-1. The moon was shining sulkily.

53-2. A burden weighed heavily on his shoulder. [Indicate the man referred to
by 'z_1'.]

53-3. He is an excellent speaker but has nothing to say. [For 'he' put 'x_1'.]

53-4. Uneasy lies the head that wears a crown.

53-5. Oranges taste good.

53-6. The wine seems good. [Use the symbol 'Θ'.]

53-7. Gossip is vice enjoyed vicariously.

§ 54. Use numerical predicates.

54-1. The temperature of the room is 68°. [Regard 'the room' as a proper
name.]

54-2. Peter is 6 feet tall.

54-3. Johnny likes to drive his car at 80 miles per hour.

54-4. The egg is as hot as the water in which it is boiled.

54-5. Some of the sequoia trees are older than some of the Egyptian pyramids.

54-6. Heard melodies are sweet, but those unheard are sweeter. [Use 'x' and 'y' for the lowest-level arguments, although a melody is of a higher type than physical objects.]

54-7. There is nothing in the world so irresistibly contagious as laughter and good humor. [Cf. remark added to 54-6 just above.]

§ 59. Classify each part of speech, using the table on pp. 352-353 and adding the respective number of this table, e.g., 'II, 2, a'. Do not classify with respect to category II, 1. For higher functions, also omit categories II, 2, and II, 3.

59-1. The present king of England was ceremonially crowned at Westminster Abbey.

59-2. There was a rain storm last night.

59-3. A fine grain development of a film is usually done slowly.

Solutions

3–1. 'Red' designates a color but red indicates danger, even though the name of danger is colored black.

3–2. The Roman numeral 'IIII' means the same as 'IV' and both designate the number 4.

3–3. 'A word' is a phrase consisting of two words, although 'every English phrase' consists of three words.

3–4. John where Jack had had 'had', had had 'had had'; 'had had' had had a better effect on the teacher.

3–5. A word that by the addition of some letters becomes 'shorter' is 'short'.

7–A–1. *a:* the fire was produced by arson. *b:* the fire was produced by an explosion.

$a \lor b$

7–A–2. *a:* there will be a better future.

$a \lor \bar{a}$

7–A–3. *a:* you sent me the book. *b:* it has been lost in the mail.

$\bar{a} \lor b$

7–A–4. *a:* we dance. *b:* we pay the piper.

$\overline{a.\bar{b}}$

7–A–5. *a:* Peter comes. *b:* I shall go.

$a \supset \bar{b}$

7–A–6. *a:* you wish to enlist in the army. *b:* you wish to enlist in the navy. *c:* you must be physically fit.

$a \lor b \supset c$

7–A–7. *a:* you wash that dress. *b:* you iron that dress. *c:* you will ruin it.

$a.b \supset c$

7–A–8. *a:* Napoleon could invade Britain. *b:* Hitler could invade Britain.

$\overline{a \lor b}$

7–A–9. *a:* John goes. *b:* Mary goes. *c:* Peter will go.

$\overline{a \lor b} \supset c$

7–A–10. *a:* you are wood. *b:* you are stones. *c:* you are men.

$\bar{a}.\bar{b}.c$

417

7–A–11. *a:* we have asked for help. *b:* we desire it.

$\bar{a}.\bar{b}$

7–A–12. *a:* the board will take action. *b:* it sees its way to do something effective towards reestablishing order. *c:* it sees its way to do something effective towards preventing further disruption.

$\overline{b \vee c} \supset \bar{a}$

7–B–1. *a:* the speech of the Prime Minister will be transmitted by Station KNX. *b:* the speech of the Prime Minister will be transmitted by Station KFI.

$a \vee b$ adjunctive

7–B–2. *a:* Plato was an empiricist. *b:* Aristotle was an empiricist.

$\overline{a \vee b}$ adjunctive

7–B–3. *a:* we recognize a change of opinion. *b:* we frankly admit a change of emphasis.

$\bar{a}.b$ adjunctive

7–B–4. *a:* the compass needle points to the north. *b:* there is some iron in the environment.

$a \vee b$ connective

7–B–5. *a:* the volume of a gas is reduced. *b:* the pressure increases.

$a \supset b$ connective

7–B–6. *a:* the good people have made the world better. *b:* they have made it duller.

$\bar{a} \supset b$ adjunctive

7–B–7. *a:* a plant is shut off from the light. *b:* it loses its green color.

$a \supset b$ connective

7–B–8. *a:* we first consider a large variety of instances. *b:* we can form a trustworthy opinion.

$a \equiv b$ connective

7–B–9. *a:* thy soul liveth. *b:* I will not leave thee.

$a \equiv b$ adjunctive

7–B–10. *a:* a number is positive. *b:* its square root is a real number.

$a \equiv b$ connective

8–1. synthetic   8–4. tautological   8–7. synthetic
8–2. tautological   8–5. synthetic   8–8. tautological
8–3. contradictory   8–6. contradictory   8–9. synthetic

11–A–1. *a:* the lilies of the field toil. *b:* they spin.

$\bar{a}.\bar{b}$
$\overline{a \vee b}$ [5b]

11–A–2. *a:* you have your cake. *b:* you eat it.

$\overline{a.b}$

$\bar{a} \vee \bar{b}$ [5a]

11–A–3. *a:* logic is dry. *b:* logic is exciting.

$\overline{a \vee b}$

$\bar{a}.b$ [5b, 1d]

11–A–4. *a:* you step on the brake. *b:* the car slows down.

$a \supset b$

$\bar{a} \vee b$ [6a]

$\overline{a.\bar{b}}$ [6b]

$\bar{b} \supset \bar{a}$ [6c]

Notice that the expression $\bar{a} \supset \bar{b}$ cannot be inferred.

11–A–5. *a:* a man is prejudiced. *b:* he will recognize the value of a liberal education.

$\bar{a} \supset b$

$a \vee b$ [6a]

$\overline{\bar{a}.\bar{b}}$ [6b]

$\bar{b} \supset a$ [6c]

11–A–6. *a:* you come with me. *b:* I shall go.

$\bar{a} \supset \bar{b}$

$b \supset a$ [6c]

11–A–7. *a:* you have the cash. *b:* you can obtain a sufficient loan. *c:* you can buy a car.

$\overline{a \vee b} \supset \bar{c}$

$a \vee b \vee \bar{c}$ [6a]

$c \supset a \vee b$ [6c]

11–A–8. *a:* Mr. A will run for the presidency (identical with: Mr. A will be the candidate). *b:* Mr. B will be the Republican candidate.

$a.(a \vee b)$

a [4e]

11–A–9. *a:* I shall get face 'six' on one die. *b:* I shall get face 'six' on the other die.

$a \vee b$

$a \vee \bar{a}.b$ [5e]

Notice that only in the second form the disjunction is exclusive. This transformation into an exclusive disjunction is used in the calculus of probability. The probability of '$a \vee b$' cannot be obtained by addition of the probabilities of 'a' and of 'b' because this disjunction is inclusive. It can be constructed, however, by the addition of the probabilities of 'a' and '$\bar{a}.b$', the latter being given by the product of the probabilities of '$\bar{a}$' and 'b'. Determine the numerical values by using the probability $\frac{1}{6}$ for 'a' and 'b', and check the result by counting among the 36 possible combinations those in which you have face 6 on at least one die.

11–A–10. *a:* women are fickle. *b:* they are slow of understanding.

$a.(\bar{a} \vee b)$
$a.\bar{a} \vee a.b$ [4a]
$a.b$ [5d]

11–A–11. *a:* there is a new thing under the sun. *b:* Copernicus is an innovator. *c:* Einstein is an innovator.

$\bar{a} \vee b.c$
$(\bar{a} \vee b).(\bar{a} \vee c)$ [4b]
$(a \supset b).(a \supset c)$ [6a]
$a \supset b.c$ [6e]

The last result can also be obtained directly from the first line by the use of 6a.

11–A–12. *a:* silence is golden. *b:* radio is a friend to the lonely.

$\bar{a} \supset b.\bar{b}$
$a \vee b.\bar{b}$ [6a, 1d]
a [5d]

11–A–13. *a:* Cleopatra was alive in 1938. *b:* she was married to Hitler. *c:* she was married to Mussolini.

$\overline{a.\bar{b}.\bar{c}}$
$\bar{a} \vee b \vee c$ [5a, 1d]
$a \supset b \vee c$ [6a]

11–A–14. *a:* Peter comes. *b:* Paul comes. *c:* we play chess.

$a \vee b \supset c$
$(a \supset c).(b \supset c)$

11–A–15. *a:* Peter comes. *b:* Paul comes. *c:* we shall play the Schubert trio tonight.

$a.b \supset c$
$a \supset (b \supset c)$ [6d]
$b \supset (a \supset c)$ [6d]
$(a \supset c) \vee (b \supset c)$ [6h]

The last form represents an unusual mode of speech. It can be explained as follows. Should only Paul come, but not Peter, we cannot infer that the trio will be played. In that case, it is Peter's coming which would have implied the playing of the trio. The 'or' between the implications leaves it open which implication is true. If none of the persons comes, both implications are true, in the adjunctive sense. When the implications are understood in the connective sense, the observed facts, at least, do not disprove the existence of one of the implications.

11–A–16. *a:* John understands Plato. *b:* John knows Greek.

$$(a \supset b) \supset \bar{a}$$
$$\bar{a} \vee b \supset \bar{a} \ [6a]$$
$$(\bar{a} \supset \bar{a}).(b \supset \bar{a}) \ [6f]$$
$$(a \vee \bar{a}).(b \supset \bar{a}) \ [6a]$$
$$b \supset \bar{a} \ [5c]$$

Additional premise: $b \supset a$

$$(b \supset \bar{a}).(b \supset a) \equiv b \supset a.\bar{a} \ [6e]$$
$$b \supset a.\bar{a}$$
$$\bar{b} \vee a.\bar{a} \ [6a]$$
$$\bar{b} \ [5d]$$

11–A–17. *a:* triangles are congruent. *b:* they have equal angles. *c:* they have equal areas.

$$(a \supset b.c).(b.c \supset a)$$
$$a \equiv b.c \ [7a]$$
$$a.b.c \vee \bar{a}.\bar{b}.\bar{c} \ [7b]$$
$$a.b.c \vee \bar{a}.(\bar{b} \vee \bar{c}) \ [5a]$$
$$a.b.c \vee \bar{a}.\bar{b} \vee \bar{a}.\bar{c} \ [4a]$$

The numbers in brackets refer to the formulas on p. 38. The last expression represents the simplest form:

11–B–1. $(a \vee \bar{c}).\bar{b} \equiv a.\bar{b} \vee \bar{c}.\bar{b} \ [4a]$

11–B–2. $a \supset (b \supset c) \equiv a.b \supset c \ [6d] \equiv \overline{a.b} \vee c \ [6a] \equiv \bar{a} \vee \bar{b} \vee c \ [5a]$

11–B–3. $a.c \supset b \equiv \overline{a.c} \vee b \ [6a] \equiv \bar{a} \vee \bar{c} \vee b \ [5a]$

11–B–4. $\overline{a.\bar{b} \vee c} \equiv \overline{a.\bar{b}}.\bar{c} \ [5a] \equiv (\bar{a} \vee b).\bar{c} \ [5a] \equiv \bar{a}.\bar{c} \vee b.\bar{c} \ [4a]$

11–B–5. $\overline{a \vee b \supset c} \equiv (a \vee b).\bar{c} \ [6b] \equiv a.\bar{c} \vee b.\bar{c} \ [4a]$

11–B–6. $(a \vee b).(\bar{a} \vee c) \equiv a.\bar{a} \vee a.c \vee b.\bar{a} \vee b.c \ [4c] \equiv a.c \vee b.\bar{a} \vee b.c \ [5d]$

11–B–7. $\overline{a.b.(c \supset d)} \equiv \bar{a} \vee \bar{b} \vee \overline{c \supset d} \ [5a] \equiv \bar{a} \vee \bar{b} \vee c.\bar{d} \ [6b]$

11–B–8. $(a \vee b).\overline{a.b} \equiv (a \vee b).(\bar{a} \vee \bar{b}) \ [5a] \equiv a.\bar{a} \vee a.\bar{b} \vee b.\bar{a} \vee b.\bar{b} \ [4c]$
$$\equiv a.\bar{b} \vee \bar{a}.b \ [5d]$$

11–B–9. $\overline{a \supset b} \supset c \equiv (a \supset b) \vee c \ [6a] \equiv \bar{a} \vee b \vee c \ [6a]$

11–B–10. $\overline{a.(b \vee c)} \equiv \bar{a} \vee \overline{b \vee c} \ [5a] \equiv \bar{a} \vee \bar{b}.\bar{c} \ [5b]$

11–B–11. $a \supset \overline{b.c} \equiv \bar{a} \vee \overline{b.c} \ [6a] \equiv \bar{a} \vee \bar{b} \vee \bar{c} \ [5a]$

11–B–12. $(a \equiv a \supset b) \equiv a.(a \supset b) \vee \bar{a}.\overline{a \supset b} \ [7b] \equiv a.(\bar{a} \vee b) \vee \bar{a}.a.\bar{b} \ [6a, 6b]$
$$\equiv a.(\bar{a} \vee b) \ [5d] \equiv a.\bar{a} \vee a.b \ [4a] \equiv a.b \ [5d]$$

11-B-13. $[\overline{a.(b \supset c)} \equiv d] \equiv (\bar{a} \vee \overline{b \supset c} \equiv d)$ [5a] $\equiv (\bar{a} \vee b.\bar{c} \equiv d)$ [6b]
$\equiv (\bar{a} \vee b.\bar{c}).d \vee \overline{\bar{a} \vee b.\bar{c}}.\bar{d}$ [7b] $\equiv \bar{a}.d \vee b.\bar{c}.d \vee a.\overline{b.\bar{c}}.\bar{d}$ [4a, 5b]
$\equiv \bar{a}.d \vee b.\bar{c}.d \vee a.\bar{d}.(\bar{b} \vee c)$ [5a] $\equiv \bar{a}.d \vee b.\bar{c}.d \vee a.\bar{d}.\bar{b} \vee a.\bar{d}.c$ [4a]

11-B-14. $\overline{c \equiv d} \equiv (c \equiv \bar{d})$ [7c] $\equiv c.\bar{d} \vee \bar{c}.d$ [7b]

11-B-15. $\overline{a \supset b.\bar{c}} \equiv a.\overline{b.\bar{c}}$ [6b] $\equiv a.(\bar{b} \vee c)$ [5a] $\equiv a.\bar{b} \vee a.c$ [4a]

11-B-16. $\overline{(a \supset b).(c \supset a)} \equiv \overline{a \supset b} \vee \overline{c \supset a}$ [5a] $\equiv a.\bar{b} \vee c.\bar{a}$ [6b]

11-B-17. $(a \equiv \overline{c \vee d}) \equiv a . \overline{c \vee d} \vee \bar{a}.(c \vee d)$ [7b] $\equiv a.\bar{c}.\bar{d} \vee \bar{a}.c \vee \bar{a}.d$ [5b, 4a]

11-B-18. $\overline{(\bar{a} \supset b \vee c) \supset a} \equiv (\bar{a} \supset b \vee c).\bar{a}$ [6b] $\equiv (a \vee b \vee c).\bar{a}$ [6a]
$\equiv a.\bar{a} \vee \bar{a}.b \vee \bar{a}.c$ [4a] $\equiv \bar{a}.b \vee \bar{a}.c$ [5d]

11-C-1. $a.\bar{b} \vee \bar{a}.b \vee \bar{a}.\bar{b}$
The expansion in major T-cases is identical with the original formula.

11-C-2. $a.b.c \vee \bar{a}.b.c \vee \bar{a}.\overline{b.c}$ [major T-cases]
$a.b.c \vee \bar{a}.b.c \vee \bar{a}.b.\bar{c} \vee \bar{a}.\bar{b}.c \vee \bar{a}.\bar{b}.\bar{c}$ [elementary T-cases]

11-C-3. $a.b.\bar{c} \vee a.\overline{b.\bar{c}} \vee \bar{a}.b.\bar{c}$ [major T-cases]
$a.b.\bar{c} \vee a.b.c \vee a.\bar{b}.\bar{c} \vee a.\bar{b}.c \vee \bar{a}.b.\bar{c}$ [elementary T-cases]

11-D-1. $\overline{\overline{a \supset b}} \equiv \overline{\overline{a.\bar{b}}}$ [6b] $\equiv a.\bar{b}$ [1d]

11-D-2. $a \vee (b \supset \bar{c}) \equiv \bar{a} \supset (b \supset \bar{c})$ [6a] $\equiv \bar{a} \supset (\bar{b} \vee \bar{c})$ [6a] $\equiv \bar{a} \supset \overline{b.c}$ [5a]

11-D-3. $a \supset (b \supset c) \equiv \bar{a} \vee \bar{b} \vee c$ [6a] $\equiv \overline{a.b} \vee c$ [5a] $\equiv a.b \supset c$ [6a]

11-D-4. $b \vee \bar{b} \supset c \equiv \overline{b \vee \bar{b}} \vee c$ [6a] $\equiv b.\bar{b} \vee c$ [5b] $\equiv c$ [5d]

11-D-5. $a \vee \bar{c} \supset b \equiv (a \supset b).(\bar{c} \supset b)$ [6f] $\equiv (\bar{a} \vee b).(c \vee b)$ [6a]

11-D-6. $a \vee (\bar{a} \supset c) \equiv \bar{a} \supset (\bar{a} \supset c)$ [6a] $\equiv \bar{a}.\bar{a} \supset c$ [6d] $\equiv \bar{a} \supset c$ [1c]

11-D-7. $a \supset b.c \equiv \bar{a} \vee b.c$ [6a] $\equiv (\bar{a} \vee b).(\bar{a} \vee c)$ [4b] $\equiv (a \supset b).(a \supset c)$ [6a]

11-D-8. $\overline{a \supset b.c} \equiv a.\overline{b.c}$ [6b] $\equiv a.(\bar{b} \vee \bar{c})$ [5a] $\equiv a.\bar{b} \vee a.\bar{c}$ [4a]
$\equiv \overline{a.\bar{b}} \supset a.\bar{c}$ [6a] $\equiv \bar{a} \vee b \supset a.\bar{c}$ [5a] $\equiv (a \supset b) \supset a.\bar{c}$ [6a]

11-D-9. $(a \equiv b) \equiv a.b \vee \bar{a}.\bar{b}$ [7b] $\equiv (a \vee \bar{a}).(b \vee \bar{a}).(a \vee \bar{b}).(b \vee \bar{b})$ [4d]
$\equiv (\bar{a} \vee b).(\bar{b} \vee a)$ [5c] $\equiv (a \supset b).(b \supset a)$ [6a]

11-D-10. $\overline{a \equiv b} \equiv \overline{a.b \vee \bar{a}.\bar{b}}$ [7b] $\equiv \overline{a.b} . \overline{\bar{a}.\bar{b}}$ [5b] $\equiv (\bar{a} \vee \bar{b}).(a \vee b)$ [5a]
$\equiv \bar{a}.a \vee \bar{a}.b \vee \bar{b}.a \vee \bar{b}.b$ [4c] $\equiv \bar{a}.b \vee a.\bar{b}$ [5d] $\equiv (a \equiv \bar{b})$ [7b]

11-D-11. $(a \equiv \bar{b}) \equiv (a \supset \bar{b}).(\bar{b} \supset a)$ [7a] $\equiv (\bar{a} \vee \bar{b}).(b \vee a)$ [6a]

11-D-12. $(a \equiv b) \supset c \equiv \bar{c} \supset \overline{a \equiv b}$ [6c] $\equiv \bar{c} \supset (a \equiv \bar{b})$ [7c]
$\equiv \bar{c} \supset (a \supset \bar{b}).(\bar{b} \supset a)$ [7a] $\equiv \bar{c} \supset (a \vee b).(\bar{a} \vee \bar{b})$ [6a]
$\equiv \bar{c} \supset (a \vee b).\overline{a.b}$ [5a]

11-D-13. $a.(a \equiv b) \equiv a(a.b \vee \bar{a}.\bar{b})$ [7b] $\equiv a.a.b \vee a.\bar{a}.b$ [4a] $\equiv a.b$ [1c, 5d]

11-D-14. $\bar{a}.b \vee (a \equiv b) \equiv \bar{a}.b \vee a.b \vee \bar{a}.\bar{b}$ [7b] $\equiv \bar{a}.(b \vee \bar{b}) \vee a.b$ [4a]
$\equiv \bar{a} \vee a.b$ [5c] $\equiv \bar{a} \vee b$ [5e] $\equiv a \supset b$ [6a]

Solutions

11–D–15. $a \vee (b \equiv c.a) \equiv a \vee b.c.a \vee \overline{b.c.a}$ [7b] $\equiv a \vee \overline{b.c.a}$ [4e]
$\equiv a \vee \bar{b}.(\bar{c} \vee \bar{a})$ [5a] $\equiv a \vee \bar{b}.\bar{c} \vee \bar{b}.\bar{a}$ [4a] $\equiv a \vee \bar{b} \vee \bar{b}.\bar{c}$ [5e]
$\equiv a \vee \bar{b}$ [4e] $\equiv b \supset a$ [6a]

15–A–1. *a:* a boy is forbidden to smoke. *b:* he will smoke. *c:* a boy is forbidden to drive a car. *d:* he will drive a car.

> $(a \supset b).(c \supset d)$
> $a.c \supset b.d$ [8j]
> $a \vee c \supset b \vee d$ [8k]

15–A–2. *a:* the earth moves into the shadow of the moon. *b:* there will be an eclipse. *c:* the stars become visible.

> $(a \supset b).(b \supset c)$
> $a \supset c$ [8l]

15–A–3. *a:* an American citizen is put into jail. *b:* he is entitled to a hearing within 24 hours.

> $a \supset b$
> $a.c \supset b$ [8g]

15–B–1. *a:* a man is 21 years old. *b:* he is older than 21. *c:* he is a citizen. *d:* he is a negro. *v:* he can vote.

> $(a \vee b).c.d \supset \bar{v}$
> $v \supset (a \vee b).c$
> ---
> $\overline{(a \vee b).c} \supset \bar{v}$ [6c]
> $\overline{(a \vee b).c}. d \supset \bar{v}$ [8g]
> $(a \vee b).c.d \vee \overline{(a \vee b).c}. d \supset \bar{v}$ [6f]
> $d.[(a \vee b).c \vee \overline{(a \vee b.c)}] \supset \bar{v}$ [4a]
> $d \supset \bar{v}$ [5c]

15–B–2. *a:* Bill is put into maze A. *b:* Bill is hungry. *c:* Bill has learned maze A. *d:* Bill is able to find food.

> $a.b \supset c \vee d$
> $c \supset d$
> a
> $\bar{d}$
> ---
> $\bar{c}$ [(1, § 15) applied to second and fourth premise]
> $\overline{a.b}$ [6c applied to first premise]
> $\bar{a} \vee \bar{b}$ [5a]
> $\bar{b}$ [11, § 15]

15–B–3. $\overline{b \vee c} \supset \bar{a}$ [6c applied to first premise]
$\bar{b}.\bar{c} \supset \bar{a}$ [5b]
$\bar{a}$ [1, § 14]

15–B–4. $\overline{b \supset c} \supset a$ [6a applied to first premise]
$b.\bar{c} \supset a$ [6b]
$b.\bar{c}$ [(11, § 15) applied to second and third premise]
a [(1, § 14) applied to second conclusion]

15–C–1. *a:* there had to be heavy taxes in this country. *b:* an American army was sent to Europe.

$$a \vee \bar{b}$$
$$\underline{b}$$
$$a \ [\text{11}, \S \ 15]$$

15–C–2. *a:* Napoleon was sick during the battle of Leipzig. *r:* the decimal system is in use in present-day England.

$$\bar{a} \supset b$$
$$\cdots$$
$$\cdots$$
$$\underline{q \supset r}$$
$$\bar{a} \supset r \ [\text{15}, \S \ 15]$$

15–C–3. *a:* this is a galvanometer. *b:* current passes through the wire. *c:* the needle is deflected.

$$a.b \supset c$$
$$\bar{c}$$
$$\underline{a}$$
$$\overline{a.b} \ [(\text{1}, \S \ 15) \ \text{applied to first and second premise}]$$
$$\bar{a} \vee \bar{b} \ [\text{5a}]$$
$$\bar{b} \ [\text{11}, \S \ 15]$$

15–C–4. *a:* the car starts. *b:* the points of the distributor are unclean. *c:* the carburetor is obstructed.

$$\bar{a} \supset b \vee c$$
$$\underline{\bar{b}.\bar{a}}$$
$$\overline{b \vee c} \supset a \ [\text{6c applied to first premise}]$$
$$\bar{b}.\bar{c} \supset a \ [\text{5b}]$$
$$\bar{b} \supset (\bar{c} \supset a) \ [\text{6d}]$$
$$\bar{c} \supset a \ [\text{1}, \S \ 14]$$
$$\bar{a} \supset c \ [\text{6c}]$$
$$c \ [\text{1}, \S \ 14]$$

15–C–5. *a:* wages are raised. *b:* prices are raised. *c:* there will be inflation. *d:* the government will have to abdicate. *e:* the government is able.

$$a.b \supset c$$
$$c \supset d$$
$$e.\bar{d}$$
$$\underline{a}$$
$$\bar{d} \ [\text{from third premise}]$$
$$\bar{c} \ [(\text{1}, \S \ 15) \ \text{applied to second and third premise}]$$
$$\overline{a.b} \ [(\text{1}, \S \ 15) \ \text{applied to first premise}]$$
$$\bar{a} \vee \bar{b} \ [\text{5a}]$$
$$\bar{b} \ [\text{11}, \S \ 15]$$

15-C-6. *a:* this salt is put into water. *b:* it is heated. *c:* it dissolves. *d:* it is salt X. *e:* it is salt Y. *f:* it will make a red flame. *g:* it will make a yellow flame.

$a.b \supset (c \supset d \vee e)$
$d \supset f$
$e \supset g$
$a.b.c$

g
$\underline{\quad g \supset \bar{f} \text{ [from the meaning of the terms]}}$
$a.b.c \supset d \vee e$ [6d]
$d \vee e$ [1, § 14]
$\bar{f}$ [(1, § 14) applied to last two premises]
$\bar{d}$ [(1, § 15) applied to second premise]
e [(11, § 15) applied to second conclusion]

15-C-7. *a:* the switch is turned on. *b:* the vacuum is perfect. *c:* there will be a tension of 10,000 volts. *d:* the glass will show a green fluorescent light.

$a.b \equiv c.d$
$\underline{a.c}$
$a.b \supset d$ [7a and 8i applied to first premise]
$a \supset (b \supset d)$ [6e]
$b \supset d$ [1, § 14]
$c.d \supset b$ [7a and 8i applied to first premise]
$c \supset (d \supset b)$ [6e]
$d \supset b$ [1, § 14]
$b \equiv d$ [7a]

17-1. *i:* insane. x_1: George III.

$i(x_1)$

17-2. *p:* president. x_1: George Washington. y_1: United States.

$p(x_1, y_1)$

17-3. *s:* separates. x_1: Bosphorus. y_1: Europe. z_1: Asia.

$s(x_1, y_1, z_1)$

17-4. *t:* as tall as. x_1: Eiffel tower. y_1: Empire State Building.

$\overline{t(x_1, y_1)}$

17-5. *s:* see. x_1: I. y_1: Donald. z_1: William.

$\overline{s(x_1, y_1) \vee s(x_1, z_1)}$

17-6. *s:* situated between. x_1: Oregon. y_1: California. z_1: Washington.

$s(x_1, y_1, z_1)$

18-A-1. *c:* crow. *b:* black.

$(x)[c(x) \supset b(x)]$

18–A–2. *b:* brother. *r:* resemble.

$$(x)(y)[b(x, y) \supset r(x, y)]$$

18–A–3. *w:* well. *e:* ends well.

$$(x)[e(x) \supset w(x)]$$

18–A–4. *d:* dog. *b:* bark. *bt:* bite.

$$(x)[d(x).b(x) \supset \overline{bt(x)}]$$

18–A–5. *m:* mother. *l:* love. *b:* boy.

$$(x)(y)[m(x, y).b(y) \supset l(x, y)]$$

18–A–6. *f:* follower. *m:* manner. *i:* imitated. z_1: Oscar Wilde.

$$(x)(y)[f(x, z_1).m(y, z_1) \supset i(x, y)]$$

18–A–7. *m:* man. *w:* woman. *wo:* working. y_1: Douglas factory. *c:* citizen. z_1: United States.

$$(x)\{[m(x) \vee w(x)].wo(x, y_1) \supset c(x, z_1)\}$$

18–A–8. *m:* man. *w:* woman. *ww:* working as welders. *r:* receive. *e:* equal. *wa:* wage.

$$(x)(y)(u)(v)[m(x).w(y).ww(x).ww(y).r(x, u).r(y, v).wa(u).wa(v) \supset e(u, v)]$$

18–B–1. *b:* black. *s:* swan.

$$(\exists x)b(x).s(x)$$

18–B–2. *t:* tower. *v:* vertical.

$$(\exists x)t(x).\overline{v(x)}$$

18–B–3. *b:* book. *g:* gave. y_1: John. z_1: Mary.

$$(\exists x)b(x).g(y_1, z_1, x)$$

18–B–4. *b:* boy. *e:* eats. *a:* apple.

$$(\exists x)(\exists y)b(x).a(y).e(x, y)$$

18–B–5. *m:* man. *c:* came. *b:* brought. *bf:* briefcase. z_1: Dr. A.

$$(\exists x)(\exists y)m(x).bf(y).c(x).b(x, z_1, y)$$

18–B–6. *r:* rotten. *s:* situated. y_1: Denmark.

$$(\exists x)r(x).s(x, y_1)$$

18–B–7. *l:* logician. *a:* admire. y_1: Aristotle.

$$(\exists x)l(x).\overline{a(x, y_1)}$$

19–1. *s:* soldier. y_1: America. *c:* citizen.

$$\overline{(x)[s(x, y_1) \supset c(x, y_1)]}$$
$$(\exists x)s(x, y_1).\overline{c(x, y_1)}$$

19–2. *p:* planet. *sl:* self-luminous.

$$(x)[p(x) \supset \overline{sl(x)}]$$
$$\overline{(\exists x)p(x).sl(x)}$$

19-3. m: mother. h: hates.

$$(x)(y)[m(x, y) \supset \overline{h(x, y)}]$$
$$\overline{(\exists x)(\exists y)m(x, y).h(x, y)}$$

19-4. f: father. b: boy.

$$\overline{(x)(y)[f(x, y) \supset \overline{b(y)}]}$$
$$(\exists x)(\exists y)f(x, y).\overline{b(y)}$$

Notice that the word 'only' is not a function, but is expressed by logical means alone.

19-5. m: man. y_1: the moon. l: living on.

$$\overline{(\exists x)m(x).l(x, y_1)}$$
$$(x)[m(x) \supset \overline{l(x, y_1)}]$$

19-6. c: cold. d: day. ac: as cold as. y_1: this day.

$$(x)[c(x).d(x).(x \neq y_1) \supset \overline{ac(x, y_1)}]$$
$$\overline{(\exists x)\, c(x).d(x).(x \neq y_1).ac(x, y_1)}$$

19-7. h: healthy. m: man. t: take. d: drug.

$$(x)(y)[h(x).m(x).d(y) \supset \overline{t(x, y)}]$$
$$\overline{(\exists x)(\exists y)h(x).m(x).d(y).t(x, y)}$$

19-8. c: clever. g: girl. ma: marries. m: man. l: love.

$$(x)(y)[c(x).g(x).m(y).\overline{l(x, y)} \supset \overline{ma(x, y)}]$$
$$\overline{(\exists x)(\exists y)c(x).g(x).m(y).\overline{l(x, y)}.ma(x, y)}$$

20-A-1. m: mammal. b: breathe through. l: lung.

$$(x)(\exists y)[m(x) \supset b(x, y).l(y)]$$

also correct (cf. 13, § 20): $(x)[m(x) \supset (\exists y)b(x, y).l(y)]$

20-A-2. s: soldier. u: uniform. w: wears.

$$(x)(\exists y)[s(x) \supset w(x, y).u(y)]$$

20-A-3. s: soldier. f: followed. o: order. g: general.

$$(\exists x)(\exists y)(z)[s(z) \supset g(x, z).o(y, x).f(z, y)]$$

20-A-4. m: man. w: woman. wk: working. z_1: Douglas factory. we: wear. b: badge.

$$(x)(\exists y)\{[m(x) \lor w(x)].wk(x, z_1) \supset we(x, y).b(y)\}$$

20-A-5. m: metal. l: lighter than. w: water.

$$(\exists x)(y)\{m(x).[w(y) \supset l(x, y)]\}$$

also correct: $(\exists x)m(x).(y)[w(y) \supset l(x, y)]$ [11a, § 25]

The condition that x and y must have equal volume is here not expressed because the linguistic formulation does not state it explicitly.

20-A-6. *m:* man. y_1: John. *h:* hate. z_1: Margot.

$(x)[m(x).(x \neq y_1) \supset h(x, z_1)]$

20-A-7. *m:* man. *t:* taller.

$(\exists x)(y)\{m(x).[m(y).(y \neq x) \supset t(x, y)]\}$
also correct: $(\exists x)m(x).(y)[m(y).(y \neq x) \supset t(x, y)]$

20-A-8. *f:* flea. *h:* has. *sm:* smaller.

$(x)(\exists y)[f(x) \supset h(x, y).f(y).sm(y, x)]$

20-A-9. *l:* living being. *d:* descend. *ps:* primitive species.

$(\exists x)(y)\{ps(x).[l(y) \supset d(y, x)]\}$

20-B-1. *s:* stone. *r:* rolling. *g:* gathers. *m:* moss.

$(x)[r(x).s(x) \supset \overline{(\exists y)m(y).g(x, y)}]$
$(x)(y)\{r(x).s(x) \supset [m(y) \supset \overline{g(x, y)}]\}$ [10, §19; 12, §20]

20-B-2. *m:* mammal. *l:* live. z_1: ocean. *w:* water animal. *e:* emerge. *b:* breathing.

$(\exists x)m(x).l(x, z_1) \supset (\exists x)w(x).[b(x) \supset e(x)]$
$(x)(\exists y)\{m(x).l(x, z_1) \supset w(y).[b(y) \supset e(y)]\}$ [10 and 13, § 20]

20-B-3. *h:* human being. *p:* planet. z_1: Earth. *l:* living on. *m:* modern. *sc:* scientist. *s:* surprised.

$(\exists x)(\exists y)h(x).p(y).(y \neq z_1).l(x, y) \supset (u)[m(u).sc(u) \supset \overline{s(u)}]$
$(x)(y)(u)\{h(x).p(y).(y \neq z_1).l(x, y) \supset [m(u).sc(u) \supset \overline{s(u)}]\}$

In the first form we could put 'x' for 'u'.

20-B-4. *at:* consist of atoms. y_1: Democritus. *ph:* great philosopher.

$(x)at(x) \supset ph(y_1)$
$(\exists x)[at(x) \supset ph(y_1)]$ [11, § 20]

20-B-5. *m:* man. *h:* hero. *w:* woman. *b:* beautiful. *mo:* movies. *a:* appeal.

$(x)[m(x) \supset h(x)].(x)[w(x) \supset b(x)] \supset (x)(y)[mo(x) \supset \overline{a(x, y)}]$
$(\exists x)(u)(v)\{[m(x) \supset h(x)].[w(x) \supset b(x)] \supset [mo(u) \supset \overline{a(u, v)}]\}$ [3, 11 and 12, § 20]

20-B-6. *f:* species of flies. *tr:* transfers. *m:* malaria germs. *s:* swamp. *st:* stagnant. *p:* pool. *dr:* dried up.

$(x)[(\exists y)m(y).tr(x, y) \supset f(x)] \supset \{(z)[s(z) \lor st(z).p(z) \supset dr(z)] \supset \overline{(\exists y)m(y)}\}$
$(\exists x)(\exists y)(\exists z)(u)\{[m(y).tr(x, y) \supset f(x)] \supset [[s(z) \lor st(z).p(z) \supset dr(z)] \supset \overline{m(u)}]\}$
[(10) and (8), (9) and (11), (12), (13), § 20; (2, § 19)]

21-1. *j:* just. y_1: Aristides.

$j(x) \supset j(y_1)$
$(x)[j(x) \supset j(y_1)]$
$(\exists x)j(x) \supset j(y_1)$

21-2. *o:* oil deposit. *s:* situated. z_1: Germany. *m:* man. *d:* developed.

$o(x).s(x, z_1) \supset (\exists y)m(y).d(y, x)$

$(x)[o(x).s(x, z_1) \supset (\exists y)m(y).d(y, x)]$

The phrase 'there were' contained in the sentence does not represent an existential operator, but means as much as 'were situated'.

21-3. *o:* oil deposit. *s:* situated. z_1: Germany.

$\overline{o(x).s(x, z_1)}$

$(x)\overline{[o(x).s(x, z_1)]}$

$\overline{(\exists x)o(x).s(x, z_1)}$

21-4. *r:* race. *g:* as good as.

$r(x).r(y) \supset g(x, y)$

$(x)(y)[r(x).r(y) \supset g(x, y)]$

21-5. *r:* rule. *nl:* native language. *k:* knows. *s:* surprised. u_1: I.

$(\exists z)nl(z, x).(y)[r(y, z) \supset k(x, y)] \supset s(u_1)$

$(x)\{(\exists z)nl(z, x).(y)[r(y, z) \supset k(x, y)] \supset s(u_1)\}$

$(\exists x)(\exists z)nl(z, x).(y)[r(y, z) \supset k(x, y)] \supset s(u_1)$

22-1. Uniform. Not interconnective. One-one function. Symmetrical. Intransitive. Irreflexive.

22-2. Nondivided field. Not interconnective. Many-many function. Mesosymmetrical. Mesotransitive. Mesoreflexive.

22-3. Uniform. Not interconnective. Many-many function. Asymmetrical. Transitive. Reflexive.

22-4. Uniform. Interconnective. Many-many function. Asymmetrical. Transitive. Reflexive.

22-5. Uniform. Not interconnective. Many-many function. Symmetrical. Mesotransitive. Irreflexive.

22-6. Divided field. Not interconnective. One-many function. Asymmetrical. Irreflexive.

22-7. Uniform. Not interconnective. Many-many function. Symmetrical. Transitive. Reflexive.

22-8. Uniform. Not interconnective. Many-many function. Asymmetrical. Transitive. Irreflexive.

22-9. Uniform. Interconnective. Many-many function. Mesosymmetrical. Transitive. Mesoreflexive.

22-10. Uniform. Not interconnective. One-one function. Asymmetrical. Intransitive. Irreflexive.

47-1. x_1 = Robert Browning. *w* = wife. *i* = ill.

$i[(\imath y)w(y, x_1)]$

47-2. x_1 = John Stuart Mill. f = father. a = author. u_1 = *Analysis of the Phenomena of the Human Mind*.

$$(\imath y)f(y, z_1) = (\imath z)a(z, u_1)$$

47-3. x_1 = Bill. tw = twin brother. lt = lent. j = jacket.

$$y_1 = (\imath y)tw(y, x_1)$$
$$z_1 = (\imath z)j(z, y_1)$$
$$lt(y_1, x_1, z_1)$$

47-4. x_1 = Thomas Hobbes. b = was born. yr = year. d = defeat.

z_1 = Armada.
$$t_1 = (\imath t)yr(t).d(z_1, t)$$
$$b(x_1, t_1)$$

47-5. p = president. y_1 = United States. z_1 = Civil War. wr = wrote. l = letter. u_1 = Mrs. Bixby. pl = take place.

$$t_1 = (\imath t)pl(z_1, t)$$
$$x_1 = (\imath x)p(x, y_1, t_1)$$
$$(\exists y)l(y).wr(x_1, y, u_1)$$

47-6. u_1 = thou. w = wear. f = father. s = serpent. st = stung. cr = crown.

$$x_1 = (\imath x)f(x, u_1)$$
$$y_1 = (\imath y)s(y).st(y, x_1)$$
$$z_1 = (\imath z)cr(z).w(x_1, z)$$
$$w(y_1, z_1)$$

47-7. m = miserable. md = medicine. h = hope.

$$(x)\{m(x) \supset (\exists y)md(y, x).h(y).(z)[md(z, x) \supset (z = y)]\}$$

The definite article is not used here for a description, but for the expression of generality.

47-8. h = hen. cm = cackle more than. e = egg. l = lay. lr = larger than.

$$(\imath x)h(x).(y)[h(y).(y \neq x) \supset cm(x, y)] \neq (\imath z)h(z).(u)(v)[e(u).e(v).$$
$$l(z, u).\overline{l(z, v)} \supset lr(u, v)]$$

48-1. x_1 = Torricelli. d = discover. a = air. ex = exert. pr = pressure.

$$p =_{Df} (y)(\exists z)[a(y) \supset ex(y, z).pr(z)]$$
$$d[x_1, (\imath v)p^*(v)]$$

48-2. x_1 = Carthage. d = be destroyed. y_1 = Rome. z_1 = Mediterranean. r = ruler. m = make.

$$v_1 = (\imath v)[d(x_1)]^*(v)$$
$$u_1 = (\imath u)r(u, z_1)$$
$$m(v_1, y_1, u_1)$$

Solutions

48-3. pm = permit. sm = smokes in. b = bus. m = man.

$$p =_{Df} (\exists x)(\exists y)m(x).b(y).sm(x, y)$$
$$\overline{pm[(\imath v)p^*(v)]}$$

Here we have used 'y' for the space argument. We could also introduce a function 'located in', abbreviated by 'in', and write 'p' in the following form:

$$p = (\exists x)(\exists y)(\exists t)m(x).b(y).sm(x, t).in(x, y, t)$$

48-4. mbl = more blessed. gv = give to.

$$g(x) =_{Df} (\exists y)gv(x, y)$$
$$r(x) =_{Df} (\exists y)gv(y, x)$$
$$(v)(u)\{[g(x)]^*(v).[r(x)]^*(u) \supset mbl(v, u, x)\}$$

We regard the Bible quotation as meaning that the giving by x is more blessed *for* x than the receiving by x, because the giving by x is at the same time a receiving by y, and thus is not more blessed for y.

48-5. sn = snow. f = fall.

$$p =_{Df} (\exists x)sn(x).f(x)$$
$$(\exists v)p^*(v)$$

Whereas in the two preceding examples the 'it' anticipates a clause describing an event, the word 'it' is used here independently with reference to the event of snowing. We regard the sentence 'it snows' as meaning 'there is snowing'. The tendency to analogy has led language to express here a bound variable in the form of a specialized variable, the pseudo-constant 'it'.

48-6. pr = produce. s = synthetic. r = rubber. y_1 = United States. mi = make independent. im = is imported from. n = natural. z_1 = East Indies.

$$p_1 =_{Df} (\exists x)s(x).r(x).pr(y_1, x)$$
$$p_2 =_{Df} (\exists x)n(x).r(x).im(x, z_1)$$
$$mi[(\imath v)p_1^*(v), y_1, (\imath u)p_2^*(u)]$$

48-7. d = was discovered in. g = gold. br = brought. I = class of all immigrants. λ = large. y_1 = California.

$$p =_{Df} (\exists x)g(x).d(x, y_1)$$
$$(\exists C)(C \subset I)(C \in \lambda).br[(\imath v)p^*(v), C, y_1]$$

49-1. x_1 = Columbus. y_1 = India. d = discovered. bl = believed.

$$(\exists v)_L [d(x_1, y_1)]^*(v).bl(x_1, v)$$

49-2. $x_1{}^{li}$ = Romeo. $y_1{}^{li}$ = Juliet. d = dead. kn = know.

$$(\exists v)_{li} [\overline{d(y_1{}^{li})}]^*(v).\overline{kn(x_1{}^{li}, v)}$$

49-3. x_1 = Socrates. r = refuse. es = escape.

$$(\exists v)_{in} [es(x_1)]^*(v).r(x_1, v)$$

49-4. x_1 = Pharaoh. K = kine. c = come up out of. r = river.

$p = _{Df} (\exists F)_{x_1} (\exists z)_{x_1} (y)[(y \in F) \equiv (y \in K).c(y, z).r(z)].(F \in 7)$
$(\exists v)_{x_1} d(x_1, v).p^*(v)$

49-5. t = teach. at = attempt. ins = inspire. ds = desire. i = iron.
c = cold. h = hammer. l = like.

$dl(y, x) = _{Df} (\exists u)_{in} ds(y, u).[l(x, y)]^*(u)$
$(x)(y)\{(\exists v)_{in} [l(x, y)]^*(v).at(x, v)$
$\overline{.(\exists w)[dl(y, x)]^*(w).ins(x, y, w) \supset (z)[(\exists r)i(r).c(r).h(z, r) \supset l(x, z)]\}}$

50-1. f = father. m = merchant. sp = speaks.

$m\{(\imath x)f[x, (\imath y)sp(y, \Theta)]\}$

50-2. r = road. in = indicate. gs = gesture. rf = referred to. g = goes to.
y_1 = Yosemite.

$x_1 = (\imath x)r(x).(\exists y)gs(y).rf(y, \Theta).in(y, x)$
$g(x_1, y_1)$

50-3. s = shalt see. z_1 = Philippi. ad = is addressed by. sp = speaks.

$x_1 = (\imath x)ad(x, \Theta)$
$y_1 = (\imath y)sp(y, \Theta)$
$s(x_1, y_1, z_1)$

50-4. l = shall leave. pl = place. sp = speaks. ad = is addressed by.

$x_1 = (\imath x)(\exists z)sp(x, \Theta, z)$
$y_1 = (\imath y)ad(y, \Theta)$
$z_1 = (\imath z)pl(z).(\exists x)sp(x, \Theta, z)$
$l(x_1, y_1, z_1)$

50-5. st = stone. ins = inscribed on. lb = is laid beneath. h = in hope for.
z_1 = Zion. ll = landlord. u_1 = the Lion.

$x_1 = (\imath x)st(x).ins(\Theta, x)$
$y_1 = (\imath y)ll(y, u_1)$
$lb(y_1, x_1).h(y_1, z_1)$

50-6. rg = rugged. e = elm. rd = rude. ff = forefather. h = hamlet.
sl = sleep beneath. sp = speaks.

$z_1 = (\imath z)h(z).rf[z, (\imath u)sp(u, \Theta)]$
$(\exists C)(x)[(x \in C) \equiv e(x).rg(x).rf(x, \Theta)].(y)[ff(y, z_1)$
$.rd(y) \supset (\exists x)(x \in C)sl(y, x)]$

51-A-1.   1: swearest. 2: hast supped.

S, R_1, E_1
$E_2 - S, R_2$

51-A-2.   1: have. 2: shall have.

S, R_1, E_1
$S, R_2 \quad - E_2$

51-A-3. 1: broke. 2: will break.

$$E_1, R_1 - S$$
$$S, R_2 - E_2$$

Note that the reference point is not kept constant.

51-A-4. 1: will have left. 2: arrive.

$$S - E_1 - R_1$$
$$R_2, E_2$$

Note that in the second clause the point of speech is neglected. We therefore omit it in the diagram.

51-A-5. 1: have lied. 2: cannot believe.

$$E_1 - S, R_1$$
$$S, R_2, E_2$$

51-A-6. 1: lied. 2: cannot believe.

$$E_1, R_1 - S$$
$$S, R_2, E_2$$

Notice that here the use of the word 'after' in combination with the time determinations induces us to use the simple past in the first clause, in contradistinction to the preceding example; the time comparison thus refers to the reference points.

51-A-7. 1: have seen. 2: have touch'd.

$$E_1 \longrightarrow S, R_1$$
$$E_2 - S, R_2$$

In this quotation from a sixteenth-century English poem by Ben Jonson the positional use of the reference point is abandoned, and the comparison refers, not to the reference points, but to the event points.

51-A-8. 1: think. 2: will bite. 3: hast trod.

$$S, R_1, E_1$$
$$S \qquad\qquad - R_2, E_2$$
$$E_3 - R_3$$

In the second clause, the positional principle compels us to interpret the simple future in the form $S - R, E$. In the third clause the point of speech is neglected.

51-A-9. 1: love. 2: believe. 3: will die. 4: shall grieve.

$$S, R_1, E_1$$
$$S, R_2, E_2$$
$$S \qquad\qquad - R_3, E_3$$
$$S \qquad\qquad\qquad\qquad - R_4, E_4$$

The positional principle requires the interpretation $S - R, E$ for the simple future. The word 'ere' pulls the reference points R_3 and R_4 apart.

51-A-10. 1: will be. 2: have found. 3: was. 4: sought.

$$S \text{——} R_1, E_1$$
$$E_2 \text{—} R_2$$

$$R_3, E_3 \text{—} S$$
$$R_4, E_4 \text{—} S$$

In the second clause the point of speech is neglected.

51-B-1. c = come. r = receive. x_1 = Peter. y_1 = Dr. Jones.

$$(\exists t)(t < t_0).c(x_1, t).r(y_1, x_1, t)$$

51-B-2. c = come. s = show. x_1 = Peter. y_1 = I. l = letter. w = written. u_1 = John.

$$(\exists t)(\exists z)(t < t_0).c(x_1, t).s(x_1, y_1, z, t).[z = (\imath z')(\exists t')w(u_1, z', t').l(z').(t' < t)]$$

51-B-3. c = comes. x_1 = Peter. y_1 = I. z_1 = John. s = see.

$$(\exists t)(\exists t')c(x_1, t).s(y_1, z_1, t').(t_0 < t' < t)$$

51-B-4. c = comes. x_1 = Peter. g = give. z_1 = you. l = letter.

$$(t)\{c(x_1, t).(t > t_0).(t')[(t_0 \leqq t' < t) \supset \overline{c(x_1, t')}] \supset (\exists y)l(y).g(x_1, y, z_1, t)\}$$

The expression containing 't' qualifies the major all-operator in such a way that the implication is stated only for the first coming of Peter after the time of speech. The conjunction 'when' differs from 'if' in that it states the existence of a t for which the implicans is true; since '$a.(a \supset b)$' is the same as '$a.b$', a when-clause can be interpreted as a logical product (cf. 51-B-3).

51-B-5. d = disciple. x_1 = Aristotle. y_1 = Plato. z_1 = Socrates.

$$(\exists t)(\exists t')d(x_1, y_1, t).d(y_1, z_1, t').(t' < t < t_0)$$

51-B-6. b = book. l = look. r = reprint. y_1 = you.

$$(\exists t')r(x_1, t').(t' > t_0).[x_1 = (\imath x)(\exists t)b(x).l(y_1, x, t).(t < t_0)]$$

51-B-7. g = gold. h = heavier. i = iron.

$$(x)(y)[g(x).i(y) \supset h(x, y, t)]$$

In this example the present tense of 'is' must be interpreted as indicating that the time is used as a free variable.

51-B-8. m = man. f = can fool.

$$(y)(\exists t)[m(x).m(y) \supset f(x, y, t)].(\exists y)(t)[m(x).m(y) \supset f(x, y, t)]$$
$$\overline{.(y)(t)[m(x).m(y) \supset f(x, y, t)]}$$

Notice that we may interpret the symbol 'f' simply as standing for 'to fool'. The modality 'can' of the problem sentence then may be regarded as expressed by the existential operators in the form of an extensional modality (cf. § 23).

51–B–9. b = bold. m = man. e = eat. o = oyster.

$$x_1 = (\imath x)m(x).(\exists y)(\exists t)e(x, y, t).o(y).(t < t_0)$$
$$.(z)(t')(u)[m(z).(z \neq x).e(z, u, t').o(u) \supset (t' > t)]$$
$$(\exists t)b(x_1, t).(t < t_0)$$

51–B–10. T = class of all time points. σ = small. c = come. ad = is addressed to. sp = spirit of. D = class of all delights.

$$x_1 = (\imath x)ad(x, \Theta) = (\imath x)sp(x, D)$$
$$(\exists C)(C \subset T).(t)[c(x_1, t) \equiv (t \in C)].\sigma(C)$$

53–1. x_1 = the moon. γ = was shining. δ = sulky.

$$(\exists f)f(x_1).\gamma(f).\delta(f)$$

53–2. b = burden. γ = weight. δ = heavy. sh = shoulder. z_1 = he.

$$(\exists x)(\exists f)(\exists y)b(x).sh(y, z_1).f(x, y).\gamma(f).\delta(f)$$

53–3. x_1 = he. γ = excellent. δ = speaker. hs = has to say.

$$(\exists f)f(x_1).\gamma(f).\delta(f).\overline{(\exists y)hs(x_1, y)}$$

It is obvious that 'excellent' is used here as an adverb, since the man is not characterized as excellent.

53–4. γ = uneasy. δ = lie. h = head. w = wear. c = crown.

$$(x)(y)(f)[h(x).w(x, y).c(y).f(x).\delta(f) \supset \gamma(f)]$$

Here we regard 'uneasy' as an adverb qualifying 'lies'. In another interpretation, it can be regarded as an adjective predicating a property of the head when it is lying. — The definite article is used here for the expression of generality.

53–5. or = orange. γ = taste. δ = good.

$$(x)[or(x) \supset (\exists f)f(x)\gamma(f).\delta(f)]$$

53–6. w = wine. g = good. rf = referred to. sp = speaks. bl = believes. cs = causes.

$$y_1 = (\imath y)sp(y, \Theta)$$
$$x_1 = (\imath x)w(x).rf(x, y_1)$$
$$p =_{Df} g(x_1) \qquad b(y_1) =_{Df} (\exists u)_L\, p^*(u).bl(y_1, u)$$
$$(\exists f)(\exists u)(\exists v)f(x_1).[f(x_1)]^*(u).[b(y_1)]^*(v).cs(u, v)$$

53–7. g = gossip. v = vice. γ = enjoy. δ = vicarious. rf = refers to.

$$sp(x, y).g(y) \supset (\exists z)(\exists f).v(z).rf(y, z).f(x, z).\gamma(f).\delta(f)$$

We have written the statement in terms of the free variables 'x' and 'y', so that the implication is stated for any person x and any gossip y. Since the wording of the statement does not use these free variables, it is better construed as an identity of the corresponding functions:

$$sp(\hat{x}, \hat{y}).g(\hat{y}) \subset (\exists z)(\exists f).v(z).rf(\hat{y}, z).f(\hat{x}, z).\gamma(f).\delta(f)$$

We use here the sign of class inclusion between functions in the same way as it is used between class names.

54–1. ϑ = temperature. x_1 = the room.

$$(\imath f)f(x_1).\vartheta(f) = 68°$$

54–2. x_1 = Peter. γ = tallness.

$$(\exists f)f(x_1).\gamma(f).(f = 6 \text{ ft.})$$

54–3. x_1 = Johnny. l = likes. c = car. b = belongs. δ = drive. σ = speed.

$$y_1 =_{Df} (\imath y)c(y).b(y, x_1)$$
$$d(x_1, y_1) = (\exists f)\delta(x_1, y_1, f).\sigma(f).(f = 80 \text{ mls/hr})$$
$$(v)\{[d(x_1, y_1)]^*(v) \supset l(x_1, v)\}$$

54–4. e = egg. w = water. γ = heat. b = is boiled in.

$$(x)(y)(f)(\exists g)[e(x).w(y).b(x, y).f(y).\gamma(f) \supset g(x).\gamma(g).(f = g)]$$

54–5. st = sequoia tree. γ = age. e = Egyptian. pr = pyramid.

$$(\exists x)(\exists y)(\exists f)(\exists g)[st(x).e(y).pr(y).f(x).g(y).\gamma(f).\gamma(g).(f > g)]$$

54–6. h = heard. m = melody. σ = sweet.

$$(x)(y)(\exists f)(\exists g)[m(x).h(x) \supset f(x).\sigma(f)].[m(y).\overline{h(y)} \supset g(y).\sigma(g).(g > f)]$$

54–7. l = laughter. g = good. h = humor. γ = contagious. δ = irresistible.

$$(x)(y)\{l(x).g(y).h(y) \supset (\exists p)(\exists q)p(x).q(y).\gamma(p).\gamma(q).\delta(p).\delta(q)$$
$$.\overline{(\exists z)(\exists r)r(z).\gamma(r).[(r \geqq p) \vee (r \geqq q)]}$$

59–1. 'the': iota-operator, logical term, III, 1, c.

'present': token-reflexive time description, I, 2, b.

'king': simple function, II, 2, a; thing-function, II, 3, a; first type, II, 4, a.

'of': logical term, I, a, α.

'England': proper name, I, 1.

'was': logical term, III, 1, c, including a time indication (contracted term).

'crown': simple function, II, 2, a; thing-function, II, 3, a; first type, II, 4, a.

'ed': time indication, I, 2, b.

'ceremonially': higher function, II, 4, b.

  'ly': logical term, III, 1, b.

'at': semi-logical term, III, 1, a, α.

'Westminster Abbey': proper name, I, 1.

'.': assertion sign, logical term, III, 3, a.

59–2. 'There was': existential operator, logical term, III, 2, b.

  'was': contracted term, including a time indication.

'a': logical term, III, 1, c.

'rain storm': complex function, II, 2, b.

  'rain': indirect part; simple function, II, 2, a; fact-function, II, 3, b; first type, II, 4, a.

  'storm': direct part; simple function, II, 2, a; fact-function, II, 3, b; first type, II, 4, a.

'last night': complex function, II, 2, b, used for a token-reflexive description of a time argument, I, 2, b.

'last': indirect part; incomplete symbol.

'night': direct part; simple function, II, 2, a; thing-function, II, 3, a; first type, II, 4, a.

'.': assertion sign, logical symbol, III, 3, a.

59-3. 'a': logical term indicating a free variable, I, a, β.

'fine grain development': complex function, II, 2, b.

'fine', 'grain': indirect part; simple function, II, 2, a; thing-function, II, 3, a; first type, II, 4, a.

'development': direct part; simple function, II, 2, a; fact-function, II, 3, b; first type, II, 4, a.

'of': logical term, III, 1, a, α.

'a': logical term indicating a free variable, I, a, β.

'film': simple function, II, 2, a; thing-function, II, 3, a; first type, II, 4, a.

'is': logical term, III, 1, c.

'usually': numerical qualifier, higher function, II, 4, b.

'ly': logical term, III, 1, b.

'done': simple function, II, 2, a; thing-function, II, 3, a; first type, II, 4, a.

'slowly': functional modifier, higher function, II, 4, b.

'ly': logical term, III, 1, b.

'.': assertion sign, logical term, III, 3, a.

Index

439

A CATALOGUE OF
SELECTED DOVER BOOKS
IN ALL FIELDS OF INTEREST

A CATALOGUE OF SELECTED DOVER
BOOKS IN ALL FIELDS OF INTEREST

CONDITIONED REFLEXES, Ivan P. Pavlov. Full translation of most complete statement of Pavlov's work; cerebral damage, conditioned reflex, experiments with dogs, sleep, similar topics of great importance. 430pp. 5⅜ x 8½. 60614-7 Pa. $4.50

NOTES ON NURSING: WHAT IT IS, AND WHAT IT IS NOT, Florence Nightingale. Outspoken writings by founder of modern nursing. When first published (1860) it played an important role in much needed revolution in nursing. Still stimulating. 140pp. 5⅜ x 8½. 22340-X Pa. $2.50

HARTER'S PICTURE ARCHIVE FOR COLLAGE AND ILLUSTRATION, Jim Harter. Over 300 authentic, rare 19th-century engravings selected by noted collagist for artists, designers, decoupeurs, etc. Machines, people, animals, etc., printed one side of page. 25 scene plates for backgrounds. 6 collages by Harter, Satty, Singer, Evans. Introduction. 192pp. 8⅞ x 11¾. 23659-5 Pa. $4.50

MANUAL OF TRADITIONAL WOOD CARVING, edited by Paul N. Hasluck. Possibly the best book in English on the craft of wood carving. Practical instructions, along with 1,146 working drawings and photographic illustrations. Formerly titled *Cassell's Wood Carving*. 576pp. 6½ x 9¼. 23489-4 Pa. $7.95

THE PRINCIPLES AND PRACTICE OF HAND OR SIMPLE TURNING, John Jacob Holtzapffel. Full coverage of basic lathe techniques— history and development, special apparatus, softwood turning, hardwood turning, metal turning. Many projects—billiard ball, works formed within a sphere, egg cups, ash trays, vases, jardiniers, others—included. 1881 edition. 800 illustrations. 592pp. 6⅛ x 9¼. 23365-0 Clothbd. $15.00

THE JOY OF HANDWEAVING, Osma Tod. Only book you need for hand weaving. Fundamentals, threads, weaves, plus numerous projects for small board-loom, two-harness, tapestry, laid-in, four-harness weaving and more. Over 160 illustrations. 2nd revised edition. 352pp. 6½ x 9¼. 23458-4 Pa. $5.00

THE BOOK OF WOOD CARVING, Charles Marshall Sayers. Still finest book for beginning student in wood sculpture. Noted teacher, craftsman discusses fundamentals, technique; gives 34 designs, over 34 projects for panels, bookends, mirrors, etc. "Absolutely first-rate"—E. J. Tangerman. 33 photos. 118pp. 7¾ x 10⅝. 23654-4 Pa. $3.00

THE PHILOSOPHY OF HISTORY, Georg W. Hegel. Great classic of Western thought develops concept that history is not chance but a rational process, the evolution of freedom. 457pp. 5⅜ x 8½. 20112-0 Pa. $4.50

LANGUAGE, TRUTH AND LOGIC, Alfred J. Ayer. Famous, clear introduction to Vienna, Cambridge schools of Logical Positivism. Role of philosophy, elimination of metaphysics, nature of analysis, etc. 160pp. 5⅜ x 8½. (Available in U.S. only) 20010-8 Pa. $1.75

A PREFACE TO LOGIC, Morris R. Cohen. Great City College teacher in renowned, easily followed exposition of formal logic, probability, values, logic and world order and similar topics; no previous background needed. 209pp. 5⅜ x 8½. 23517-3 Pa. $3.50

REASON AND NATURE, Morris R. Cohen. Brilliant analysis of reason and its multitudinous ramifications by charismatic teacher. Interdisciplinary, synthesizing work widely praised when it first appeared in 1931. Second (1953) edition. Indexes. 496pp. 5⅜ x 8½. 23633-1 Pa. $6.00

AN ESSAY CONCERNING HUMAN UNDERSTANDING, John Locke. The only complete edition of enormously important classic, with authoritative editorial material by A. C. Fraser. Total of 1176pp. 5⅜ x 8½. 20530-4, 20531-2 Pa., Two-vol. set $14.00

HANDBOOK OF MATHEMATICAL FUNCTIONS WITH FORMULAS, GRAPHS, AND MATHEMATICAL TABLES, edited by Milton Abramowitz and Irene A. Stegun. Vast compendium: 29 sets of tables, some to as high as 20 places. 1,046pp. 8 x 10½. 61272-4 Pa. $12.50

MATHEMATICS FOR THE PHYSICAL SCIENCES, Herbert S. Wilf. Highly acclaimed work offers clear presentations of vector spaces and matrices, orthogonal functions, roots of polynomial equations, conformal mapping, calculus of variations, etc. Knowledge of theory of functions of real and complex variables is assumed. Exercises and solutions. Index. 284pp. 5⅝ x 8¼. 63635-6 Pa. $4.50

THE PRINCIPLE OF RELATIVITY, Albert Einstein et al. Eleven most important original papers on special and general theories. Seven by Einstein, two by Lorentz, one each by Minkowski and Weyl. All translated, unabridged. 216pp. 5⅜ x 8½. 60081-5 Pa. $3.00

THERMODYNAMICS, Enrico Fermi. A classic of modern science. Clear, organized treatment of systems, first and second laws, entropy, thermodynamic potentials, gaseous reactions, dilute solutions, entropy constant. No math beyond calculus required. Problems. 160pp. 5⅜ x 8½. 60361-X Pa. $2.75

ELEMENTARY MECHANICS OF FLUIDS, Hunter Rouse. Classic undergraduate text widely considered to be far better than many later books. Ranges from fluid velocity and acceleration to role of compressibility in fluid motion. Numerous examples, questions, problems. 224 illustrations. 376pp. 5⅝ x 8¼. 63699-2 Pa. $5.00

DRAWINGS OF WILLIAM BLAKE, William Blake. 92 plates from Book of Job, *Divine Comedy, Paradise Lost,* visionary heads, mythological figures, Laocoon, etc. Selection, introduction, commentary by Sir Geoffrey Keynes. 178pp. 8⅛ x 11. 22303-5 Pa. $4.00

ENGRAVINGS OF HOGARTH, William Hogarth. 101 of Hogarth's greatest works: *Rake's Progress, Harlot's Progress, Illustrations for Hudibras, Before and After, Beer Street and Gin Lane,* many more. Full commentary. 256pp. 11 x 13¾. 22479-1 Pa. $7.95

DAUMIER: 120 GREAT LITHOGRAPHS, Honore Daumier. Wide-ranging collection of lithographs by the greatest caricaturist of the 19th century. Concentrates on eternally popular series on lawyers, on married life, on liberated women, etc. Selection, introduction, and notes on plates by Charles F. Ramus. Total of 158pp. 9⅜ x 12¼. 23512-2 Pa. $5.50

DRAWINGS OF MUCHA, Alphonse Maria Mucha. Work reveals draftsman of highest caliber: studies for famous posters and paintings, renderings for book illustrations and ads, etc. 70 works, 9 in color; including 6 items not drawings. Introduction. List of illustrations. 72pp. 9⅜ x 12¼. (Available in U.S. only) 23672-2 Pa. $4.00

GIOVANNI BATTISTA PIRANESI: DRAWINGS IN THE PIERPONT MORGAN LIBRARY, Giovanni Battista Piranesi. For first time ever all of Morgan Library's collection, world's largest. 167 illustrations of rare Piranesi drawings—archeological, architectural, decorative and visionary. Essay, detailed list of drawings, chronology, captions. Edited by Felice Stampfle. 144pp. 9⅜ x 12¼. 23714-1 Pa. $7.50

NEW YORK ETCHINGS (1905-1949), John Sloan. All of important American artist's N.Y. life etchings. 67 works include some of his best art; also lively historical record—Greenwich Village, tenement scenes. Edited by Sloan's widow. Introduction and captions. 79pp. 8⅜ x 11¼. 23651-X Pa. $4.00

CHINESE PAINTING AND CALLIGRAPHY: A PICTORIAL SURVEY, Wan-go Weng. 69 fine examples from John M. Crawford's matchless private collection: landscapes, birds, flowers, human figures, etc., plus calligraphy. Every basic form included: hanging scrolls, handscrolls, album leaves, fans, etc. 109 illustrations. Introduction. Captions. 192pp. 8⅞ x 11¾. 23707-9 Pa. $7.95

DRAWINGS OF REMBRANDT, edited by Seymour Slive. Updated Lippmann, Hofstede de Groot edition, with definitive scholarly apparatus. All portraits, biblical sketches, landscapes, nudes, Oriental figures, classical studies, together with selection of work by followers. 550 illustrations. Total of 630pp. 9⅛ x 12¼. 21485-0, 21486-9 Pa., Two-vol. set $14.00

THE DISASTERS OF WAR, Francisco Goya. 83 etchings record horrors of Napoleonic wars in Spain and war in general. Reprint of 1st edition, plus 3 additional plates. Introduction by Philip Hofer. 97pp. 9⅜ x 8¼. 21872-4 Pa. $3.75

HISTORY OF BACTERIOLOGY, William Bulloch. The only comprehensive history of bacteriology from the beginnings through the 19th century. Special emphasis is given to biography-Leeuwenhoek, etc. Brief accounts of 350 bacteriologists form a separate section. No clearer, fuller study, suitable to scientists and general readers, has yet been written. 52 illustrations. 448pp. 5⅝ x 8¼.   23761-3 Pa. $6.50

THE COMPLETE NONSENSE OF EDWARD LEAR, Edward Lear. All nonsense limericks, zany alphabets, Owl and Pussycat, songs, nonsense botany, etc., illustrated by Lear. Total of 321pp. 5⅜ x 8½. (Available in U.S. only)   20167-8 Pa. $3.00

INGENIOUS MATHEMATICAL PROBLEMS AND METHODS, Louis A. Graham. Sophisticated material from Graham Dial, applied and pure; stresses solution methods. Logic, number theory, networks, inversions, etc. 237pp. 5⅜ x 8½.   20545-2 Pa. $3.50

BEST MATHEMATICAL PUZZLES OF SAM LOYD, edited by Martin Gardner. Bizarre, original, whimsical puzzles by America's greatest puzzler. From fabulously rare Cyclopedia, including famous 14-15 puzzles, the Horse of a Different Color, 115 more. Elementary math. 150 illustrations. 167pp. 5⅜ x 8½.   20498-7 Pa. $2.50

THE BASIS OF COMBINATION IN CHESS, J. du Mont. Easy-to-follow, instructive book on elements of combination play, with chapters on each piece and every powerful combination team—two knights, bishop and knight, rook and bishop, etc. 250 diagrams. 218pp. 5⅜ x 8½. (Available in U.S. only)   23644-7 Pa. $3.50

MODERN CHESS STRATEGY, Ludek Pachman. The use of the queen, the active king, exchanges, pawn play, the center, weak squares, etc. Section on rook alone worth price of the book. Stress on the moderns. Often considered the most important book on strategy. 314pp. 5⅜ x 8½.
20290-9 Pa. $3.50

LASKER'S MANUAL OF CHESS, Dr. Emanuel Lasker. Great world champion offers very thorough coverage of all aspects of chess. Combinations, position play, openings, end game, aesthetics of chess, philosophy of struggle, much more. Filled with analyzed games. 390pp. 5⅜ x 8½.
20640-8 Pa. $4.00

500 MASTER GAMES OF CHESS, S. Tartakower, J. du Mont. Vast collection of great chess games from 1798-1938, with much material nowhere else readily available. Fully annotated, arranged by opening for easier study. 664pp. 5⅜ x 8½.   23208-5 Pa. $6.00

A GUIDE TO CHESS ENDINGS, Dr. Max Euwe, David Hooper. One of the finest modern works on chess endings. Thorough analysis of the most frequently encountered endings by former world champion. 331 examples, each with diagram. 248pp. 5⅜ x 8½.   23332-4 Pa. $3.50

SECOND PIATIGORSKY CUP, edited by Isaac Kashdan. One of the greatest tournament books ever produced in the English language. All 90 games of the 1966 tournament, annotated by players, most annotated by both players. Features Petrosian, Spassky, Fischer, Larsen, six others. 228pp. 5⅜ x 8½.   23572-6 Pa. $3.50

ENCYCLOPEDIA OF CARD TRICKS, revised and edited by Jean Hugard. How to perform over 600 card tricks, devised by the world's greatest magicians: impromptus, spelling tricks, key cards, using special packs, much, much more. Additional chapter on card technique. 66 illustrations. 402pp. 5⅜ x 8½. (Available in U.S. only)   21252-1 Pa. $3.95

MAGIC: STAGE ILLUSIONS, SPECIAL EFFECTS AND TRICK PHOTOGRAPHY, Albert A. Hopkins, Henry R. Evans. One of the great classics; fullest, most authorative explanation of vanishing lady, levitations, scores of other great stage effects. Also small magic, automata, stunts. 446 illustrations. 556pp. 5⅜ x 8½.   23344-8 Pa. $5.00

THE SECRETS OF HOUDINI, J. C. Cannell. Classic study of Houdini's incredible magic, exposing closely-kept professional secrets and revealing, in general terms, the whole art of stage magic. 67 illustrations. 279pp. 5⅜ x 8½.   22913-0 Pa. $3.00

HOFFMANN'S MODERN MAGIC, Professor Hoffmann. One of the best, and best-known, magicians' manuals of the past century. Hundreds of tricks from card tricks and simple sleight of hand to elaborate illusions involving construction of complicated machinery. 332 illustrations. 563pp. 5⅜ x 8½.   23623-4 Pa. $6.00

MADAME PRUNIER'S FISH COOKERY BOOK, Mme. S. B. Prunier. More than 1000 recipes from world famous Prunier's of Paris and London, specially adapted here for American kitchen. Grilled tournedos with anchovy butter, Lobster a la Bordelaise, Prunier's prized desserts, more. Glossary. 340pp. 5⅜ x 8½. (Available in U.S. only)   22679-4 Pa. $3.00

FRENCH COUNTRY COOKING FOR AMERICANS, Louis Diat. 500 easy-to-make, authentic provincial recipes compiled by former head chef at New York's Fitz-Carlton Hotel: onion soup, lamb stew, potato pie, more. 309pp. 5⅜ x 8½.   23665-X Pa. $3.95

SAUCES, FRENCH AND FAMOUS, Louis Diat. Complete book gives over 200 specific recipes: bechamel, Bordelaise, hollandaise, Cumberland, apricot, etc. Author was one of this century's finest chefs, originator of vichyssoise and many other dishes. Index. 156pp. 5⅜ x 8.
23663-3 Pa. $2.50

TOLL HOUSE TRIED AND TRUE RECIPES, Ruth Graves Wakefield. Authentic recipes from the famous Mass. restaurant: popovers, veal and ham loaf, Toll House baked beans, chocolate cake crumb pudding, much more. Many helpful hints. Nearly 700 recipes. Index. 376pp. 5⅜ x 8½.
23560-2 Pa. $4.00

THE AMERICAN SENATOR, Anthony Trollope. Little known, long unavailable Trollope novel on a grand scale. Here are humorous comment on American vs. English culture, and stunning portrayal of a heroine/villainess. Superb evocation of Victorian village life. 561pp. 5⅜ x 8½.
23801-6 Pa. $6.00

WAS IT MURDER? James Hilton. The author of *Lost Horizon* and *Goodbye, Mr. Chips* wrote one detective novel (under a pen-name) which was quickly forgotten and virtually lost, even at the height of Hilton's fame. This edition brings it back—a finely crafted public school puzzle resplendent with Hilton's stylish atmosphere. A thoroughly English thriller by the creator of Shangri-la. 252pp. 5⅜ x 8. (Available in U.S. only)
23774-5 Pa. $3.00

CENTRAL PARK: A PHOTOGRAPHIC GUIDE, Victor Laredo and Henry Hope Reed. 121 superb photographs show dramatic views of Central Park: Bethesda Fountain, Cleopatra's Needle, Sheep Meadow, the Blockhouse, plus people engaged in many park activities: ice skating, bike riding, etc. Captions by former Curator of Central Park, Henry Hope Reed, provide historical view, changes, etc. Also photos of N.Y. landmarks on park's periphery. 96pp. 8½ x 11. 23750-8 Pa. $4.50

NANTUCKET IN THE NINETEENTH CENTURY, Clay Lancaster. 180 rare photographs, stereographs, maps, drawings and floor plans recreate unique American island society. Authentic scenes of shipwreck, lighthouses, streets, homes are arranged in geographic sequence to provide walking-tour guide to old Nantucket existing today. Introduction, captions. 160pp. 8⅞ x 11¾. 23747-8 Pa. $6.95

STONE AND MAN: A PHOTOGRAPHIC EXPLORATION, Andreas Feininger. 106 photographs by *Life* photographer Feininger portray man's deep passion for stone through the ages. Stonehenge-like megaliths, fortified towns, sculpted marble and crumbling tenements show textures, beauties, fascination. 128pp. 9¼ x 10¾. 23756-7 Pa. $5.95

CIRCLES, A MATHEMATICAL VIEW, D. Pedoe. Fundamental aspects of college geometry, non-Euclidean geometry, and other branches of mathematics: representing circle by point. Poincare model, isoperimetric property, etc. Stimulating recreational reading. 66 figures. 96pp. 5⅝ x 8¼.
63698-4 Pa. $2.75

THE DISCOVERY OF NEPTUNE, Morton Grosser. Dramatic scientific history of the investigations leading up to the actual discovery of the eighth planet of our solar system. Lucid, well-researched book by well-known historian of science. 172pp. 5⅜ x 8½. 23726-5 Pa. $3.00

THE DEVIL'S DICTIONARY. Ambrose Bierce. Barbed, bitter, brilliant witticisms in the form of a dictionary. Best, most ferocious satire America has produced. 145pp. 5⅜ x 8½. 20487-1 Pa. $1.75

PRINCIPLES OF ORCHESTRATION, Nikolay Rimsky-Korsakov. Great classical orchestrator provides fundamentals of tonal resonance, progression of parts, voice and orchestra, tutti effects, much else in major document. 330pp. of musical excerpts. 489pp. 6½ x 9¼.   21266-1 Pa. $6.00

TRISTAN UND ISOLDE, Richard Wagner. Full orchestral score with complete instrumentation. Do not confuse with piano reduction. Commentary by Felix Mottl, great Wagnerian conductor and scholar. Study score. 655pp. 8⅛ x 11.   22915-7 Pa. $12.50

REQUIEM IN FULL SCORE, Giuseppe Verdi. Immensely popular with choral groups and music lovers. Republication of edition published by C. F. Peters, Leipzig, n. d. German frontmaker in English translation. Glossary. Text in Latin. Study score. 204pp. 9⅜ x 12¼.
23682-X Pa. $6.00

COMPLETE CHAMBER MUSIC FOR STRINGS, Felix Mendelssohn. All of Mendelssohn's chamber music: Octet, 2 Quintets, 6 Quartets, and Four Pieces for String Quartet. (Nothing with piano is included). Complete works edition (1874-7). Study score. 283 pp. 9⅜ x 12¼.
23679-X Pa. $6.95

POPULAR SONGS OF NINETEENTH-CENTURY AMERICA, edited by Richard Jackson. 64 most important songs: "Old Oaken Bucket," "Arkansas Traveler," "Yellow Rose of Texas," etc. Authentic original sheet music, full introduction and commentaries. 290pp. 9 x 12.   23270-0 Pa. $6.00

COLLECTED PIANO WORKS, Scott Joplin. Edited by Vera Brodsky Lawrence. Practically all of Joplin's piano works—rags, two-steps, marches, waltzes, etc., 51 works in all. Extensive introduction by Rudi Blesh. Total of 345pp. 9 x 12.   23106-2 Pa. $13.50

BASIC PRINCIPLES OF CLASSICAL BALLET, Agrippina Vaganova. Great Russian theoretician, teacher explains methods for teaching classical ballet; incorporates best from French, Italian, Russian schools. 118 illustrations. 175pp. 5⅜ x 8½.   22036-2 Pa. $2.00

CHINESE CHARACTERS, L. Wieger. Rich analysis of 2300 characters according to traditional systems into primitives. Historical-semantic analysis to phonetics (Classical Mandarin) and radicals. 820pp. 6⅛ x 9¼.
21321-8 Pa. $8.95

EGYPTIAN LANGUAGE: EASY LESSONS IN EGYPTIAN HIERO-GLYPHICS, E. A. Wallis Budge. Foremost Egyptologist offers Egyptian grammar, explanation of hieroglyphics, many reading texts, dictionary of symbols. 246pp. 5 x 7½. (Available in U.S. only)
21394-3 Clothbd. $7.50

AN ETYMOLOGICAL DICTIONARY OF MODERN ENGLISH, Ernest Weekley. Richest, fullest work, by foremost British lexicographer. Detailed word histories. Inexhaustible. Do not confuse this with Concise Etymological Dictionary, which is abridged. Total of 856pp. 6½ x 9¼.
21873-2, 21874-0 Pa., Two-vol. set $10.00

CATALOGUE OF DOVER BOOKS

TONE POEMS, SERIES II: TILL EULENSPIEGELS LUSTIGE STREICHE, ALSO SPRACH ZARATHUSTRA, AND EIN HELDEN-LEBEN, Richard Strauss. Three important orchestral works, including very popular *Till Eulenspiegel's Marry Pranks*, reproduced in full score from original editions. Study score. 315pp. 9⅜ x 12¼. (Available in U.S. only)
23755-9 Pa. $7.50

TONE POEMS, SERIES I: DON JUAN, TOD UND VERKLARUNG AND DON QUIXOTE, Richard Strauss. Three of the most often performed and recorded works in entire orchestral repertoire, reproduced in full score from original editions. Study score. 286pp. 9⅜ x 12¼. (Available in U.S. only)
23754-0 Pa. $7.50

11 LATE STRING QUARTETS, Franz Joseph Haydn. The form which Haydn defined and "brought to perfection." (*Grove's*). 11 string quartets in complete score, his last and his best. The first in a projected series of the complete Haydn string quartets. Reliable modern Eulenberg edition, otherwise difficult to obtain. 320pp. 8⅜ x 11¼. (Available in U.S. only)
23753-2 Pa. $6.95

FOURTH, FIFTH AND SIXTH SYMPHONIES IN FULL SCORE, Peter Ilyitch Tchaikovsky. Complete orchestral scores of Symphony No. 4 in F Minor, Op. 36; Symphony No. 5 in E Minor, Op. 64; Symphony No. 6 in B Minor, "Pathetique," Op. 74. Bretikopf & Hartel eds. Study score. 480pp. 9⅜ x 12¼. 23861-X Pa. $10.95

THE MARRIAGE OF FIGARO: COMPLETE SCORE, Wolfgang A. Mozart. Finest comic opera ever written. Full score, not to be confused with piano renderings. Peters edition. Study score. 448pp. 9⅜ x 12¼. (Available in U.S. only)
23751-6 Pa. $11.95

"IMAGE" ON THE ART AND EVOLUTION OF THE FILM, edited by Marshall Deutelbaum. Pioneering book brings together for first time 38 groundbreaking articles on early silent films from *Image* and 263 illustrations newly shot from rare prints in the collection of the International Museum of Photography. A landmark work. Index. 256pp. 8¼ x 11.
23777-X Pa. $8.95

AROUND-THE-WORLD COOKY BOOK, Lois Lintner Sumption and Marguerite Lintner Ashbrook. 373 cooky and frosting recipes from 28 countries (America, Austria, China, Russia, Italy, etc.) include Viennese kisses, rice wafers, London strips, lady fingers, hony, sugar spice, maple cookies, etc. Clear instructions. All tested. 38 drawings. 182pp. 5⅜ x 8.
23802-4 Pa. $2.50

THE ART NOUVEAU STYLE, edited by Roberta Waddell. 579 rare photographs, not available elsewhere, of works in jewelry, metalwork, glass, ceramics, textiles, architecture and furniture by 175 artists—Mucha, Seguy, Lalique, Tiffany, Gaudin, Hohlwein, Saarinen, and many others. 288pp. 8⅜ x 11¼.
23515-7 Pa. $6.95

MUSHROOMS, EDIBLE AND OTHERWISE, Miron E. Hard. Profusely illustrated, very useful guide to over 500 species of mushrooms growing in the Midwest and East. Nomenclature updated to 1976. 505 illustrations. 628pp. 6½ x 9¼.   23309-X Pa. $7.95

AN ILLUSTRATED FLORA OF THE NORTHERN UNITED STATES AND CANADA, Nathaniel L. Britton, Addison Brown. Encyclopedic work covers 4666 species, ferns on up. Everything. Full botanical information, illustration for each. This earlier edition is preferred by many to more recent revisions. 1913 edition. Over 4000 illustrations, total of 2087pp. 6⅛ x 9¼.   22642-5, 22643-3, 22644-1 Pa., Three-vol. set $24.00

MANUAL OF THE GRASSES OF THE UNITED STATES, A. S. Hitchcock, U.S. Dept. of Agriculture. The basic study of American grasses, both indigenous and escapes, cultivated and wild. Over 1400 species. Full descriptions, information. Over 1100 maps, illustrations. Total of 1051pp. 5⅜ x 8½.   22717-0, 22718-9 Pa., Two-vol. set $12.00

THE CACTACEAE,, Nathaniel L. Britton, John N. Rose. Exhaustive, definitive. Every cactus in the world. Full botanical descriptions. Thorough statement of nomenclatures, habitat, detailed finding keys. The one book needed by every cactus enthusiast. Over 1275 illustrations. Total of 1080pp. 8 x 10¼.   21191-6, 21192-4 Clothbd., Two-vol. set $35.00

AMERICAN MEDICINAL PLANTS, Charles F. Millspaugh. Full descriptions, 180 plants covered: history; physical description; methods of preparation with all chemical constituents extracted; all claimed curative or adverse effects. 180 full-page plates. Classification table. 804pp. 6½ x 9¼.
                                                    23034-1 Pa. $10.00

A MODERN HERBAL, Margaret Grieve. Much the fullest, most exact, most useful compilation of herbal material. Gigantic alphabetical encyclopedia, from aconite to zedoary, gives botanical information, medical properties, folklore, economic uses, and much else. Indispensable to serious reader. 161 illustrations. 888pp. 6½ x 9¼. (Available in U.S. only)
                                    22798-7, 22799-5 Pa., Two-vol. set $11.00

THE HERBAL or GENERAL HISTORY OF PLANTS, John Gerard. The 1633 edition revised and enlarged by Thomas Johnson. Containing almost 2850 plant descriptions and 2705 superb illustrations, Gerard's *Herbal* is a monumental work, the book all modern English herbals are derived from, the one herbal every serious enthusiast should have in its entirety. Original editions are worth perhaps $750. 1678pp. 8½ x 12¼.
                                                    23147-X Clothbd. $50.00

MANUAL OF THE TREES OF NORTH AMERICA, Charles S. Sargent. The basic survey of every native tree and tree-like shrub, 717 species in all. Extremely full descriptions, information on habitat, growth, locales, economics, etc. Necessary to every serious tree lover. Over 100 finding keys. 783 illustrations. Total of 986pp. 5⅜ x 8½.
                                    20277-1, 20278-X Pa., Two-vol. set $10.00

THE COMPLETE BOOK OF DOLL MAKING AND COLLECTING, Catherine Christopher. Instructions, patterns for dozens of dolls, from rag doll on up to elaborate, historically accurate figures. Mould faces, sew clothing, make doll houses, etc. Also collecting information. Many illustrations. 288pp. 6 x 9.   22066-4 Pa. $4.00

THE DAGUERREOTYPE IN AMERICA, Beaumont Newhall. Wonderful portraits, 1850's townscapes, landscapes; full text plus 104 photographs. The basic book. Enlarged 1976 edition. 272pp. 8¼ x 11¼.
23322-7 Pa. $6.00

CRAFTSMAN HOMES, Gustav Stickley. 296 architectural drawings, floor plans, and photographs illustrate 40 different kinds of "Mission-style" homes from *The Craftsman* (1901-16), voice of American style of simplicity and organic harmony. Thorough coverage of Craftsman idea in text and picture, now collector's item. 224pp. 8⅛ x 11.   23791-5 Pa. $6.00

PEWTER-WORKING: INSTRUCTIONS AND PROJECTS, Burl N. Osborn. & Gordon O. Wilber. Introduction to pewter-working for amateur craftsman. History and characteristics of pewter; tools, materials, step-by-step instructions. Photos, line drawings, diagrams. Total of 160pp. 7⅞ x 10¾.   23786-9 Pa. $3.50

THE GREAT CHICAGO FIRE, edited by David Lowe. 10 dramatic, eye-witness accounts of the 1871 disaster, including one of the aftermath and rebuilding, plus 70 contemporary photographs and illustrations of the ruins—courthouse, Palmer House, Great Central Depot, etc. Introduction by David Lowe. 87pp. 8¼ x 11.   23771-0 Pa. $4.00

SILHOUETTES: A PICTORIAL ARCHIVE OF VARIED ILLUSTRATIONS, edited by Carol Belanger Grafton. Over 600 silhouettes from the 18th to 20th centuries include profiles and full figures of men and women, children, birds and animals, groups and scenes, nature, ships, an alphabet. Dozens of uses for commercial artists and craftspeople. 144pp. 8⅜ x 11¼.
23781-8 Pa. $4.00

ANIMALS: 1,419 COPYRIGHT-FREE ILLUSTRATIONS OF MAMMALS, BIRDS, FISH, INSECTS, ETC., edited by Jim Harter. Clear wood engravings present, in extremely lifelike poses, over 1,000 species of animals. One of the most extensive copyright-free pictorial sourcebooks of its kind. Captions. Index. 284pp. 9 x 12.   23766-4 Pa. $7.50

INDIAN DESIGNS FROM ANCIENT ECUADOR, Frederick W. Shaffer. 282 original designs by pre-Columbian Indians of Ecuador (500-1500 A.D.). Designs include people, mammals, birds, reptiles, fish, plants, heads, geometric designs. Use as is or alter for advertising, textiles, leathercraft, etc. Introduction. 95pp. 8¾ x 11¼.   23764-8 Pa. $3.50

SZIGETI ON THE VIOLIN, Joseph Szigeti. Genial, loosely structured tour by premier violinist, featuring a pleasant mixture of reminiscenes, insights into great music and musicians, innumerable tips for practicing violinists. 385 musical passages. 256pp. 5⅝ x 8¼.   23763-X Pa. $3.50

CATALOGUE OF DOVER BOOKS

A MAYA GRAMMAR, Alfred M. Tozzer. Practical, useful English-language grammar by the Harvard anthropologist who was one of the three greatest American scholars in the area of Maya culture. Phonetics, grammatical processes, syntax, more. 301pp. 5⅜ x 8½.   23465-7 Pa. $4.00

THE JOURNAL OF HENRY D. THOREAU, edited by Bradford Torrey, F. H. Allen. Complete reprinting of 14 volumes, 1837-61, over two million words; the sourcebooks for *Walden*, etc. Definitive. All original sketches, plus 75 photographs. Introduction by Walter Harding. Total of 1804pp. 8½ x 12¼.   20312-3, 20313-1 Clothbd., Two-vol. set $50.00

CLASSIC GHOST STORIES, Charles Dickens and others. 18 wonderful stories you've wanted to reread: "The Monkey's Paw," "The House and the Brain," "The Upper Berth," "The Signalman," "Dracula's Guest," "The Tapestried Chamber," etc. Dickens, Scott, Mary Shelley, Stoker, etc. 330pp. 5⅜ x 8½.   20735-8 Pa. $3.50

SEVEN SCIENCE FICTION NOVELS, H. G. Wells. Full novels. *First Men in the Moon, Island of Dr. Moreau, War of the Worlds, Food of the Gods, Invisible Man, Time Machine, In the Days of the Comet.* A basic science-fiction library. 1015pp. 5⅜ x 8½. (Available in U.S. only)   20264-X Clothbd. $8.95

ARMADALE, Wilkie Collins. Third great mystery novel by the author of *The Woman in White* and *The Moonstone.* Ingeniously plotted narrative shows an exceptional command of character, incident and mood. Original magazine version with 40 illustrations. 597pp. 5⅜ x 8½.   23429-0 Pa. $5.00

MASTERS OF MYSTERY, H. Douglas Thomson. The first book in English (1931) devoted to history and aesthetics of detective story. Poe, Doyle, LeFanu, Dickens, many others, up to 1930. New introduction and notes by E. F. Bleiler. 288pp. 5⅜ x 8½. (Available in U.S. only)   23606-4 Pa. $4.00

FLATLAND, E. A. Abbott. Science-fiction classic explores life of 2-D being in 3-D world. Read also as introduction to thought about hyperspace. Introduction by Banesh Hoffmann. 16 illustrations. 103pp. 5⅜ x 8½.   20001-9 Pa. $1.50

THREE SUPERNATURAL NOVELS OF THE VICTORIAN PERIOD, edited, with an introduction, by E. F. Bleiler. Reprinted complete and unabridged, three great classics of the supernatural: *The Haunted Hotel* by Wilkie Collins, *The Haunted House at Latchford* by Mrs. J. H. Riddell, and *The Lost Stradivarious* by J. Meade Falkner. 325pp. 5⅜ x 8½.   22571-2 Pa. $4.00

AYESHA: THE RETURN OF "SHE," H. Rider Haggard. Virtuoso sequel featuring the great mythic creation, Ayesha, in an adventure that is fully as good as the first book, *She.* Original magazine version, with 47 original illustrations by Maurice Greiffenhagen. 189pp. 6½ x 9¼.   23649-8 Pa. $3.00

THE COMPLETE WOODCUTS OF ALBRECHT DURER, edited by Dr. W. Kurth. 346 in all: "Old Testament," "St. Jerome," "Passion," "Life of Virgin," Apocalypse," many others. Introduction by Campbell Dodgson. 285pp. 8½ x 12¼.   21097-9 Pa. $6.95

DRAWINGS OF ALBRECHT DURER, edited by Heinrich Wolfflin. 81 plates show development from youth to full style. Many favorites; many new. Introduction by Alfred Werner. 96pp. 8⅛ x 11.   22352-3 Pa. $4.00

THE HUMAN FIGURE, Albrecht Dürer. Experiments in various techniques—stereometric, progressive proportional, and others. Also life studies that rank among finest ever done. Complete reprinting of *Dresden Sketchbook*. 170 plates. 355pp. 8⅜ x 11¼.   21042-1 Pa. $6.95

OF THE JUST SHAPING OF LETTERS, Albrecht Dürer. Renaissance artist explains design of Roman majuscules by geometry, also Gothic lower and capitals. Grolier Club edition. 43pp. 7⅞ x 10¾   21306-4 Pa. $2.50

TEN BOOKS ON ARCHITECTURE, Vitruvius. The most important book ever written on architecture. Early Roman aesthetics, technology, classical orders, site selection, all other aspects. Stands behind everything since. Morgan translation. 331pp. 5⅜ x 8½.   20645-9 Pa. $3.75

THE FOUR BOOKS OF ARCHITECTURE, Andrea Palladio. 16th-century classic responsible for Palladian movement and style. Covers classical architectural remains, Renaissance revivals, classical orders, etc. 1738 Ware English edition. Introduction by A. Placzek. 216 plates. 110pp. of text. 9½ x 12¾.   21308-0 Pa. $7.50

HCRIZONS, Norman Bel Geddes. Great industrialist stage designer, "father of streamlining," on application of aesthetics to transportation, amusement, architecture, etc. 1932 prophetic account; function, theory, specific projects. 222 illustrations. 312pp. 7⅞ x 10¾.   23514-9 Pa. $6.95

FRANK LLOYD WRIGHT'S FALLINGWATER, Donald Hoffmann. Full, illustrated story of conception and building of Wright's masterwork at Bear Run, Pa. 100 photographs of site, construction, and details of completed structure. 112pp. 9¼ x 10.   23671-4 Pa. $5.00

THE ELEMENTS OF DRAWING, John Ruskin. Timeless classic by great Viltorian; starts with basic ideas, works through more difficult. Many practical exercises. 48 illustrations. Introduction by Lawrence Campbell. 228pp. 5⅜ x 8½.   22730-8 Pa. $2.75

GIST OF ART, John Sloan. Greatest modern American teacher, Art Students League, offers innumerable hints, instructions, guided comments to help you in painting. Not a formal course. 46 illustrations. Introduction by Helen Sloan. 200pp. 5⅜ x 8½.   23435-5 Pa. $3.50

THE SENSE OF BEAUTY, George Santayana. Masterfully written discussion of nature of beauty, materials of beauty, form, expression; art, literature, social sciences all involved. 168pp. 5⅜ x 8½.   20238-0 Pa. $2.50

ON THE IMPROVEMENT OF THE UNDERSTANDING, Benedict Spinoza. Also contains *Ethics, Correspondence,* all in excellent R. Elwes translation. Basic works on entry to philosophy, pantheism, exchange of ideas with great contemporaries. 402pp. 5⅜ x 8½.   20250-X Pa. $3.75

THE TRAGIC SENSE OF LIFE, Miguel de Unamuno. Acknowledged masterpiece of existential literature, one of most important books of 20th century. Introduction by Madariaga. 367pp. 5⅜ x 8½.
20257-7 Pa. $3.50

THE GUIDE FOR THE PERPLEXED, Moses Maimonides. Great classic of medieval Judaism attempts to reconcile revealed religion (Pentateuch, commentaries) with Aristotelian philosophy. Important historically, still relevant in problems. Unabridged Friedlander translation. Total of 473pp. 5⅜ x 8½.   20351-4 Pa. $5.00

THE I CHING (THE BOOK OF CHANGES), translated by James Legge. Complete translation of basic text plus appendices by Confucius, and Chinese commentary of most penetrating divination manual ever prepared. Indispensable to study of early Oriental civilizations, to modern inquiring reader. 448pp. 5⅜ x 8½.   21062-6 Pa. $4.00

THE EGYPTIAN BOOK OF THE DEAD, E. A. Wallis Budge. Complete reproduction of Ani's papyrus, finest ever found. Full hieroglyphic text, interlinear transliteration, word for word translation, smooth translation. Basic work, for Egyptology, for modern study of psychic matters. Total of 533pp. 6½ x 9¼. (Available in U.S. only)   21866-X Pa. $4.95

THE GODS OF THE EGYPTIANS, E. A. Wallis Budge. Never excelled for richness, fullness: all gods, goddesses, demons, mythical figures of Ancient Egypt; their legends, rites, incarnations, variations, powers, etc. Many hieroglyphic texts cited. Over 225 illustrations, plus 6 color plates. Total of 988pp. 6⅛ x 9¼. (Available in U.S. only)
22055-9, 22056-7 Pa., Two-vol. set $12.00

THE ENGLISH AND SCOTTISH POPULAR BALLADS, Francis J. Child. Monumental, still unsuperseded; all known variants of Child ballads, commentary on origins, literary references, Continental parallels, other features. Added: papers by G. L. Kittredge, W. M. Hart. Total of 2761pp. 6½ x 9¼.
21409-5, 21410-9, 21411-7, 21412-5, 21413-3 Pa., Five-vol. set $37.50

CORAL GARDENS AND THEIR MAGIC, Bronsilaw Malinowski. Classic study of the methods of tilling the soil and of agricultural rites in the Trobriand Islands of Melanesia. Author is one of the most important figures in the field of modern social anthropology. 143 illustrations. Indexes. Total of 911pp. of text. 5⅝ x 8¼. (Available in U.S. only)
23597-1 Pa. $12.95

GEOMETRY, RELATIVITY AND THE FOURTH DIMENSION, Rudolf Rucker. Exposition of fourth dimension, means of visualization, concepts of relativity as Flatland characters continue adventures. Popular, easily followed yet accurate, profound. 141 illustrations. 133pp. 5⅜ x 8½.
23400-2 Pa. $2.75

THE ORIGIN OF LIFE, A. I. Oparin. Modern classic in biochemistry, the first rigorous examination of possible evolution of life from nitrocarbon compounds. Non-technical, easily followed. Total of 295pp. 5⅜ x 8½.
60213-3 Pa. $4.00

THE CURVES OF LIFE, Theodore A. Cook. Examination of shells, leaves, horns, human body, art, etc., in *"the* classic reference on how the golden ratio applies to spirals and helices in nature . . . . "—Martin Gardner. 426 illustrations. Total of 512pp. 5⅜ x 8½.
23701-X Pa. $5.95

PLANETS, STARS AND GALAXIES, A. E. Fanning. Comprehensive introductory survey: the sun, solar system, stars, galaxies, universe, cosmology; quasars, radio stars, etc. 24pp. of photographs. 189pp. 5⅜ x 8½. (Available in U.S. only)
21680-2 Pa. $3.00

THE THIRTEEN BOOKS OF EUCLID'S ELEMENTS, translated with introduction and commentary by Sir Thomas L. Heath. Definitive edition. Textual and linguistic notes, mathematical analysis, 2500 years of critical commentary. Do not confuse with abridged school editions. Total of 1414pp. 5⅜ x 8½.
60088-2, 60089-0, 60090-4 Pa., Three-vol. set $18.00

DIALOGUES CONCERNING TWO NEW SCIENCES, Galileo Galilei. Encompassing 30 years of experiment and thought, these dialogues deal with geometric demonstrations of fracture of solid bodies, cohesion, leverage, speed of light and sound, pendulums, falling bodies, accelerated motion, etc. 300pp. 5⅜ x 8½.
60099-8 Pa. $4.00

Prices subject to change without notice.

Available at your book dealer or write for free catalogue to Dept. GI, Dover Publications, Inc., 180 Varick St., N.Y., N.Y. 10014. Dover publishes more than 175 books each year on science, elementary and advanced mathematics, biology, music, art, literary history, social sciences and other areas.